Reshaping the World

Reshaping the World
DEBATES ON MESOAMERICAN COSMOLOGIES

EDITED BY
Ana Díaz

UNIVERSITY PRESS OF COLORADO
Louisville

© 2020 by University Press of Colorado

Published by University Press of Colorado
245 Century Circle, Suite 202
Louisville, Colorado 80027

All rights reserved
Manufactured in the United States of America

 The University Press of Colorado is a proud member of the Association of University Presses.

The University Press of Colorado is a cooperative publishing enterprise supported, in part, by Adams State University, Colorado State University, Fort Lewis College, Metropolitan State University of Denver, Regis University, University of Colorado, University of Northern Colorado, University of Wyoming, Utah State University, and Western Colorado University.

∞ This paper meets the requirements of the ANSI/NISO Z39.48–1992 (Permanence of Paper)

ISBN: 978-1-60732-943-5 (hardcover)
ISBN: 978-1-64642-005-6 (paperback)
ISBN: 978-1-60732-953-4 (ebook)
https://doi.org/10.5876/9781607329534

Library of Congress Cataloging-in-Publication Data

Names: Díaz, Ana (Ethnologist), editor.
Title: Reshaping the world : debates on Mesoamerican cosmologies / edited by Ana Díaz.
Description: Louisville : University Press of Colorado, 2019. | Includes bibliographical references and index.
Identifiers: LCCN 2019032321 (print) | LCCN 2019032322 (ebook) | ISBN 9781607329435 (cloth) | ISBN 9781646420056 (paper) | ISBN 9781607329534 (ebook)
Subjects: LCSH: Indian cosmology. | Indians of Mexico—Religion. | Indians of Central America—Religion. | Indians of Mexico—Rites and ceremonies. | Indians of Central America—Rites and ceremonies.
Classification: LCC F1219.3.C65 R47 2019 (print) | LCC F1219.3.C65 (ebook) | DDC 299.792—dc23
LC record available at https://lccn.loc.gov/2019032321
LC ebook record available at https://lccn.loc.gov/2019032322

Cover illustration: *Muwieri*, Museo Nacional de Antropología, Mexico City (reproduction authorized by the Instituto Nacional de Antropología e Historia, Secretaría de Cultura).

To Toke Sellner Reunert (1971–2016), a dear friend and brilliant colleague, whom we lost during the final editing of this volume. May you travel safely on new and unknown paths of the above and the below.

Contents

List of figures	ix
Preface	xiii
Acknowledgments	xv

Introduction: Rethinking the Mesoamerican Cosmos
  Ana Díaz 3

PART I. RECOGNITION: ON DESCRIBING OTHERS' WORLDS

Chapter 1. Colliding Universes: A Reconsideration of the Structure of the Precolumbian Mesoamerican Cosmos
  *Jesper Nielsen and
  Toke Sellner Reunert †* 31

Chapter 2. Incorporating Mesoamerican Cosmology within a Global History of Religion: Some Considerations on the Work of Lorenzo Pignoria
  Sergio Botta 70

Chapter 3. Dissecting the Sky: Discursive Translations in Mexican Colonial Cosmographies
  Ana Díaz 100

Part II. Inventiveness: Reshaping Experience in Colonial Cosmologies

Chapter 4. The Colonial Encounter: Transformations of Indigenous Yucatec Conceptions of *K'uh*
Gabrielle Vail — 141

Chapter 5. Zapotec Travels in Time and Space: The Correlation between the 260-Day Cycle and a Multilevel Cosmological Model
David Tavárez — 180

Chapter 6. A Cosmology of Water: The Universe According to the Ch'orti' Maya
Kerry Hull — 209

Part III. Complexity: Breaking Paradigms on Cosmological Conceptions

Chapter 7. Distance and Power in Classic Maya Texts
Alexandre Tokovinine — 251

Chapter 8. The Sky, the Night, and the Number Nine: Considerations of the Nahua Vision of the Universe
Katarzyna Mikulska — 282

Chapter 9. Creating and Destroying the Upper Part of the Cosmos: A New Approach to the Study of Wixarika Cosmology
Johannes Neurath — 319

About the Authors — 341
Index — 345

Figures

1.1. The multilayered structure of the Aztec cosmos, *Codex Vaticanus A* — 34
1.2. Cosmogram, *Codex Fejérváry-Mayer* — 39
1.3. Aztec bone rasp (*omichicahuaztli*) — 40
1.4. Aztec stone relief known as the *Stone of the Heavens* — 41
1.5. Pedro de Gante preaching to Nahuas — 49
1.6. Diego de Valadés's "great chain of being": heaven, earth, and hell — 50
1.7. The first two scenes of the sacred narrative of the *Selden Roll* — 54
1.8. Konrad von Megenberg, showing the heavenly realm divided into layers — 55
1.9. Roman gods in a layered universe — 56
2.1. Homoyoca, implicitly compared to Jupiter via an image redrawn rom *Codex Vaticanus A* — 80
2.2. *Codex Vaticanus A*, 1v — 81
2.3. Osiris and Isis — 83
2.4. Pignoria's "messenger" of Citlallatonac — 84
2.5. Piognoria redrew the central image of *Codex Vaticanus A* 7v — 85
2.6. Quetzalcoatl, adapted by Pignoria — 86
2.7. Quetzalcoatl in *Codex Vaticanus A* — 87
2.8. The Constantine' medallion — 89
2.9. *Hoc signo victor eris* — 90

3.1. The Mexican cosmos, *Codex Vaticanus A*   101
3.2. Cosmographical reconstruction of the double-pyramid structure   102
3.3. The three levels of creation as a *continuum*   111
3.4. 12-Wind descends from the sky; 9-Wind descends from the sky   115
3.5. Reconstruction of the Nahua underworld   116
3.6. Two figures stand on a starry band, *Codex Vindobonensis*, 13   117
3.7. The Altar of Venus, a squared altar with starry iconography   118
3.8. A round altar with sky and solar iconography   119
3.9. Descent of 9-Wind to Earth, *Selden Roll*   121
3.10. Band-goddesses dressed with embroided skirts   122
3.11. Vessel with iconographical motifs that resemble an embroided skirt   126
3.12. The solar sky, destiny of warriors dwho died in combat   128
4.1. Examples of *k'uh* in the Maya codices   146
4.2. Ceremony involving sanctifying deity effigies, *Codex Madrid*   147
4.3. Yearbearer ceremony (*Dresden Codex*)   148
4.4. Itzamna wearing K'awil's nose and named as Bolon (Nine) Yax   151
4.5. Yearbearer ceremonies (*Madrid Codex*)   152
4.6. The four Pawahtuns and Chaak planting   155
4.7. The ascent of the crocodilian   157
4.8. God L and K'awil   160
4.9. Mural from Tulum showing Chaak and Chak Chel   163
4.10. The creator deities, Chak Chel and Itzamna   164
4.11. Possible depiction of Kawak yearbearer ceremonies   166

4.12. Hisin falling into the underworld; Hun Ahaw ascendant   167
5.1. Zapotec diagram of cosmoloical space   187
5.2. Notes on the first three days in a divinatory count   189
5.3. Zapotec cosmological diagram: nine levels above and nine below earth   190
5.4. The first two trecenas in Book 85-1   192
5.5. Annotation concerning the sixth feast in Trecena 1   194
6.1. Ch'orti'-speaking areas, Guatemala   210
6.2. Cornfields in the Ch'orti' area   211
6.3. A ceremony performed in a Ch'orti' hamlet by a Mam-speaking priest   212
6.4. A Ch'orti' home altar, hamlet of Amatillo above Jocotán, Guatemala   216
6.5. The five colored lakes in Ch'orti' cosmology   218
6.6. The Paddler Gods being "bathed" in watery clouds   233
6.7. Two Paddler Gods "bathing"; the glyphic compound reading *yati*   233
7.1. Classic-period sites in the southern Maya lowlands   252
7.2. Inscriptions referring to places   254
7.3. Inscriptions using the term tz'ul ("foreigner")   256
7.4. Sites with families in quadripartite lists or identifying with cardinal directions   257
7.5. Deep-time places   260
7.6. Shared origins: sites associated with chi-T316 place name or Teotihuacan   264
7.7. Early Classic representations of traveling   267
7.8. Traveling in Cancuen narratives   269
8.1. Night sky   284
8.2. Starry sky   285

8.3. Bone musical instrument or *omichicahuaztli*	286
8.4. Clear daytime sky	287
8.5. Cloudy sky	288
8.6. The interior of the earth, or the World of the Dead	289
8.7. The Venus star	291
9.1. Wooden pyramid from Te'akata	327
9.2. *Imumui*, a stair for Father Sun's ascent	327
9.3. The Taimarita hill, with its natural tiers	329
9.4. A *xiriki* altar with thatched roof, Xawiepa ceremonial center	331
9.5. Huichol embroided textile, with motif of bicephalic eagle and peyote	333
9.6. A shaman's *muwieri*	333

Preface

As shown in the image of the cover of this volume, the conceptions of the cosmos in America varied notably from those articulated through Christian cosmography and theology. The feathered stick reproduced on the cover of this book (a *muwieri* [see chapter 9]) is one impersonator of the sky. But his/her identity could not be discovered by sixteenth-century friars, nor by us, without a proper introduction on Wixarika's thought.

The purpose of this volume is to question the configuration of the Mesoamerican cosmos as a unitary, fixed, ahistorical, pan-Mesoamerican structure. Most of the contributors to this volume offer critical readings of archaeological and historical sources in order to destabilize the dominant precolumbian cosmological model of 13/9 upper/lower layers. Three main points of departure are common to our attempts to understand Mesoamerican cosmological models: (1) a reevaluation of precolumbian *vis-à-vis* colonial sources, (2) a critical revision of the idea of a multilayered universe composed of 13 and nine vertically arranged layers or levels, and (3) the proposal of new cosmographic alternatives to understand the structure of the Mesoamerican cosmos.

The present volume is the result of a continued series of inter- and cross-disciplinary discussions and exchanges of ideas. Initial studies that critically discussed previous ideas on the Mesoamerican cosmological models had been published by Katarzyna Mikulska (2008), Jesper Nielsen and Toke Sellner Reunert (2009), and Ana Díaz (2009). To our surprise, we had arrived at very similar conclusions, although working independently and drawing upon different kinds of sources and data. In 2011 we met in Mexico City at the Museo Nacional de Antropología, in the conference "Cosmologías indígenas, nuevas aproximaciones" organized by Ana Díaz. There, Katarzyna Mikulska, Jesper Nielsen, Toke Sellner Reunert, Berenice Alcántara, Vanya Valdovinos, and Ana Díaz joined in a discussion that eventually formed the backbone of a book.

In 2015, I thus edited a first version of a book, collecting these new and more detailed studies, all written in Spanish and published by the UNAM (see Díaz 2015). The volume was conceived of as a textbook for students of Mesoamerican studies and was received with great interest by the Mexican academic community. After further fruitful discussions, meetings, and exchanges of ideas and data, some of the authors decided to integrate our reflections into another volume, written for a wider international public, and to invite other specialists to collaborate in the conversation.

REFERENCES

Díaz, Ana. 2009. "La primera lámina del Códice Vaticano A. ¿Un modelo para justificar la topografía celestial de la antigüedad pagana indígena?" *Anales del Instituto de Investigaciones Estéticas* 31 (95): 5–44.

Díaz, Ana, ed. 2015. *Cielos e inframundos; una revisión a la Cosmología Mesoamericana.* México: Universidad Nacional Autónoma de México Instituto de Investigaciones Históricas, Fideicomiso Felipe Teixidor y Montserrat Alfau de Teixidor.

Mikulska, Katarzyna. 2008. "El concepto de *ilhuicatl* en la cosmovisión nahua y sus representaciones gráficas en códices." *Revista Española de Antropología Americana* 38 (2): 151–171.

Nielsen, Jesper, and Toke Sellner Reunert. 2009. "Dante's Heritage. Questioning the Multi-Layered Model of the Mesoamerican Universe." *Antiquity* 83 (2): 399–413.

Acknowledgments

This book would never have been possible without the generous support of a number of colleagues and friends. First of all, I want to thank Diana Magaloni, who was the director of the Museo Nacional de Antropología in 2011 and who opened the doors of the eminent institution for the meeting. She was also my teacher and mentor, and showed me the clues of the profession with her outstanding example.

I also wish to thank the contributors to this volume, especially Jesper Nielsen and Gabrielle Vail, who trusted in this project from its early phases and supported me with their advice in the process of completing this project. My gratitude to Alexandre Tokovinine, David Tavárez, Sergio Botta, and Kerry Hull, who accepted to join to our team, and for working in record time. To Cecelia Klein, whose comments helped the book achieve its current configuration, and whose article on weaving the cosmos became a framework for my work as an art historian.

I also want to thank María Capetillo and Susana Kolb, who were our support in the process of translating and revising the English edition. Thanks to Carlos Mondragón, Paul Liffman, and Alfonso Vite for helping with important final details. A special thanks to Vanya Valdovinos for joining our colloquium in 2011, when she was a graduate student, and for present her thoughtful approach to rock-art examples of the impersonators of the Sky. Thanks also to the anonymous reviewers of the volume, whose cogent comments led to improvements throughout the chapters.

Finally, a great thanks to the colleagues, teachers, and students who are not mentioned here by name but who were present in all phases of the project.

Reshaping the World

Introduction

Rethinking the Mesoamerican Cosmos

ANA DÍAZ

Among the many spectacular pieces of monumental sculpture and carved reliefs preserved from the Late Postclassic now on display in the Sala Mexica in Mexico City's Museo Nacional de Antropología is a little-known and easy-to-miss rectangular carved-stone panel. It shows two warrior figures and a bird descending down a narrow band that is flanked by two bands of starry symbols arranged in what appear to be layers (see figure 1.4). The title on the display board reads *Lápida de los Cielos* ("Stone of the Skies"). To provide further explanation, a second, larger board next to the stone panel shows a scene from the late sixteenth-century *Codex Vaticanus A* representing a multilayered universe with an Aztec creator god seated in the uppermost layer of the sky, and below the earth the nine layers of the hell-like underworld of Mictlan. The accompanying text briefly describes the Mexica cosmos stating that there were 13 layers in the sky and nine in the underworld. Although the stone panel does not show 13 layers, nor portray any difference among the layers (in contrast to the *Codex Vaticanus A*), the visitor is naturally meant to conclude that the warriors and the bird are traversing the heavenly spheres—hence the name given to this unique representation, which can also be translated as "Stone of the Heavens." If the visitor continues his or her itinerary to the ruins of the former main temple of the Mexica, the *Templo Mayor*, also in the heart of Mexico City, a display board will confirm this cosmological model and describe how the

DOI: 10.5876/9781607329534.c000

temple was located at the center of the universe, connecting the thirteen and nine layers, again clearly referring to the *Codex Vaticanus A*. Similarly, most, if not all, textbooks on ancient Mesoamerica and the Aztecs describe the same general version of the precolumbian cosmos; indeed, scholarly consensus seems to have been achieved on this particular issue. This consensus, however, is relatively recent.

Descriptions of the form and functioning of the indigenous cosmos have appeared in several works written since colonial times, often in a fragmented and/or contradictory fashion, and it is nearly impossible to find two sources offering the same exact version of the cosmological structure. It was not until the beginning of the twentieth century that Eduard Seler (1902–1923, 1907) was able to provide a more coherent interpretation and reconstruction of the Mexica cosmos by integrating contradictory data from a great variety of sources, among which the representation of the cosmos in the *Codex Vaticanus A* stands out. Based on this specific illustration, he formulated an explanatory model that conceives of the sky as a series of 13 levels and the underworld as composed of nine levels.

While Seler himself emphasized that Mexica cosmology was far more complex than this layered structure would seem to suggest (see Nielsen and Sellner Reunert 2015, chapter 2, this volume), the model's pedagogical efficiency and its similarity with classic Eurasian cosmographies undoubtedly eased the model's acceptance and its institutionalization as the fundamental cosmological structure shared by all Mesoamericans before and after the conquest. Seler's ideas have since been supplemented by an impressive list of renowned scholars such as Alfred Tozzer (1907), J. Eric S. Thompson (1934, 1954, 1970), Walter Krickeberg (1950), William Holland (1961, 1963), Alfonso Caso (1967), Miguel León-Portilla (1966 [1956], 1994 [1968]), Alfredo López Austin (1984, 1994, 2001, 2016), and Eduardo Matos Moctezuma (1987, 2008, 2013). Although the model of the 13 skies and nine underworlds has been slightly modified over time, it remains the basic model of the Mesoamerican cosmos, and has been formalized by scholars for more than a century to the point that it has become part of a set of standard truths about ancient Mesoamerica, and thus has been rarely questioned or challenged until recent contributions by Klein (1982), Knab (1991, 2004), Nielsen and Sellner Reunert (2009) Díaz (2009), Alcántara Rojas (2011), Valdovinos (2011), Mikulska (2008, 2015), and Neurath (2015).

Some of these works were presented and discussed at the Ciclo de Conferencias "Cosmologías indígenas, nuevas aproximaciones," which took place in the Museo Nacional de Antropología in Mexico City at 2011. It was during those two intense days of discussions and fruitful exchange of ideas and data that the idea of a book on the topic began to emerge, deriving from the collective volume that preceded the present edition (Díaz, ed. 2015). After years of continued fruitful discussions, meetings, and exchanging of ideas and data, we decided to integrate our reflections into a single volume and to invite other specialists to collaborate in the conversation. Common to our renewed attempts to understand Mesoamerican cosmological models were three main points of departure: (1) A reevaluation of precolumbian *vis-à-vis* colonial sources, (2) a critical revision of the idea of a multilayered universe composed of 13 and nine vertically arranged layers or levels, and (3) the proposal of new cosmographic alternatives to understand the composition of the Mesoamerican cosmos. This is essentially what this book is about: offering a new approach to a classic subject in Mesoamerican studies, namely the indigenous conception of the cosmos. Specifically, we are interested in reviewing the relationship between the spaces of sky, earth, and underworld.

By reanalyzing and recontextualizing the relevant sources, most of the contributing authors discuss and challenge the commonly accepted notion that these spaces were conceived of as fixed, static structures of superimposed levels unrelated to and unaffected by historical events and human actions. Instead, we propose that cosmological spaces were, and are, dynamic elements shaped, defined, and redefined throughout the course of history. The chapters in this volume thus aim to show that indigenous cosmographies could be subdivided and organized in complex and diverse arrangements, and that their constituent parts were constructed through modular articulations in orderly succession: 13 spaces, nine spaces, four cardinal points and a center, four or five color patterns, supernatural entities and phenomena, calendrical signs, and two daily qualities, that is, day and night.[1] These diverse arrangements are not fixed models, but rather components in a dynamic interplay, which cannot be adequately understood if the cosmological discourse is reduced to a superposition of nine and 13 levels.

Therefore, the segmentation or compartmentalization of the sky and the underworld cannot be reduced to the opposition of schematic horizontal and vertical arrangements. Indeed, these otherworlds are inhabited by a number of familiar phenomena or features, such as mountains, springs, palaces, roads, caves, trees, animals, and ancestors, which are also present on earth and can be recognized in the experienced environments of Mesoamerican communities.[2] All these elements, reflecting a shared daily reality, were, and in many areas continue to be, replicated in an almost specular fashion onto other cosmological levels. Through this type of process complex historical and regional cosmologies were, and are, generated.

In order to deemphasize and question some of the analytical categories that tend to predispose our understanding of these otherworldly spaces, the authors offer detailed analyses of the varied cosmological traditions in Mesoamerica, drawing upon a wide range of sources and data, including precolumbian texts and imagery, archaeological excavations, and ethnohistorical documents from the early colonial period, as well as ethnographic descriptions of different contemporary groups, focusing on the Maya, Nahua, Zapotec, and Wixáritari. Furthermore, several contributions offer historiographical background regarding the central cosmological concepts and models from medieval and renaissance Europe, thus extending the scope of our observations beyond indigenous forms of thought.

The contributing authors of this volume agree with the idea that documentary sources have their own independent cultural life, becoming part of different orders of discourse throughout history. One of our principal interests is thus to emphasize the historical contexts under which the cosmological models that have come down to us were envisioned and formulated. Our intention is not to propose another universalist model of Mesoamerican cosmology, but rather to call attention to the likely multitude of cosmographic repertoires or models that operated simultaneously as a result of historical circumstances and regional variations.

MESOAMERICAN COSMOS: FIXED STRUCTURES OR MUTABLE SPACES?

For late sixteenth-century missionaries—such as the compiler of the *Codex Vaticanus A*, Fray Pedro de los Ríos—and for late nineteenth- and early twentieth-century Mesoamericanists, the Mesoamerican universe consisted of three basic, clearly defined, and differentiated spaces. According to them, the sky was conceived of as a well-lit, diaphanous place, the underworld as a dark, ominous region to which the majority of the dead are confined, and the earth as an intermediate, neutral space inhabited by men. This arrangement seems to coincide with the cosmological structure shared by all the socalled ancient civilizations, such as the Egyptian, the Sumerian, and ancient Greek, suggesting that they were all different versions of a universal cosmological structure (Jung 1984; Eliade 1981; Seler 1907; Olivier 2010; Matos Moctezuma 2010).[3]

However, this structure is only one of many possible ways of presenting what could very well have been a much more dynamic and less static cosmological model, as evidenced by ethnographies from various regions of Oceania,

Siberia, and the Amazonian lowlands, which *destabilize the ontological premises of classical studies of literate societies and civilizations* (Wagner 1991; Coppet and Iteanu 1995; Gell 1995; Mondragón 2015). Indeed, the notion that there is one sky (heaven) that remains the same—a perennial, immutable space—is mainly compatible with and derived from a Western tradition and Aristotelian metaphysics (Aristotle 1987). Several of the contributions to this volume identify and discuss aspects of Mesoamerican cosmology, specifically the earth, sky, and underworld, as highly complex spaces with territorial limits and qualities that do not seem to be wholly compatible with the Euro-Christian vision of the universe and its constituents.

First, several authors have suggested that rather than three clearly divided, vertically arranged spaces, ancient Mesoamerican peoples recognized an opposition between "above" and "below," or perhaps rather between "here" (the earth) and "there" (the night/day sky/dream space/the remote past) (see for example Knab 1991; Mikulska 2015; Neurath 2015). This might not be understood as a static binary-opposition pattern, but as a conception that incorporates movement and change as part of the ontological machinery. Mikulska, Tokovinine and Neurath analyze this topic in detail in chapters 7, 8, and 9 of this volume, respectively. The apparent contradictions of the colonial sources—which refer to a heaven divided into sometimes nine, sometimes 13, sometimes seven or five partitions—make sense if we span our approach to incorporate multiple levels of semantic interaction, assuming that the cosmos can be ordered and approached in different ways depending on the context in which the agents relate to it. We seem to be faced with a fragmentation of the cosmological geography into a range of possibilities: two, seven, nine, or 13 regions, all depending on the historical moment and the specific cultural or even ritual context in which they are presented and named (in this volume, see Díaz, chapter 3; Hull, chapter 6; Mikulska, chapter 8; Nielsen and Sellner Reunert, chapter 1; Neurath, chapter 9; Tokovinine, chapter 7; Vail, chapter 4).

Second, the boundaries between these spaces question the image of discrete, separated areas. For example, both the sky and the underworld are home to a variety of beings (supernatural entities including gods, ancestors, and souls) who do not seem to have unique and fixed spheres of action. The geographical composition of these spheres also reveals a common organization of the places, which resembles the distribution of spaces on earth (palaces, gardens, springs, rivers) as once proposed by Knab in his ethnographic approach to the Nahuas of Puebla (1991, 2004). This argument is taken far in the chapter written by Tokovinine (chapter 7, this volume), who reveals the interweaving connections between linear and temporal distances in historical Maya narratives in precolumbian texts. In another example, Tavárez (chapter 5, this volume) shows how ancient concepts of time, arranged in the Zapotec *biyé* (a count of 260 days), set in motion a pattern of days that walk along three houses (earth, sky, and the underworld), distributing time among the vertical layers of the cosmos. Tavárez proposes that the movement of days by the houses of the cosmos can be taken as evidence of the division of the upper and lower spaces in nine fixed levels above and nine below Earth.

Botta (chspter 2), Díaz (chapter 3), Neurath (chapter 9), and Nielsen and Sellner Reunert (chapter 1) analyze the conformation of colonial Amerindian cosmological programs (Mexican and Maya) that took as reference modern metaphysical conceptions. The imposition of foreign taxonomies and discursive forms to understand and articulate regional knowledges and experiences produced universal explanations that usually took for granted the voices of the "subjects of study." To show some examples, precolumbian groups of the southern Maya lowlands seemed to emphasize the measures of time spent in travel to identify the nature of places (Tokovinine, chapter 7, this volume); Neurath discusses the constitution of the body as one of the main paradigms that affect the Wixarika social cosmos and its mutable composition (chapter 9, this volume). The organic quality of the otherworlds is underlined in the contributions of Díaz (chapter 3) and Mikulska (chapter 8) of this volume, and the complexity of the taxonomical orders observed in Maya calendrical practices—in operation before and after the contact with Christian theology—are discussed in the essays by Vail (chapter 4) and Hull (chapter 6) in this volume. These works show a fluid incorporation of principles, elements, and characters that update the complexity of American traditional chronological systems.

Finally, most of the contributions show that the analysis of colonial sources could be very useful for understanding the processes of translation, adaptation, and reconfiguration of the world, as they reveal the complex strategies used by the colonial Maya, Zapotec, or Mexica to incorporate ideas and practices that were congruent with their contemporary paradigms. Indeed this exercise shows an extraordinary example of the way in which the opposition between purely precolumbian and Christian cosmologies breaks down.

These descriptions call for a new way of conceptualizing the Mesoamerican cosmos in its different contexts, especially considering that in precolumbian times the arrangement of space (cosmological and earthly territory) was guided by patterns that possibly have not been recognized because they do not follow the discursive forms of classic and Judeo-Christian cosmologies, even when they incorporate Christian elements, which were resignified.

BACKGROUND AND METHODOLOGICAL CONSIDERATIONS

The approach to the study of Mesoamerican cosmology presented in this volume and its underlying methodological and theoretical considerations are part of a broader framework within the field of Mesoamerican and early colonial studies. Several of the ideas and interpretations presented in the following chapters, as well the methodological approach to this topic, follow earlier, important contributions to the field dealing sometimes with the Americas as a whole, but mainly with either central Mexico or the Maya region. Specifically, we draw upon earlier interpretations of the transcultural processes that took place from the colonial period onwards and that lie at the heart of questions regarding cosmology in Mesoamerica. In methodological terms, we approach our sources critically, acknowledging the important part played by the native indigenous population in early colonial society in creating them, as well as their variability.

Recently, a growing awareness of the close collaboration and mutual influence between Mesoamerican and Euro-Christian individuals, religious concepts, cultural practices, and languages has changed previous views (e.g., Ricard 1966; Bricker 1981; Levin and Navarrete 2007; Schroeder 2010; Wolf, Connors, and Waldman 2011; Rappaport and Cummins 2012) about the period and the formation of postconquest Mesoamerican colonial societies. In this volume, the contributions that are focused on the early colonial period and on the sources produced in this dynamic time period follow in the path of a series of highly influential works that have reinterpreted the contact period and the colonial sources, asking new questions of old issues and challenging long-held views. Some of the most important include the seminal works by James Lockhart (1992) on the Nahuas after the conquest, those by Nancy Farriss (1984), Inga Clendinnen (1987), and Matthew Restall (1997) on the Maya in the aftermath of the Spanish invasion, as well as a series of works offering new perspectives on the roles of indigenous populations during the conquest (see Restall 1998, 2003; Matthew and Oudijk 2007; Restall and Asselbergs 2007).

A number of studies have focused on the religious beliefs and practices of the postconquest period, including the reinterpretation and appropriation of Euro-Christian ideas by indigenous populations, as well as indigenous roles in the production of religious images in books and on the walls of churches and monasteries in New Spain (e.g., Mignolo 1995; Peterson 1993, 1995; Schroeder and Poole 2007; and several contributions in Cecil and Pugh 2009; Olivier 2010). Furthermore, an increasing number of articles and books have been published specifically on the topic of the syncretistic process of intermingling of Old and New World religious ideas[4]—that is, the adoption and remodeling of Christian themes and concepts that led to the creation of what can be called Mexican Catholicism(s), including what is sometimes referred to as Nahua Christianity and Maya Catholicism (Christensen 2013).

The issue of conversion and the question of the transmission of religious ideas and concepts across language boundaries have been closely examined. It was first treated in depth in Louise Burkhart's classic book *The Slippery Earth: Nahua–Christian Dialogue in Sixteenth-Century Mexico* (Burkhart 1989). This work is a milestone in the research on the evangelization process in New Spain (see also Dehouve 2004; Díaz Balsera 2005). It deals with the highly complex situation of transferring Christian concepts onto an indigenous language, a subject also investigated by other scholars (Anderson 1993; Christensen 2013; Kirkhusmo 2007, 2009; Alcántara Rojas 2008). Jaime Lara's two impressive volumes (Lara 2004, 2008) on the architecture, art, and liturgy in early colonial New Spain are recent landmarks in the research on the complex cultural interactions and influences that together formed the basis for the production of religious buildings, texts, and images in colonial Mexico.

However, for a better understanding of transcultural processes it is important to emphasize that reinterpretation of Christian ideas in indigenous cosmologies was the result of a "struggle for recognition," not just a form of cultural hybridization (Dean and Leibsohn 2003; also see Botta, chapter 2, this volume). The work of David Tavárez shows an interesting example of how the original practices, such as those related to the use of the *biyé*, were affected not directly by the introduction of Christian concepts but by the transformation of the society in colonial times (Tavárez 2011). He investigates, for example, how the introduction of new technologies (such as alphabetic writing) and historical data (such as the introduction of Christian characters into originally indigenous stories) were incorporated to reproduce Zapotec cultural forms (Tavárez 2007).

We know that tapestries showing the most fundamental Christian concepts were annotated with glosses in native languages (using Latin letters) to instill in them the divine doctrine, and the invention of new liturgic genres also allowed for potential overlaps and reinterpretations between the two sets of religious images, discursive forms, and terminologies (Báez 2005; Schwaller 2006; see Nielsen and Sellner Reunert, chapter 1, this volume and Botta, chapter 2, this volume). The importance of images in the conversion process, as well as the indigenous population's contribution to the production of religious images, has been discussed in depth by Constantino Reyes-Valerio (1978, 2000), among others. Recent works on material and visual culture offer new perspectives for understanding transcultural dialogue that transcends the notion of hybridism (Dean and Leibsohn 2003), as well as conceptions of absence and resistance (Mundy and Leibsohn 2012).

The religious transcultural dialogue evident in the colonial visual and material culture by which Christianity was taught, leads to the question of European influence in early colonial sources on Mesoamerican cultures and languages, and thus to a questioning of the use of these sources in our attempts to reconstruct precolumbian beliefs and practices. Constance Cortez, for instance described colonial art as an example of a cultural discourse (Cortez 2002a), using the Yucatec Maya scribe and artist Gaspar Antonio Chi, best known for his collaboration with Bishop Diego de Landa, as an example of an individual who could create works "that could be visually and mentally accessed by both cultures" (Cortez 2002b, 200).

In the colonial imagery, we indeed see how Mesoamericans strived to incorporate the old, sometimes very literally, within the new, a process that the friars often allowed to take place. Quiñones Keber previously noted that the Mesoamerican cosmologies reproduced in colonial sources were restructured for European consumption as consciously constructed cultural objects (Quiñones Keber 1995). Similarly, studies by Victoria Bricker and Helga-Maria Miram (Bricker and Miram 2002), as well as those by Erik Velásquez García (2009) and Timothy Knowlton (2009), have examined the Maya books of *Chilam Balam* in detail and demonstrated the influence of Euro-Christian culture, including imagery, in these pivotal sources on Maya religion, mythology, and cosmology. In the central Mexico region, George Kubler and Charles Gibson (1951) followed the transformations of the Mexican calendrical graphic discourse as a result of the introduction of Christian figurative cycles, a topic also studied by Betty Ann Brown (1977), Susan Spitler (2005), Ana Díaz (2011) and Anthony Aveni (2012).

A number of the best-known and widely used colonial sources, such as Bernardino de Sahagún's *Florentine Codex*, including their accompanying images, have been widely used as a visual "window" into the precolumbian past, in the sense that they have been thought to convey genuine precolumbian concepts and practices. However, it has become increasingly clear that a considerable part of these images are complex in character and often draw heavily upon Western, Euro-Christian templates. Sahagún's *Florentine Codex* has been well studied in this regard by authors such as Edmonson (1974), Klor de Alva, Nicholson, and Quiñones (1988), Schwaller (2003), and in particular by Pablo Escalante (1999, 2003, 2008) and Diana Magaloni (2003, 2004, 2011). Escalante has demonstrated the frequency with which the indigenous illustrators of the *Florentine Codex* found templates and models for their representations of precolumbian Aztec culture in European books (e.g., the Bible and volumes on natural history such as the *Hortus Sanitatis*) that were kept in the monastery schools such as the College of Santa Cruz in Tlaltelolco (Escalante 1999, 2003, 2008; Bremer 2003; Boone 2003). Diana Magaloni revealed the ways in which indigenous painters of the *Florentine Codex* adopted certain European visual repertoires, techniques, and pigments in order to create new images, which were coherent with their own past traditions as well as with the new world that was taking form (Magaloni 2003, 2004, 2011; see also Wolf and Connors 2011).

In addition, recent studies related to the continuity and transformation of sacred space, including landscape and the built environment, further suggest that "native observances" were far from forgotten, but formed an essential component of the emerging colonial religious practices and beliefs (e.g., Solari 2013). Thus, the construction of Christian churches and monasteries were often constructed on sites that in precolumbian times had formed part of a sacred geography (Hermann 2015). Prime examples of such Christian appropriations of precolumbian sacred sites are Chalma in Morelos and numerous sites in in northern Yucatan, as pointed out by Ralph Roys in 1952. Thus, rededicating the sacred site allowed for continuity in the beliefs or the *genius loci* associated with the site itself and its relationship with the surrounding landscape.

In sum, it is these active and ongoing processes of appropriation, borrowing, restructuring, and seeking compromises, and the resulting hybrid forms, that are at the center of the present volume. The contributions all take into account the roles that the indigenous population in early colonial society played in creating (as authors and artists) the sources, as well as the inhomogeneity and variability of the sources upon which we draw (in terms of time, place, and ethnicity, for instance). Thus, we try to avoid the kind of overarching generalizations that lead to reductionist simplifications or universalistic interpretations. Finally, in this volume we have all strived to familiarize ourselves with Euro-Christian traditions, be they religious, literary, artistic, or scientific (while admitting that we still have much to learn); something that we as Mesoamericanists sometimes tend to forget in our focus on the precolumbian past and in our efforts to reconstruct and understand precolumbian ways of thinking and believing.

Creativity and Reinvention

Due to our interest in showing how the conceptions of cosmology have transformed over time, we must necessarily deal with concepts such as history, tradition, and creativity. It is not our intention to write a detailed analysis of the discussions held around these terms, but it is fundamental to clarify our perspective regarding these. All the cases analyzed in the present volume are located in history. Knowledge, even sometimes ancient knowledge, permeates a given culture from one generation to the next, but we cannot assume that the Maya from Classic-period Naranjo are the same as the Maya from eighteenth-century Yucatan. While it is possible that they shared a set of common ideas and practices that transformed through time, especially after the contact with Europeans, this explanation assumes cultural uniformity in Mesoamerica and refers to a fundamental or structural continuity of thought across several generations: even when it accepts changes through time, the original nucleus survives. In this volume, rather than perceiving shared ideas and practices as a static substrate in continuous resistance to external agents, we propose to think about this process as a series of emic strategies that allowed for the adoption and integration of new concepts and ideas as historical, religious, and sociopolitical conditions changed over time. This took place not only after contact with the Old World, but—as archaeological and historical sources of Maya, Nahua, Mixtecs, and other groups indicate—change, adaptation, and creation can be found in this part of the continent dating back to the Preclassic. Therefore, more than visualizing a common nucleus affected by the introduction of new ideas, we prefer to use the metaphor of a flexible net, where the designs are created by the combination of strings of different colors manipulated by human hands. The designs reproduce patterns according to tradition (the correct way of doing things), but there is always space for creativeness and personal decisions. This flexibility affects even the cosmological conceptions.

In other words, more than mere syncretism or hybridization, understood as a collision between two cultures that interact and absorb each other, we should perhaps conceive of forms of knowledge that are flexible, active, and procedural. They make possible the integration of new knowledge, be it external or internal to the community, in order to update previous knowledge and experience to their specific conditions of life. Thus, the emphasis is not on cultural influence, but on historical suitability. While recognizing the stability and endurance of a substratum of religious beliefs, including cosmic models, over long periods of time, we must strive to refine our analysis by framing such phenomena in their specific historical context. Approaching the sources—precolumbian as well as colonial and present day—from such a perspective, it becomes apparent that descriptions and representations of the cosmos and its constituents are best conceived of as emerging, unstable products, and as the results of creative action in different places and historical moments. It is important to emphasize once again, that none of the contributors seeks to propose a new, all-encompassing "great model" that suggests a worldview shared by all Mesoamerican cultures.

Although several fundamental aspects were undoubtedly held in common, there was probably never one, stable pan-Mesoamerican vision of the universe from the Preclassic to the Postclassic. By creatively combining insights from previous historical and anthropological studies of Mesoamerican history with the analysis of Amerindian and Euro-Christian text and imagery, archaeological remains, and linguistics, this volume hopes to widen the analytical scope to reflect upon and cast new light on how the Mesoamerican cosmos was structured and restructured through history, but probably never so profoundly as in the aftermath of the Spanish conquest.

THE MESOAMERICAN COSMOS REVISITED: CONTRIBUTIONS OF THE VOLUME

The present volume is arranged in nine chapters, which are distributed in three symmetrical sections. While the chapters have been arranged according to key concepts, they have many more themes in common, and can therefore be connected in multiple ways.

I. Recognition: On Describing Others´ Worlds

Jesper Nielsen and Toke Sellner Reunert (chapter 1) open the first section by approaching the overall problem of reconstructions of precolumbian visions of the cosmos that rely mainly on colonial sources, but without thorough, critical analyses of the historical context in which they were produced. The authors suggest that a generalization has taken place regarding the idea of a multilayered Mesoamerican universe, showing that this cosmic structure with 13 layers in the sky and nine or seven in the underworld has been inferred primarily from postcolumbian central Mexican sources and not from precolumbian evidence such as Maya hieroglyphic texts or iconography. Second, and more important, the present coauthors elaborate on their already-published hypothesis (2009) that the notion of a multilayered universe was not a Mesoamerican concept in the first place. They thus suggest that this cosmic model was introduced into the area in the sixteenth century, and that it ultimately derives from a European vision of the cosmos, referred to as the Dantean worldview, which was widely accepted in southern Europe at the time of the conquest.

The contribution of Sergio Botta continues this discussion. In the second chapter, Botta introduces the *aggionta* (addition) of the Italian antiquarian Lorenzo Pignoria to the mythographic Renaissance work of Vincenzo Cartari—the *Imagini delli dei de gl'antichi*—which is one of the most original attempts to shape a global religious comparative methodology in the modern age. The *aggionta*, published for the first time in 1615, adds illustrations and descriptions of gods and idols from Mexico, Japan, and India to Cartari's work. As for the Mesoamerican gods, Pignoria collected some images from the famous *Codex Vaticanus A* and invented a syncretic cosmology that combined indigenous and Christian features. Embodying Mesoamerican cosmology in a global representation of idolatry, Pignoria sketched a peculiar historical development of religion, which functioned mainly for colonial purposes. As argued by Botta here, it was through such comparative reflection, unique in its genre, that Pignoria contributed to building a modern and global language of religion.

The third chapter, by Ana Díaz, contrasts the visual discourses of the Mexican cosmos in colonial and precolonial sources, focusing in the configuration of the sky. The chapter offers an examination of the content, context, and production process of the early colonial sources that describe the Mexican cosmography, focusing on the reconstruction strategies used by the documents' authors to give shape to a completely new indigenous cosmological conception. First, Díaz analyzes the descriptions of the cosmos written and depicted in early colonial sources, exploring the contradictions present in them, then she explores the possibility of identifying the original images that may have served as their model. Second, by examining prehispanic visual sources and focusing on Citlalinicue—a main character in the foundational narratives or cosmogonies, who is recognized as one aspect of the sky—Díaz argues for the conception of a cosmos as a living entity. The sky emerges as a body and a threshold: as an entity who assumes different aspects or qualities, revealing a multiple and dynamic constitution. In résumé, Citlalincue is present in most of cosmogonical narratives as a main character, but she seems to be disarticulated and displaced in order to build a cosmographical fixed structure with her iconographical elements.

II. Inventiveness: Reshaping Experience in Colonial Cosmologies

The fourth contribution, by Gabrielle Vail, documents how conceptions of Maya supernatural entities changed between the fourteenth and the eighteenth centuries and influenced scholarly interpretations of prehispanic cosmology.

The analysis led Vail to explore colonial conceptions of the Maya underworld (Metnal or Xib'alb'a), drawing comparisons with prehispanic sources and Christian conceptions. The chapter examines the concepts of *ángel* (angel) and *diablo* (devil), introduced by Landa and included in later Yucatec language texts such as the Books of *Chilam Balam*. Through the analysis of different series of entities, Vail concludes that Mayan deities were considered to have existed in multiple realms, but after the introduction of Christianity, it became imperative to separate three realms and assign specific supernatural entities to each space, a view that is more compatible with a Christian worldview.

In the fifth chapter of the volume, David Tavárez offers an interpretation of the connection between time (*biyé*) and space in Zapotec cosmological conceptions. This essay provides the first analysis of a correlation between the feasts in the 260-day ritual calendar and various cosmological locations, as articulated in the corpus of Zapotec texts. The northern Zapotec manuals established a spatiotemporal continuum that linked time in the 260-day calendar with locations in a three-tiered cosmos, in which Tavárez identifies a further nine levels above and nine below Earth. Therefore, this text contrasts with the position of other authors of the volume, in that Tavárez seeks to demonstrate that the notion of a cosmos structured in nine upper and nine lower layers was a model that remained from prehispanic times to the eighteenth century. This chapter is fundamental to understanding that in the American continent, conceptions of time and space were deeply intermingled, in contrast to the cosmological explanations of the world produced in Europe and ancient Near East.

The last chapter of this section, chapter 6 by Kerry Hull, moves from the colonial period to contemporary time. It is an ethnographic work of synthesis about the worldview of the Ch´orti´. The work was included in this section because it dialogues with the previous two chapters, showing the way in which Maya communities adapted and updated knowledge from different cultural contexts, which is in continuous transformation in response to specific historical phenomena. Hull shows how, over more than a hundred years, there has been a steep decline in native ceremonialism and traditional religious rites, which could be attributed to three factors: (1) missionary work by the Catholic church and by evangelical Christian groups, (2) a climate of fear surrounding the labeling of traditional practices as "witchcraft" (*brujería*), which often results in the murder of the practitioner, and (3) a "folklorization" of traditional Ch'orti' beliefs by the Guatemalan government.

The text offers a reconstruction of the cosmology of the Ch'orti', a Maya group of southern Guatemala who have in their mythology a vast array of otherworld beings who live, travel, or manifest themselves in water. Therefore, in many circumstances water is feared and avoided since it is conceptually linked to the presence of supernaturals—some good, but most malicious. In Ch'orti' thought, water is the primary portal between realms and, as a consequence, it operates as a facilitator of negative events in the lives of the Ch'orti'. In addition, water figures prominently in the cosmogonic structure of the heavens as well as the earth. Creation and global destruction in Ch'orti' mythology are acts of separating, manipulating, and organizing water. Furthermore, certain astral bodies are closely associated with water, both as bringers of life-giving rains, but also as conduits for evil spirits who afflict humans on earth. Finally, there are other cultural conceptions associated with cosmological phenomena in Ch'orti' society that create a causal link between events in the sky and resulting effects in the daily affairs of the populace.

III. Complexity: Breaking Paradigms of Cosmological Conceptions

As announced by the subtitle, this section offers paradigmatic cases that were brought into light when the authors tried to follow new guides to understanding alternative cosmologies as presented in the sources. The section opens with chapter 7, Alexander Tokovinine's contribution, the only one that focuses exclusively on the precolumbian period, and therefore, one of the richest diamonds of this compilation.

Tokovinine explores Classic Maya narratives, highlighting travels to distant places as crucial life-changing events in personal biographies and in histories of entire dynasties. The chapter begins by addressing the problem of defining spatial, temporal, and social distances from the perspective of Classic Maya inscriptions. This work reveals that the narratives of origins and pilgrimage evoke distances on different scales, from places in deep time and historic locations, beyond the confines of the Classic Maya world, to the nearby royal courts. Thus, Tokovinine shows the connection between geosocial paths and nets, emphasizing the importance of working with primary sources to understand the principles of operation that connect cosmological conceptions with deep history and lived experience.

The eighth chapter, by Katarzyna Mikulska, contributes to our understanding of the *imago mundi* as expressed by the ancient inhabitants of central Mexico in the Late Postclassic and early colonial periods. Based on the analysis of the otherworld or the supernatural spheres, Mikulska suggests that there was no clear distinction between "up" and "down" that could correlate with the fundamental dialectical cosmological scheme associated with heaven and hell in Euro-Christian thinking. For the Nahua, the difference seems instead to be based on the opposition of the night sky, the region of divine and primordial time, with the day sky, region of human and structured time. In this conception, the world of the dead ancestors, an eternal source of life and fertility, can be situated in the night sky as well as in the interior of the earth, a time-space of dreams and the unconscious. As such, it can be referred to with the number "nine," which leads to the second issue analyzed in this chapter: the possibility that numbers had semantic value and as such were used to name places and divinities.

The ninth and last work of the volume, by Johannes Neurath, posits a criticism of the tradition of studying indigenous worldviews as collective representations about nature. It questions the homogeneity of conceptions that supposedly are shared by the members of a society. Instead, it advocates for a treatment of Amerindian *Lebenswelten*. As pointed out by the author: "the simple idea to reconstruct the Huichol view of the world is problematic. Neither should cosmology be understood as a system of concepts about nature. Important aspects of the cosmos are made during ritual, so creation is an ongoing event that is not separable from ritual action." In the study of *wixarika* worldview, one could start from a geographical contrast between "above" and "below," between the *semidesert* in the east called Wirikuta and the fertile coast plains in the west. However, this elaborate system of analogies exhibits important asymmetries. By taking into account the analysis of relations and practices, Neurath seeks to demonstrate how important parts of the cosmos are not considered a natural given, but a product of ritual action. These parts are always unstable, and their existence ephemeral. Ritual action does not only create, it also destroys.

In sum, throughout its chapters this volume seeks to encourage colleagues and students to approach the subject of the Mesoamerican cosmos through new analytical, empirical, and methodological perspectives, so as to make the most of data from images, written texts, archaeological objects, and oral traditions that can provide information on these intriguing otherworldly spaces commonly identified as the three realms of the world.

NOTES

1. A concept that seems to be very useful to explain this dynamic is a type of arrangement identified by James Lockhart (1992) as cellular or modular organization. It implies an orderly, symmetrical, numerical succession of parts forming larger units.

2. See the examples of the cosmological configuration described by Tim Knab (1991, 2004) and Pedro Pitarch (1996) in their fieldwork in Nahua and Maya communities. A similar dynamic can be found in colonial descriptions such as the *Popol Wuj*.

3. Eduard Seler was influenced by the mythical-astronomical ideas of the "lunar school of mythological interpretation" led by another famous, contemporary German scholar, Ernst Siecke (Nicholson 1990).

4. For early forays into the discussion of religious syncretism in Mesoamerica, see Madsen (1960) and Thompson (1960).

REFERENCES

Alcántara Rojas, Berenice. 2008. "Cantos para bailar un cristianismo reinventado: La nahuatlización del discurso de evangelización en la Psalmodia Christiana de Fray Bernardino de Sahagún." PhD diss., Universidad Nacional Autónoma de México.

Alcántara Rojas, Berenice. 2011. "Ilhuicac, tlalticpac y mictlan: Cosmologías nahuas y cristianas en el discurso de evangelización del siglo XVI." Paper presented at the Ciclo de Conferencias "Cosmologías indígenas, nuevas aproximaciones." Museo Nacional de Antropología, Mexico, January 27.

Anderson, Arthur J. O. 1993. *Bernardino de Sahagún's Psalmodia Christiana (Christian Psalmody)*. Salt Lake City: University of Utah Press.

Aristotle. 1987. "Physical Treatises." In *The Works of Aristotle*, vol. 1, ed. Robert Maynard Hutchins, 259–498. Great Books of the Western World. London: Encyclopaedia Britannica.

Aveni, Anthony. 2012. *Circling the Square. How the Conquest Altered the Shape of Time in Mesoamerica*. Philadelphia, PA: American Philosophical Society.

Báez, Linda. 2005. *Mnemosine Novohispánica. Retórica e imágenes en el siglo XVI*. México, Instituto de Investigaciones Estéticas, UNAM.

Baird, Ellen T. 1993. *The Drawings of Sahagún's* Primeros Memoriales: *Structure and Style*. Norman: University of Oklahoma Press.

Boone, Elizabeth Hill. 2003. "The Multilingual, Bi-visual World of Sahagún's Mexico." In *Sahagún at 500: Essays on the Quincentenary of the Birth of Fr. Bernardino de Sahagún*, ed. John F. Schwaller, 137–166. Berkeley, CA: Academy of American Franciscan History.

Bremer, Thomas S. 2003. "Sahagún and the Imagination of Inter-Religious Dialogues." In *Sahagún at 500: Essays on the Quincentenary of the Birth of Fr. Bernardino de Sahagún*, ed. John F. Schwaller, 11–29. Berkeley, CA: Academy of American Franciscan History.

Bricker, Victoria R. 1981. *The Indian Christ, the Indian King: The Historical Substrate of Maya Myth and Ritual*. Austin: University of Texas Press.

Bricker, Victoria, and Helga M. Miram. 2002. *An Encounter of Two Worlds: The Book of Chilam Balam of Kaua*. New Orleans, LA: Tulane University–Middle American Research Institute.

Brown, Betty Ann. 1977. "European Influences in Early Colonial Descriptions and Illustrations of the Mexican Monthly Calendar." PhD diss., University of New Mexico, Albuquerque.

Burkhart, Louise. 1989. *The Slippery Earth: Nahua-Christian Moral Dialogue in Sixteenth-Century Mexico*. Tucson: University of Arizona Press.

Caso, Alfonso. 1967. *Los calendarios prehispánicos*. Mexico: Universidad Nacional Autónoma de México, Instituto de Investigaciones Históricas.

Cecil, Leslie G., and Timothy W. Pugh, eds. 2009. *Maya Worldviews at Conquest*. Boulder: University Press of Colorado.

Christensen, Mark. 2013. *Nahua and Maya Catholicisms: Text and Religion in Colonial Central Mexico and Yucatan*. Stanford, CA: Stanford University Press.

Clendinnen, Inga. 1987. *Ambivalent Conquests: Maya and Spaniard in Yucatan, 1517–1570*. Cambridge: Cambridge University Press.

Códice Vaticano A. 1996. Edited by Ferdinand Anders, Maarten Jansen, and Luis Reyes García. Graz: Akademische Druck- und Verlagsanstalt, and Mexcio: Sociedad Estatal Quinto Centenario, Fondo de Cultura Económica.

Coppet, Daniel, and André Iteanu, eds. 1995. *Cosmos and Society in Oceania*. Oxford and Washington, DC: Berg.

Cortez, Constance. 2002a. "The Indigenous Presence in the Colonial Visual Culture of Mexico and the Southwest." In *Telling the Santa Clara Story: Sesquicentennial Voices*, ed. Russell K. Skowronek, 28–44. Santa Clara, CA: Santa Clara University/City of Santa Clara.

Cortez, Constance. 2002b. "New Dance, Old Xius: The 'Xiu Family Tree' and Maya Cultural Continuity after European Contact." In *Heart of Creation: The Mesoamerican World and the Legacy of Linda Schele*, ed. Andrea Stone, 201–215. Tuscaloosa: University of Alabama Press.

Dean, Carolyn, and Dana Leibsohn. 2003. "Hybridity and Its Discontents: Considering Visual Culture in Colonial Spanish America." *Colonial Latin American Review* 12 (1): 5–35.

Dehouve, Danièle. 2004. *L'évangélisation des Aztèques ou le pécheur universel*. Paris: Maisonneuve & Larose.

Díaz, Ana. 2009. "La primera lámina del Códice Vaticano A. ¿Un modelo para justificar la topografía celestial de la antigüedad pagana indígena?" *Anales del Instituto de Investigaciones Estéticas* 31 (95): 5–44.

Díaz, Ana. 2011. "Las formas del tiempo. Tradiciones cosmográficas en los documentos calendáricos indígenas del México central." PhD diss. Mexico: Universidad Nacional Autónoma de México, Instituto de Investigaciones.

Díaz, Ana. 2015. "La pirámide, la falda y una jicarita llena de maíz tostado: Una crítica a la teoría de los niveles del cielo mesoamericanos." In *Cielos e inframundos; una revisión a la Cosmología Mesoamericana*, coordinated by Ana Díaz, 65–107. México: Universidad Nacional Autónoma de México Instituto de Investigaciones Históricas, Fideicomiso Felipe Teixidor y Montserrat Alfau de Teixidor.

Díaz, Ana, ed. 2015. *Cielos e inframundos: una revisión de las cosmologías mesoamericanas*. Mexico: Universidad Nacional Autónoma de México, Instituto de Investigaciones Históricas, Fideicomiso Felipe Teixidor y Montserrat Alfau de Teixidor.

Díaz Balsera, Viviana. 2005. *The Pyramid under the Cross: Franciscan Discourses of Evangelization and the Nahua Christian Subject in Sixteenth-Century Mexico*. Tucson: University of Arizona Press.

Edmonson, Munro S., ed. 1974. *Sixteenth-Century Mexico: The Work of Sahagún*. Albuquerque: School of American Research, University of New Mexico.

Eliade, Mircea. 1981. *Tratado de historia de las religiones*. Mexico: Era.

Escalante Gonzalbo, Pablo. 1999. "Los animales del Códice florentino en el espejo de la tradición occidental." *Arqueología Mexicana* VI (36): 52–59.

Escalante Gonzalbo, Pablo. 2003. "The Painters of Sahagún's Manuscripts: Mediators between Two Worlds." In *Sahagún at 500: Essays on the Quincentenary of the Birth of Fr. Bernardino de Sahagún*, ed. John F. Schwaller, 167–191. Berkeley, CA: Academy of American Franciscan History.

Escalante Gonzalbo, Pablo, ed. 2008. *El arte cristiano-indígena del siglo XVI novohispano y sus modelos europeos*. Cuernavaca, Mexico: Centro de Investigación y Docencia en Humanidades del Estado de Morelos.

Farriss, Nancy M. 1984. *Maya Society under Colonial Rule: The Collective Enterprise of Survival*. Princeton, NJ: Princeton University Press.

Gell, Alfred. 1995. "Closure and Multiplication: An Essay in Polynesian Cosmology and Ritual." In *Cosmos and Society in Oceania*, ed. Daniel Coppet and André Iteanu, 21–56. Oxford, Washington DC: Berg.

Griffiths, Nicholas, and Fernando Cervantes. 1999. *Spiritual Encounters: Interactions between Christianity and Native Religions in Colonial America*. Lincoln: University of Nebraska Press.

Hermann, Manuel. 2015. *Configuracoines territoriales en la Mixteca*. Vol. I. Estudios de historia y antropología. México: CIESAS, Publicaciones de la casa Chata.

Holland, William. 1961. "Relaciones entre la religión tzotzil contemporánea y la maya Antigua." *Anales del Instituto Nacional de Antropología e Historia* 13: 113–131.

Holland, William. 1962. "Highland Maya Native Medicine: A Study of Cultural Change." PhD diss., University of Arizona, Tucson.

Holland, William. 1963. *Medicina maya en los Altos de Chiapas*. Mexico: Instituto Nacional Indigenista.

Jáuregui, Jesús. 2008. "Quo vadis, Mesoamérica? Primera parte." *Antropología: Boletín Oficial del Instituto Nacional de Antropología e Historia* 82: 3–31.

Jung, Carl. 1984. *El hombre y sus símbolos*. Barcelona: Luis de Caralt.

Kirkhusmo Pharo, Lars. 2007. "The Concept of 'Religion' in Mesoamerican Languages." *Numen* 54 (1): 28–70.

Kirkhusmo Pharo, Lars. 2009. "Translating Non-Denominational Concepts in Describing a Religious System: A Semantic Analysis of Colonial Dictionaries in Nahuatl and Yucatec." *Historiographia Linguistica* 36 (2/3): 345–360.

Klein, Cecelia. 1982. "Woven Heaven, Tangled Earth: A Weaver's Paradigm of the Mesoamerican Cosmos." In *Ethnoastronomy and Archaeoastronomy in the American Tropics*, ed. Anthony F. Aveni, and Gary Urton, 1–35. Annals of the New York Academy of Sciences 385. New York: New York Academy of Sciences.

Klor de Alva, José Jorge. 1993. "Aztec Spirituality and Nahuatized Christianity." In *South and Meso-American Native Spirituality: From the Cult of the Feathered Serpent to the Theology of Liberation*, ed. Gary H. Gossen, 173–197. New York: Crossroad Publishing Company.

Klor de Alva, José Jorge, Henry B. Nicholson, and Eloise Quiñones Keber, eds. 1988. *The Work of Bernardino de Sahagún: Pioneer Ethnographer of Sixteenth-Century Aztec Mexico*. New York: Institute for Mesoamerican Studies, University at Albany.

Knab, Tim. 1991. "Geografía del inframundo." *Estudios de Cultura Náhuatl* 21: 31–58.

Knab, Tim. 2004. *The Dialog of Earth and Sky*. Tucson: University of Arizona Press.

Knowlton, Timothy W. 2009. "Composition and Artistry in a Classical Yucatec Maya Creation Myth: Prehispanic Ritual Narratives and Their Colonial Transmission." *The Maya and Their Sacred Narratives: Text and Context in Maya Mythologies*, ed. Geneviève Fort, Raphaël Gardiol, Sebastian Matteo, and Christopher Helmke. *Acta Mesoamericana* 20: 111–121.

Knowlton, Timothy W. 2012. *Maya Creation Myths: Words and Worlds of the Chilam Balam*. Boulder: University Press of Colorado.

Knowlton, Timothy W., and Gabrielle Vail. 2010. "Hybrid Cosmologies in Mesoamerica: A Reevaluation of the Yax Cheel Cab, a Maya World Tree." *Ethnohistory* 57 (4): 709–739.

Krickeberg, Walter. 1950. "Bauform und Weltbild im Alten Mexiko." *Paideuma* 4: 295–333.

Kubler, George, and Charles Gibson. 1951. "The Tovar Calendar." *Memoires of the Connecticut Academy of Arts & Sciences*, Vol XI. New Haven, CT: Yale University Press.

Lara, Gerardo. 2016. *¿Ignorancia invencible? Superstición e idolatría ante el Provisorato de Indios y Chinos del Arzobispado de México en el siglo XVIII*. Mexico: Instituto de Investigaciones Históricas.

Lara, Jaime. 2004. *City, Temple, Stage: Eschatological and Liturgical Theatrics in New Spain*. Notre Dame, IN: University of Notre Dame Press.

Lara, Jaime. 2008. *Christian Texts for Aztecs: Art and Liturgy in Colonial Mexico*. Notre Dame, IN: University of Notre Dame Press.

León-Portilla, Miguel. 1966 [1956]. *La filosofía náhuatl estudiada en sus fuentes*. Mexico: Universidad Nacional Autónoma de México.

León-Portilla, Miguel. 1994 [1968]. *Tiempo y realidad en el pensamiento maya*. Mexico: Universidad Nacional Autónoma de México.

Levin, Danna, and Federico Navarrete, eds. 2007. *Indios, mestizos, españoles: Interculturalidad e historiografía en la Nueva España*. Mexico: Universidad Autónoma Metropolitana-Azcapotzalco, Universidad Nacional Autónoma de México, Instituto de Investigaciones Históricas.

Lockhart, James. 1992. *The Nahuas after the Conquest: A Social and Cultural History of the Indians of Central Mexico, Sixteenth through Eighteenth Centuries*. Stanford: Stanford University Press.

López Austin, Alfredo. 1984 [1980]. *Cuerpo humano e ideología*. Mexico: Universidad Nacional Autónoma de México, Instituto de Investigaciones Antropológicas.

López Austin, Alfredo. 1994. *Tamoanchan y Tlalocan*. Mexico: Fondo de Cultura Económica.

López Austin, Alfredo. 2001. "El núcleo duro, la cosmovisión y la tradición mesoamericana." In *Cosmovisión, ritual e identidad de los pueblos indígenas de México*, ed. Johanna Broda and Félix Báez-Jorge, 47–65. Mexico: Consejo Nacional para la Cultura y las Artes, Fondo de Cultura Económica.

López Austin, Alfredo. 2016. "La cosmovisión de la tradición Mesoamericana." Part 2. *Arqueología Mexicana*, Edición especial 69.

Madsen, William. 1960. "Christo-Paganism." In *Nativism and Syncretism*, ed. Margaret A. L. Harrison and Robert Wauchope, 105–176. New Orleans, LA: Tulane University, Middle American Research Institute.

Magaloni Kerpel, Diana. 2003. "Visualizing the Nahua/Christian Dialogue: Images of the Conquest in Sahagún's Florentine Codex and their Sources." In *Sahagún at 500: Essays on the Quincentenary of the Birth of Fr. Bernardino de Sahagún*, ed. John F. Schwaller, 193–221. Berkeley, CA: Academy of American Franciscan History.

Magaloni Kerpel, Diana. 2004. "Images of the Beginning: The Painted Story of the Conquest of México in Book XII of the Florentine Codex." PhD diss. Yale University, New Haven, CT.

Magaloni Kerpel, Diana. 2011. "Painters of the New World: The Process of Making the Florentine Codex." In *Colors between Two Worlds: The Florentine Codex of Bernardino de Sahagún*, ed. Gerhard Wolf and Joseph Connors, 47–76. Milan: Max Plank Institute/Villa I Tatti, and Harvard University Center for Italian Renaissance Studies.

Matos Moctezuma, Eduardo. 1987. "Symbolism of the Templo Mayor." In *The Aztec Templo Mayor*, ed. Elizabeth H. Boone, 185–210. Washington, DC: Dumbarton Oaks Research Library and Collection.

Matos Moctezuma, Eduardo. 2008 [1975]. *Muerte a filo de obsidiana: Los nahuas frente a la muerte*. Mexico: Fondo de Cultura Económica.

Matos Moctezuma, Eduardo. 2010. *La muerte entre los mexicas*. México: Tusquets.

Matos Moctezuma, Eduardo. 2013. "La muerte entre los mexicas: Expresión particular de una realidad universal." *Arqueología Mexicana* 52: 8–33.

Matthew, Laura E., and Michel R. Oudijk, eds. 2007. *Indian Conquistadors: Indigenous Allies in the Conquest of Mesoamerica*. Norman: University of Oklahoma Press.

Mignolo, Walter D. 1995. *The Darker Side of the Renaissance: Literacy, Territoriality, and Colonization*. Ann Arbor: University of Michigan Press.

Mikulska, Katarzyna. 2008. "El concepto de ilhuicatl en la cosmovisión nahua y sus representaciones gráficas en códices." *Revista Española de Antropología Americana* 38 (2): 151–171.

Mikulska, Katarzyna. 2015. "Los cielos, los rumbos y los números: Aportes sobre la visión nahua del universo." In *Cielos e inframundos: Una revisión a la Cosmología Mesoamericana*, ed. Ana Díaz, 109–173. México: Instituto de Investigaciones Históricas, Fideicomiso Felipe Teixidor y Montserrat Alfau de Teixidor.

Mondragón, Carlos. 2015. *Un entramado de islas: Persona, medio ambiente y cambio climático en el Pacífico occidental*. Mexico: El Colegio de México, Centro de Estudios de Asia y África.

Mundy, Barbara, and Dana Leibsohn. 2012. "History from Things: Indigenous Objects and Colonial Latin America." *World History Connected*. http://worldhistoryconnected.press.illinois.edu/9.2/forum_mundy.html. Accessed February 23, 2017.

Neurath, Johannes. 2015. "La escalera del padres sol y nuestra madre joven águila." *En Cielos e inframundos: Una revisión de las cosmologías mesoamericanas*, ed. Ana Díaz, 201–215. México: Instituto de Investigaciones Históricas, Fideicomiso Felipe Teixidor y Montserrat Alfau de Teixidor.

Nicholson, Henry B. 1990. "Gesammelte Abhandlungen zur Amerikanischen Sprach und Altertumskunde: A Historical Review." In *Collected Works in Mesoamerican Linguistics and Archaeology*, vol. 1, ed. Frank E. Comparato, xiii–xvi. Culver City, CA: Labyrinthos.

Nielsen, Jesper, and Toke Sellner Reunert. 2009. "Dante's Heritage. Questioning the Multi-Layered Model of the Mesoamerican Universe." *Antiquity* 83 (2): 399–413.

Nielsen, Jesper, and Toke Sellner Reunert. 2015. "Estratos, regions e híbridos: Una reconsideración de la cosmología mesoamericana." In *Cielos e Inframundos: Una revision de las cosmologías mesoamericanas*, ed. Ana Díaz, 25–64. Mexico: Universidad Nacional Autónoma de México, Instituto de Investigaciones Históricas/Fideicomiso Felipe Teixidor y Montserrat Alfau de Teixidor.

Olivier, Guilhem. 2010. "El panteón mexica a la luz del politeismo grecolatino: El ejemplo de la obra de fray Bernardino de Sahagún." *Studi e Materiali di Storia delle Religioni* 76 (2): 389–410.

Peterson, Jeanette F. 1993. *The Paradise Garden Murals of Malinalco: Utopia and Empire in Sixteenth-Century Mexico*. Austin: University of Texas Press.

Peterson, Jeanette F. 1995. "Synthesis and Survival. The Native Presence in Sixteenth-Century Murals of New Spain." *Phoebus* 7: 14–36.

Pitarch, Pedro. 1996. *Ch'ulel: Una etnografía de las almas tzetzales*. Mexico: Fondo de Cultura Económica.

Quiñones Keber, Eloise. 1995. "Collecting Cultures: A Mexican Manuscript in the Vatican Library." In *Reframing the Renaissance: Visual Culture in Europe and Latin America, 1450–1650*, ed. Claire J. Farago, 229–242. New Haven, CT: Yale University Press.

Rappaport, Joanne, and Tom Cummins. 2012. *Beyond the Lettered City: Indigenous Literacies in the Andes*. Durham, NC: Duke University Press.

Restall, Matthew. 1997. *The Maya World: Yucatec Culture and Society, 1550–1850*. Stanford, CA: Stanford University Press.

Restall, Matthew. 1998. *Maya Conquistador*. Boston, MA: Beacon Press.

Restall, Matthew. 2003. *Seven Myths of the Spanish Conquest*. New York: Oxford University Press.

Restall, Matthew, and Florine Asselbergs. 2007. *Invading Guatemala: Spanish, Nahua, and Maya Accounts of the Conquest Wars*. University Park: Pennsylvania State University Press.

Reyes-Valerio, Constantino. 1978. *Arte indocristiano: Escultura del siglo XVI en México*. Mexico: Secretaría de Educación Pública, Instituto Nacional de Antropología e Historia.

Reyes-Valerio, Constantino. 2000. *Arte indocristiano*. Mexico: Instituto Nacional de Antropología e Historia.

Ricard, Robert. 1966. *The Spiritual Conquest of Mexico*. Berkeley: University of California Press.

Roys, Ralph L. 1952. "Conquest Sites and the Subsequent Destruction of Maya Architecture in the Interior of Northern Yucatan." *Contributions to American Anthropology and History* XI (51): 130–182.

Schroeder, Susan, ed. 2010. *The Conquest All Over Again: Nahuas and Zapotecs Thinking, Writing and Painting Spanish Colonialism*. Estbourne, UK: Sussex Academic Press.

Schroeder, Susan, and Stafford Poole, eds. 2007. *Religion in New Spain*. Albuquerque: University of New Mexico Press.

Schwaller, John F., ed. 2003. *Sahagún at 500: Essays on the Quincentenary of the Birth of Fr. Bernardino de Sahagún*. Berkeley, CA: Academy of American Franciscan History.

Schwaller, John F. 2006. "The *Ilhuica* of the Nahua: Is Heaven Just a Place?" *The Americas* 62 (3): 391–412.

Seler, Eduard. 1902–1923. *Gesammelte Abhandlungen zur Amerikanischen Sprach und Altertumskunde*. Berlin: A. Asher.

Seler, Eduard. 1907. "Mythus und Religion der Alten Mexikaner." In *Gesammelte Abhandlungen zur Amerikanischen Sprach und Altertumskunde*, vol. IV: 33–38. Fondo Museo Nacional de México (1851–1964), vol. 262, exp. 100800, fojas 121–180. Mexico: Archivo Histórico del Museo Nacional de Antropología.

Seler, Eduard. 1990. *Collected Works in Mesoamerican Linguistics and Archaeology*, 5 vols. Culver City, CA: Labyrinthos.

Solari, Amara. 2013. *Maya Ideologies of the Sacred: The Transfiguration of Space in Colonial Yucatan*. Austin: University of Texas Press.

Spitler, Susan. 2005. "Colonial Mexican Calendar Wheels: Cultural Translation and the Problem of 'Authenticity.'" In *Painted Books and Indigenous Knowledge in Mesoamerica*, ed. Elizabeth Hill Booone, 271–288. New Orleans, LA: Middle America Research Institute.

Tavárez, David. 2007. "Canciones Nicachi: textos rituales zapoteca y conocimientos rituales clásicos posteriores en Oaxaca colonial." *Reporte de FAMSI*. http://www.famsi.org/reports/02050es/section04.htm#endnotes.

Tavárez, David. 2011. *The Invisible War: Indigenous Devotions, Discipline, and Dissent in Colonial Mexico*. Stanford, CA. Stanford University Press.

Thompson, Donald E. 1960. "Maya Paganism and Christianity." In *Nativism and Syncretism*, ed. Margaret A. L. Harrison, and Robert Wauchope, 1–35. New Orleans, LA: Tulane University, Middle American Research Institute.

Thompson, J. Eric S. 1934. "Sky Bearers, Colors, and Directions in Maya and Mexican Religion." In *Contributions to American Archaeology* 10, vol. II, 209–242. Washington, DC: Carnegie Institution of Washington.

Thompson, J. Eric S. 1954. *The Rise and Fall of Maya Civilization*. Norman: University of Oklahoma Press.

Thompson, J. Eric S. 1970. *Maya History and Religion*. Norman: University of Oklahoma Press.

Tozzer, Alfred. 1907. *A Comparative Study of the Mayas and the Lacandones*. New York: McMillan Company.

Valdovinos, Vanya. 2011. "Bok´ya: Una serpiente de agua." Paper presented at the Ciclo de Conferencias "Cosmologías indígenas, nuevas aproximaciones." Museo Nacional de Antropología, Mexico, January 27.

Velásquez García, Erik. 2009. "Imagen, texto y contexto ceremonial del 'Ritual de los Ángeles': Viejos problemas y nuevas respuestas sobre la narrativa sagrada en los libros de Chilam Balam." *Acta Mesoamericana* 20: The Maya and Their Sacred Narratives: Text and Context in Maya Mythologies: 55–74.

Viveiros de Castro, Eduardo. 2004. "Perspectivismo y multinaturalismo en la América indígena." In *Tierra adentro: Territorio indígena y percepción del entorno*, ed. Alexandre Surrallés and Pedro García Hierro, 37–80. Copenhague: International Work Group for Indigenous Affairs.

Wagner, Roy. 1991. "The Fractal Person." In *Big Men and Great Men: Personifications of Power in Melanesia*, ed. Maurice Godelier, and Marylin Strathern, 159–173. Cambridge: Cambridge University Press.

Wake, Eleanor. 2010. *Framing the Sacred: The Indian Churches of Early Colonial Mexico*. Norman: University of Oklahoma Press.

Wolf, Gerhard, Joseph Connors, and Louis Waldman, eds. 2011. *Colors between Two Worlds: The Florentine Codex of Bernardino de Sahagún*. Milan: Max Plank Institute, Villa I Tatti, Harvard University Center for Italian Renaissance Studies.

Wood, Stephanie. 2003. *Transcending Conquest: Nahua Views of Spanish Colonial Mexico*. Norman: University of Oklahoma Press.

Part I
Recognition

On Describing Others' Worlds

1

Colliding Universes

A Reconsideration of the Structure of the Precolumbian Mesoamerican Cosmos

Jesper Nielsen and
Toke Sellner Reunert

> The same Indian collaborators who helped Sahagún to Indianize the Christian message were responsible for Christianizing the Indian past. (Escalante 2003, 186)

A recurring methodological discussion in Mesoamerican and Andean research has centered on the use of ethnohistorical or ethnographic analogies or the "direct historical approach" (e.g., Kubler 1973; Willey 1973; Nicholson 1976; Jansen 1988; Quilter 1996). Although the authors of this chapter find this approach both inevitable and productive, and while we do not believe that potential disjunctions in form and meaning should cause us to exclude the use of analogies, the present study emphasizes how important it is to investigate and trace the development of any cultural element, belief or otherwise, in time and space with extreme care. We suggest that a generalization has taken place regarding the idea of a multilayered Mesoamerican universe, showing that this cosmic structure has been inferred primarily from postcolumbian central Mexican sources and not from precolumbian evidence such as Maya hieroglyphic texts or iconography. Second, we elaborate on our already-published hypothesis that the notion of a multilayered universe (beyond the three tiers of heaven, earth, and underworld) was not a Mesoamerican concept in the first place, but was only introduced into the area in the sixteenth century, and that it ultimately derives from European visions of the cosmos (Nielsen and Sellner

DOI: 10.5876/9781607329534.c001

Reunert 2009; see also recent contributions on the subject by Díaz 2009; 2011, 158–200; 2015; Díaz and Alcántara 2011). Our point here is *not* to say that this multilayered cosmology, which in our view was a result of European influence, was necessarily caused by a direct influence from the writings of Dante. The influence may very well have been indirect, since the Dantean worldview was widely accepted in most of Europe at the time of the conquest. In this chapter we further examine under which circumstances and by whom the sixteenth-century sources on Mesoamerican religion and cosmography were composed. We begin, however, with a look at two sources that present different yet clearly related visions of the Mesoamerican cosmos.

LAYERS OR REGIONS: THE TOPOGRAPHY OF THE OTHERWORLD

In the mid-sixteenth-century K'iche' epic, the *Popol Vuh*, we can read the narrative of the Hero Twins and their experiences in six houses or caves in *Xib'alb'a*, the dreadful underworld. In each of these houses the twins have to face and deal with a specific test or danger. Thus, the second house is called "Blade House" and is filled with sharp cutting blades and the fourth house is the "House of Jaguars" (Christenson 2003, 163–174). In the description it would appear that all houses are arranged on the same level, that is, the houses are all situated at "ground level" in *Xib'alb'a*. If we compare the houses and their characteristics to the layers of the Aztec underworld according to the *Codex Vaticanus A* (dated to between 1566 and 1589; see Glass and Robertson 1975, 186), we find some significant similarities between the houses of the *Popol Vuh* and the correspondent layers in *Vaticanus A* (Miller and Taube 1993, 177–178). In the *Codex Vaticanus A*, the first layer beneath the earth is called *Apanohuayan* ("Water Passage"), while the second is named *Tepetl Imonamiquiyan* ("Where the Hills Clash Together") (Nicholson 1971, 406–408, fig. 7, table 2) (also see figure 1.1). In the *Popol Vuh* the Hero Twins begin their journey to Xib'alb'a by passing through great river canyons, which can indeed be described as narrow spaces where the rocks come together, and they cross the big rivers of Pus and Blood (Christenson 2003, 160). In *Vaticanus A* we also encounter layers named *Itztepetl* ("Obsidian Knife Hill"), *Itzehecayan* ("Place of the Obsidian-Bladed Wind"), and *Temiminaloyan* ("Where Someone Is Shot with Arrows"), which all can be compared to the "Blade House" in the K'iche' epic in the sense that they are places where blades inflict pain on humans. The "Place of the Obsidian-Bladed Wind" finds its parallel in the "House of Cold" where the Hero Twins are met by a constant storm of hail (Christenson 2003, 169).

The Aztecs called the eighth layer of the underworld *Teyollocualoyan* ("Where Someone's Heart Is Eaten") and this is illustrated by a coyote or jaguar devouring a human heart. The obvious K'iche' parallel to this is the "House of Jaguars" and the house of man-eating bats. It would thus seem that while the K'iche' highland Maya associated these tests and frightening beings with a group of horizontally arranged houses, the contemporaneous Aztecs identified them with a series of vertically arranged layers.

Is this simply an example of two alternative Mesoamerican concepts? Or can the difference best be explained by other historical developments and cultural mechanisms? As already indicated, we show that it cannot be verified that the vertical multilayered structure is an indigenous precolumbian model of the cosmos. All known sources that make clear references to such a multilayered cosmological structure are from the central Mexican highlands and all are of postcolumbian origin. We therefore suggest that the concept of a multilayered universe came only with the Spanish intruders—more specifically the Franciscans and Dominicans—and with what we here call a Dantean worldview with nine layers in both heaven and the underworld. Yet today, the idea of a multilayered Mesoamerican cosmos has become extremely widespread and accepted in the field; we briefly discuss how this particular image of the universe has been described and applied in the scholarly literature on the Aztecs and the ancient Maya.

PREVIOUS RESEARCH AND THE CONSENSUS ON MESOAMERICAN COSMOLOGY

It is fair to say that most Mesoamericanists of today have been taught along the same lines when it comes to the worldview and cosmology of the ancient Maya and Aztecs. The majority of textbooks available describe how the spheres of the heaven and the underworld were divided into a number of layers (the exact numbers varying depending on the source used). Susan Toby Evans states: "To Mesoamericans, the cosmos had multiple levels. The Aztecs saw nine levels of the underworld and 13 of the upper world, with earth as the first level of each" (Toby 2004, 292; see also Sharer and Traxler 2006, 730–732; Wagner 2006, 286–287; Stuart 2011, 87). Mary Miller and Karl Taube note that "at the time of the Conquest, most Central Mexican people believed in the cosmographical scheme of nine levels of the Underworld, with 13 levels of upper world," and they refer to the *Codex Vaticanus A* as the source "where the 9–13 scheme receives its most explicit and ample representation." However, they also add that "the Maya certainly perceived layers of both Underworld and

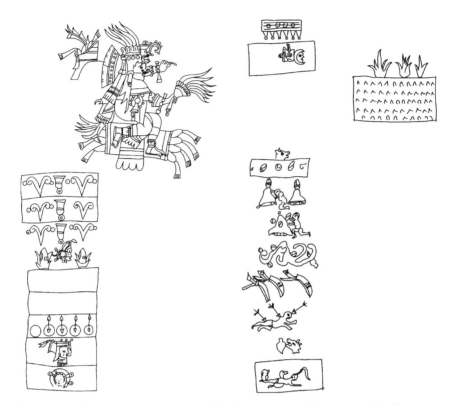

FIGURE 1.1. *The multilayered structure of the Aztec cosmos as it is represented in the Codex Vaticanus A (fols. 1v–2r), sixteenth century. (Drawing by Victor Medina.) The drawing visualizes the idea of a layered cosmos, with ten heavenly layers to the left and the remaining three appearing to the right. Below, to the right, are the nine levels of the underworld. The uppermost level of the sky is represented by the deity Tonacatecuhtli, an aged god of creation and as such an ideal parallel to the Christian god.*

upper world but the notion of nine levels of the Underworld is not specific or universal for the Maya, nor is it for either the Mixtecs or the Zapotecs" (Miller and Taube 1993, 177–178). Miller and Taube's note of caution is important, as it turns out that it is almost exclusively *Codex Vaticanus A* that is referred to, if any source is mentioned at all, in the broad synthesizing statements of a Mesoamerican multilayered cosmography.

If we want to understand how the *Vaticanus A* image came to dominate scholars' ideas and perception of Mesoamerican cosmography we need to begin with Eduard Seler, the "dean" of Mesoamerican studies. It was Seler

(1849–1922) who presented the first lengthy discussion of central Mexican cosmography, including the belief in the existence of several layers in the heaven and in the underworld. In his brilliant analysis Seler was careful to explain that a variety of ways to perceive the layout of the underworld seems to have existed (Seler 1996, 3–23). For example, he wrote that "in the same way as the earth, this flat, two-dimensional shield [of the underworld] was cut up into nine regions corresponding to the four cardinal directions. Thus, the underworld was divided into nine regions" (Seler 1996, 9). He also noted that some texts speak of nine heavens rather than thirteen (Seler 1996, 9, 17–19), and in describing the underworld and heavenly layers depicted in the *Codex Vaticanus A* scene in great detail, he pointed out that "the meaning of the several heavenly spheres is not as clear as those of the underworld. The task of filling in the thirteen of them has apparently bothered the old philosophers even more than that of characterizing the nine hells completely" (Seler 1996, 15). Furthermore, Seler stressed that at certain points the illustration seems to have been influenced by a European style—in the way the moon is represented, for example (Seler 1996, 15). Despite Seler's nuanced presentation of the potentially very different arrangements of the underworld according to local Aztec schools and traditions, the *Vaticanus A* scheme has since become the standard way of understanding and interpreting ancient central Mexican cosmography (e.g., León-Portilla 1956, 122–128; López Austin 1988, I, 52–61; 1997). Apparently, the image of a clear and recognizable vertical structure of the universe was favored to other contrasting and unillustrated versions. In 1961 Walter Krickeberg, another well-known German Americanist, wrote of the cosmology of the Postclassic cultures:

> According to the most ancient version, the earth is a flat disc sandwiched between the bases of two immense step-pyramids . . . The stream which flows round the earth and round the bases of the two pyramids is called Chicunauhapan, 'nine stream', because there are nine heavens and nine underworlds. The Aztecs replaced this stepped universe with a layered one and increased the number of heavens to thirteen. In the step hypothesis the highest heaven and the deepest hell lie at the apex of each pyramid; in the horizontal layer hypothesis the two constitute the top and bottom layers. The older idea is the more logical, it fits better with the idea that the sun climbs and descends a pyramid each day (and does the reverse every night) . . . The Aztecs regarded each of the thirteen upper layers as the seat of a particular deity or natural phenomenon, and each of the nine nether layers as a region in which a succession of terrors awaited the sun and the souls of the dead. (Krickeberg et al. 1968, 39–40)

A problem with Krickeberg's presentation is the lack of references to sources. It is unclear what "the most ancient version" is, and Krickeberg's reasoning that the step-version was replaced by a layer-version by the Aztecs is equally problematic. What can be concluded, however, is that his analysis contributed to establish the generalized and unquestioned "fact" that the Aztecs believed in a cosmic structure arranged vertically in layers of nine and 13.

If we turn to the Maya area, we have no precolumbian or early colonial sources that depict or describe the universe in a way similar to that of the *Codex Vaticanus A*. In his famous *Relación de las cosas de Yucatán* (c. 1566), Bishop Diego de Landa does not mention layers (Tozzer 1941), and it would seem that the idea of a 9–13 layer model among the Classic Maya was first introduced by J. Eric S. Thompson. In 1934 he thus suggested that "*it is possible* that the various groups of Maya direction gods, such as Bacabs, Chacs and Pauahtuns, were considered to be on different celestial or terrestrial planes" (Thompson 1934, 216; emphasis added). Hence, directional gods were first transferred to vertical layers on a purely speculative basis. Some years later, in his synthesis of the Yucatec Maya religious worldview, Thompson's good friend and colleague Ralph Roys followed up on the same idea: "The Maya cosmos, like that of the Mexicans, seems to have consisted of thirteen heavens and nine hells. In the native literature we frequently read of a group of thirteen gods called Oxlahun-ti-ku and another of nine gods known as Bolon-ti-ku. They are evidently the deities of the thirteen heavens and nine hells; especially since the Bolon-ti-ku appear to be of a malevolent character" (Roys 1943, 73; see also Anders 1963, 74–85). Roys based his conclusion on evidence from the *Book of Chilam Balam from Chumayel*, and we shall return to this late colonial and heavily Christianized source below. In 1954 Thompson wrote of the Maya: "They *appear to have believed* that the sky was divided into thirteen compartments, in each of which certain gods resided. These *may have been thought to* be arranged as that number of horizontal layers one above the other, or as steps, six ascending on the east and six descending on the west . . . *There seems no reason to doubt* that the Maya, like the Aztec, believed that there were nine underworlds" (Thompson 1954, 225–226, emphasis added). The wording has changed slightly compared to 1934, but later, in his *Maya History and Religion*, Thompson (1970) left no room for doubt: "there *are* thirteen layers of the skies . . . just as there *are* nine layers of the underworld" (Thompson 1970, 280; emphasis added). Thus, Thompson applied the *Codex Vaticanus A* and *Chilam Balam of Chumayel* visions of the universe to the precolumbian Maya, assuming that such basic structures were shared by all Mesoamerican cultures across time and space. The idea of a multilayered cosmos became so deeply embedded in Mesoamerican research that it was automatically accepted as a genuine precolumbian concept. According to Alfonso Villa Rojas:

> it is necessary to consider the ideas corresponding to its [the universe's] vertical structure with its thirteen heavens and nine lower levels, so prevalent in Central America. Archaeologists have pointed out the antiquity of these concepts, since the glyphs of the nine gods of the underworld as well as certain allusions to the deities of the upper levels appear in Classic inscriptions . . . This vertical dimension, with its two divisions of nine and thirteen floors continued to thrive in the interpretation of cosmic phenomena far into the eighteenth century. (Villa Rojas 1988, 135)

As already noted, Maya hieroglyphic inscriptions do not, however, provide the necessary evidence of a multilayered universe.

Soon, the cosmic model influenced interpretations of Mesoamerican sacred architecture and city-planning. Cecilia Klein, drawing on Krickeberg's research, refers to the existence of five or seven layers of the sky in Postclassic central Mexican sources, and notes that temple pyramids in this region often have this number of layers (Klein 1975, 80). As already pointed out, other sources mention other numbers of layers, and it would seem difficult to correlate any specific numbers with temple architecture in such generalizing terms. Rudolph van Zantwijk (1981) offered a detailed interpretation of the Templo Mayor based on the layered cosmology, drawing on the same general sources (see also Matos Moctezuma 1987, 186–189; Mendoza 1977, 67–72), and again it was assumed that the regions were stacked rather than arranged horizontally. The alleged nine layers of the underworld are also argued to be represented architecturally in the site layout of Teotihuacan (Sugiyama 1993, 121) and specific Maya temple pyramids, such as the Temple of the Inscriptions of Palenque and the K'uk'ulkan pyramid in Chichen Itza, and in the nine steps leading to the southern subterranean chambers in Palenque's palace (Carlson 1981, 154). Obviously, such patterns in the use of recurring numbers is significant, but it must be emphasized that many Maya temple pyramids or stairs to "underworldly" spaces do not follow this scheme. Also, it is not our point here to deny any association between the underworld and the number nine. Thus, the so-called Twin Pyramid groups of Tikal, often thought to be architectural cosmograms, include buildings with nine doorways on their south sides (Ashmore 1992, 174–176), and are perhaps emulating the series of horizontally arranged houses or caves of the underworld similar to those of the *Popol Vuh*.

To briefly reiterate, we have so far not encountered any secure reference to a multilayered universe among the precolumbian Maya, and it is equally problematic to provide empirical evidence that the Late Postclassic Aztecs and Nahua-speaking groups subscribed to the multilayered model. In order to understand how the Mesoamerican and European cosmic views apparently got conflated, as we argue they did, we need a basic understanding of the fundamentals of each of these two distinct systems, as well as the cultural context in which they were confronted.

MESOAMERICAN VISIONS OF THE UNIVERSE

As we turn to Mesoamerica it readily becomes apparent that, for instance among the colonial and modern Mayas, a four-part image of the universe is by far the most commonly represented (e.g., Villa Rojas 1988, 122–134). This horizontal structure, based on a central middle point and the four world directions, can be traced further back into precolumbian Mesoamerica, from the Postclassic to the Classic period, and even as far back as to the Preclassic and Olmec culture (e.g., Nicholson 1971, 403–406; Freidel, Schele, and Parker 1993, 71–73, 128–131; Miller and Taube 1993, 77–78; 186–187; Estrada-Belli 2006, 59–63; Headrick 2007, 159–162; Astor-Aguilera 2010; Hopkins and Josserand 2012; Stuart 2011, 78–87). One of the best-known representations of this horizontally divided cosmos is found on the first page of the *Codex Fejérváry Mayer* (figure 1.2), where the god of fire, Xiuhtecuhtli, is shown standing at the center of the earth acting as "the master of the four directions" and, as noted by Seler, "at each of the four terminals of the world two gods are designated as protectors and lords, which [together with Xiuhtecuhtli] produces the number nine" (Seler 1996, 7–8; see also Boone 2007, 114–117). The gods and trees of the four directions are framed by a so-called formée cross with the arms of the cross oriented toward the four cardinal directions, but counting the intercardinal loops and the center we reach the number nine, just as we have nine gods. Thus, we clearly have nine sections or divisions of the cosmos. A comparable image is found on pages 75–76 of the Maya *Codex Madrid*.

Elizabeth Boone noted this "widespread emphasis on the quadripartite nature of the universe" (Boone 2007, 174). Still, the idea of a multilayered cosmos appears to be governing her interpretation of the cosmogony on pages 39–40 of the *Codex Borgia*. Boone thus describes the sacrifice of the night sun, which happens nine times over: "The ninefold sacrifice makes sense, of course, because nine is the number of the layers of the underworld and of the Lords of the Night. We are probably witnessing the trials of the sun once it sets into the underworld" (Boone 2007, 199, fig. 114). A close look at the lower half of the *Borgia* scene does not, however, reveal any obvious indication of verticality and layers. More

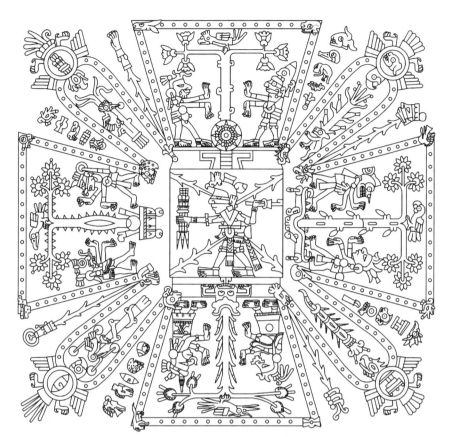

FIGURE 1.2. *The cosmogram on the first page of the* Codex Fejérváry-Mayer. *The four cardinal directions and four intercardinal directions, with the god Xiuhtecuhtli in the center, suggest nine divisions of the cosmos. (Late Postclassic; drawing by Victor Medina.)*

likely, we see a horizontally structured scene situated within the underworld, as indicated by the crocodilian earth god framing nine gods partaking in a collective and simultaneous sacrifice of the personified night sun. Indeed, precolumbian iconography and inscriptions do not seem to hold any secure references to a multilayered universe. Deities, supernatural beings, and places, in particular among the Classic Maya, often appear with numerals, of which 1, 3, 5, 7, 9, and 13 are common (e.g., Christenson 2003; Freidel, Schele and Parker 1993; Miller and Taube 1993, 151–152; Carlson 1997) but there is currently no evidence that any of these numbers, and more specifically 7, 9, and 13, can be associated with vertical layers rather than sections or regions of the underworld or heaven.[1]

We would also like to bring attention to a couple of representations of the cosmos from the Late Postclassic Aztec culture that we have not considered in our previous research. First, two incised-bone rasps—one the so-called Hueso de Culhuacan (figure 1.3), the other an unprovenanced piece on display in the Sala Mexica of the Museo Nacional de Antropología in Mexico City—show a single sky band with stars, the sun, and possibly also Venus at the top of the scenes. At the bottom of the composition of the former is the earth monster, or Tlaltecuhtli, and shown between the cosmic regions are warriors and other allusions to warfare (see von Winning 1969, 275, figs. 396–397; Mikulska 2008; 2010, 143, figs. 8a–b). A comparable scene is found on a turquoise mosaic shield from the Mixteca region, showing a curved mountain representing the earthly level (Pasztory 1983, 276–277, plates 294–295; Pohl, Fields, and Lyall 2012, 39, fig. 22). Thus, a clear pattern of opposition is evident in the juxtapositions of the starry sky band with its sun disk and the earth. In these examples there are no indications of additional layers in heaven or the underworld. Also on display in the Museo Nacional de Antropología is a fascinating Mexica stone relief, unfortunately badly broken,

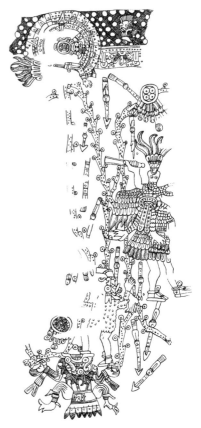

FIGURE 1.3. *Aztec bone rasp (*omichicahuaztli*) from Culhuacan, with a starry sky band and the sun across the top and the earth monster at bottom. (Museo Nacional de Antropología in Mexico City, cat. no. 11.0.07003; drawing by Jesper Nielsen.)*

that evidently represents the same kind of scenery as those just mentioned. This is the so-called *Lápida de los Cielos* (figure 1.4). Here we find two vertical panels flanking a central panel depicting a bird and what seem to be two descending warriors. The two outer panels show a row of at least seven layers of alternating stars and "Venus" symbols, each separated by a narrow band, and it could be argued that this is evidence of several heavenly layers (hence the

FIGURE 1.4. *Aztec stone relief known as the* Stone of the Heavens *(Lápida de los Cielos). The two faces and two edges are shown. The left and right outer columns on the face at left show stacked star and Venus symbols, perhaps indicating a infinite and deep sky rather than multitiered levels of the universe. (Postclassic, Central Mexico. Museo Nacional de Antropología in Mexico City. Digital Archive of the Collections of the National Museum of Anthropology. Reproduction authorized by the Instituto Nacional de Antropología e Historia. Secretaría de Cultura-INAH-MTM-MNA-CANON.)*

designation of the sculpture), perhaps similar to the layer-like roof structures with deceased warriors in the *Codex Borgia* (p. 33) (see also Díaz, chapter 3, this volume, and Mikulska, chapter 8, this volume). However, we suggest that the unique representation of undifferentiated layers on the *Lápida de los Cielos* is best understood as a generic way of representing the heavens, comparable to the so-called sky-bands that frame the image of Palenque's famous king K'inich Janaab Pakal on the sarcophagus lid discovered in the Temple of the Inscriptions (Martin and Grube 2008, 167). The Maya sky-band consists of a segmented band with repeated references to different celestial bodies, including the sun, moon, and stars (Miller and Taube 1993, 154–155). The layers of the *Lápida de los Cielos* can also be compared to the way in which the sea is represented graphically in Mesoamerican iconography as horizontally stacked undulating waves, which does not imply the notion of a multilayered ocean.

Thus, we do not regard the "layers" of the *Lápida de los Cielos* as evidence of a multilayered universe but simply as a way of depicting the deep and seemingly endless night sky, almost as a mirror image to the sea. As we see below, an image of the heavens comparable to the one of the *Lápida de los Cielos* occurs in the postcolumbian Mixtec *Selden Roll*, where stacked "layers" of stars are also penetrated or traversed by a passage.

We have already indicated that the depiction of the Mesoamerican cosmos that is most commonly referred to is the famous illustration of the layered universe in the *Codex Vaticanus A* (folios 1v–2r). Other early colonial sources—for example, the *Historia de los Mexicanos por sus pinturas* and the *Histoire du Mechique* (Garibay 1965, 69–70, 102–103, 110)—also make reference to either nine or 13 heavens and the gods associated with them, thus seemingly indicating a cosmography similar to the one depicted in the *Vaticanus A* (see also *Florentine Codex*, book 6, Dibble and Anderson 1969, 176, 187, 202, 206; *Florentine Codex*, book 10, Dibble and Anderson 1961, 168–169, *Codex Chimalpopoca*, Bierhorst 1992, 30, 149). However, if we compare the *Vaticanus A* representation with Sahagún's most lengthy and detailed description of the Aztec underworld, we find some notable discrepancies: the latter has no reference to layers, and the text rather provides us with the impression of moving through a continuous horizontal landscape. If we consult the appendix of Book 3 of the *Florentine Codex* (Anderson and Dibble 1978, 41–44), we find that:

> This is what the natives thought, the old men and the rulers: that all who died went to [one of] three places when they died. The first place was there in the place of the dead [Mictlan] . . . And there to the place of the dead went all those who died on earth, who only died of sickness: the rulers, the commoners . . . [and when they had placed the dead and a number of goods in the grave, they said]:
>
> "Here is wherewith thou wilt pass where the mountains come together. And here is wherewith thou wilt pass by the road which the serpent watcheth. And here is wherewith thou wilt pass by the blue lizard, the *xochitonal*. And here is wherewith thou wilt travel the eight deserts. And here is wherewith thou wilt cross the eight hills. Here is wherewith thou wilt pass the place of the obsidian-bladed winds." . . . And also they caused him to take with him a little dog, a yellow one . . . It was said that it would take [the dead one] across the place of the nine rivers in the place of the dead. And when there was arrival with Mictlan tecutli, he gave him the various things with which they had adorned the dead here on earth . . . And there in the nine places of the dead, in that place there was complete disappearance.

In this description there is no reference to layers or to a downward movement from one place or layer to another. The journey of the dead to the final destination, described as the Ninth Place, could equally well be interpreted as a journey through regions situated at the same level, and terminating at a central location. Not all of the dead, however, would arrive at Mictlan; others would go to Tlalocan and others to the heavenly home of the sun god (Anderson and Dibble 1978, 47–49).

Again, we find no indication of a multitude of layers, but rather a basic contrast between "down" and "up." Another of Sahagún's texts, the *Primeros Memoriales* (Sahagún 1997, 177–178) contains an additional description of the underworld, and a reference to *chicunavatenco*, "the ninth underworld," but once more, this location is not explicitly described as being situated below other underworlds rather than being placed next to or in sequence with other regions of the underworld.

If we turn to the myths describing journeys to the underworld, any reference to cosmic layers beyond the basic three tiers are once more absent. Thus, Quetzalcoatl travels to Mictlan in search of the bones of the previous human race, apparently without encountering obstacles in any number of layers (e.g., *Leyenda de los Soles*; see Bierhorst 1992). A much later source, Hernando Ruiz de Alarcón's *Treatise on the Heathen Superstitions*, written in 1629, also provides important information. Summarizing some of Ruiz de Alarcón's observations, J. Richard Andrews and Ross Hassig conclude that "the native view of the supernatural world persisted in the concepts of a celestial realm, Topan ('Above-us'), also called Chiucnauhtopan ('Nine-Topan') and Chiucnauhtlanepaniuhcan ('Nine-layering-place'), has nine levels, although these are not discussed separately, nor are internal distinctions drawn" (Andrews and Hassig 1984, 21). The obvious key term here is Chiucnauhtlanepaniuhcan and, interestingly, Ruiz de Alarcón originally translated this as "a las nueve juntas o emparejamientos" (Andrews and Hassig 1984, 224). Andrews and Hassig consider this a misinterpretation, hence their translation "Nine-layering-place." However, there is no part of the word that can be translated as "layering" or "layer," and *nepaniuh* is a form of the verb *nepanoa*, meaning "juntarse una cosa con otra" (Karttunen 1992, 169). Thus, the term simply designates a place where nine entities are united or fitted together (Una Canger, personal communication 2007; see also Haly 1992, 289). With this in mind, there is no reason to regard Ruiz de Alarcón's original translation as erroneous. On the contrary, Andrews and Hassig's interpretation of the term may be colored by the expectations generated by the prevailing understanding of Mesoamerican cosmology, and

Chiucnauhtlanepaniuhcan could thus equally well refer to a horizontal division of the heavens (see Mikulska, chapter 8, this volume).²

As already mentioned, none of our primary early colonial sources on the Maya, Diego de Landa's *Relación* and the K'iche' epic *Popol Vuh*, mentions layers (beyond the earth's surface, a raised sky realm, and an underworld), whereas the importance of the four cardinal directions and the centre is described or referred to in several instances (e.g., Tozzer 1941, 135–139; Christenson 2003, 65). In other, and much later, Yucatec Maya sources we do, however, find references that suggest a layered structure of the cosmos (for a discussion of early eighteenth-century Zapotec sources, see Tavárez, chapter 5, this volume). In the *Book of Chilam Balam of Chumayel*, the "13-God" (Oxlahun-ti-ku) and the "9-God" (Bolon-ti-ku) are mentioned in the context of "the thirteen *taz* [layers] of the heavens, *taz* covering such things as blankets spread out one above the other" (Thompson 1970, 195). We are also introduced to the "the seventh stratum of the earth" and "the four layers of stars" (Roys 1933, 31–32, 99–101). It would thus seem that the link between the 13 and nine gods and cosmic layers derive from the *Chumayel* manuscript, and that Thompson and Roys assumed, as discussed previously, that a similar belief existed in precolumbian times. However, it is important to stress that by the time the *Chilam Balam of Chumayel* was written in the final decades of the eighteenth century, Yucatec Maya culture had long been influenced by Christianity and European concepts and genres, as is evident throughout large parts of the book (Mignolo 1995, 204–207; Knowlton 2009). A similar influence from European cosmology is apparent in the *Book of Chilam Balam of Kaua*, which includes a geocentric, spherical model of the universe that may have been "brought to Yucatán in Rodrigo Zamorano's *Cronología y reportorio de la razón de los tiempos*, which was published in 1585" (Bricker and Miram 2002, 92, see also 12–15, 33–35; Díaz and Álcantara 2011). Bricker and Miram also raise the issue of the possible European origin of the multilayered universe: "The problem with these assumptions is that *tas* and *yal* refer not only to the 'layers' of the sky in these Books of Chilam Balam but also to the celestial spheres of the European universe in the *Kaua*. It is therefore possible that the references to *oxlahun tas caan* and *oxlahun yal caan* represent a syncretism of European and Maya concepts of the universe, not a survival of the Maya worldview into colonial times" (Bricker and Miram 2002, 36; see also Graham 2009). In his recent book on creation myths in the Books of Chilam Balam, Timothy Knowlton also pointed out that "plane" or "division" would be a more appropriate translation of *tas* than "layer," as "it does not necessarily imply a simple, vertically stacked cosmos" (Knowlton 2010, 72).

Our point here is that the sources describing a multilayered Mesoamerican universe are all of postcolumbian origin and are almost certainly influenced by Euro-Christian concepts regarding the structure of the cosmos. In the few extensive descriptions of journeys to the underworld (which involves several features that can be compared to the layers of the *Vaticanus A* scene and the houses of the *Popol Vuh*), the dead do not pass through layers. Most myths and rituals, both pre- and postcolumbian, on the other hand, emphasize the cardinal directions and a horizontal structure, and we suggest that in precolumbian times the oft-mentioned groups of nine and 13 deities could in fact have resided over horizontally arranged regions rather than layers. Altar 3 from the Classic Maya site of Altar de los Reyes (Campeche, Mexico) bears a hieroglyphic text referring to the "divine earth or lands, the 13 [earth or lands]" in combination with 13 emblem glyphs associated with known Maya polities, strongly suggesting that the Maya of this region acknowledged 13 subdivisions or regions of the earth (Grube 2003). What stands out is the clear horizontal emphasis coupled with the number 13, and the geography and layout of the underworld may well have mirrored a similar structure.

To support this hypothesis, we refer briefly to a fascinating description of the "geography of the underworld" from present-day San Miguel Tzinacapan (Puebla, Mexico). In Timothy Knab's accounts from this Nahua-speaking community we find detailed descriptions of the directions and regions of the underworld, not a series of layers (Knab 1991, 1993, 2009). According to the *curanderos*, or shamans, of the village, the underworld, here called "Talocan" and in Spanish ambiguously referred to both as "Most Holy Earth" and "inferno," in many ways mirrors the earthly level and its organization. The underworld is organized according to the four cardinal directions and can be reached through four "mouths" or caves (Knab 1993, 57–63). One of Knab's informants relates that inside the underworld there "are fourteen of everything . . . Thirteen outside the center for us, and one of everything inside for the Lords" (Knab 1993, 63). In Knab's own words: "el número especifica la cercanía de un lugar al centro de Talocan y su importancia relativa, trece es el más cercano y el catorce está en el centro de Talocan. Como todos los rasgos geográficos existen en el centro del inframundo hay solamente trece entre el centro y los cuatro lados" (Knab 1991, 47). In Sahagún's *Primeros Memoriales*, the numbers associated with the geography of the underworld were eight and nine (eight deserts, eight hills, nine rivers), while among the San Miguel Tzinacapan shamans of today it is rather the number 14 that has gained importance in the division of the underworldly terrain, and we learn of 14 rivers, lakes, hills, fields, roads, county seats, and so on in the underworld (Knab 1991, 47). Basically,

Talocan thus mirrors the land of the living: "There is everything there that is on the earth" (Knab 1993, 61).

In general, the rituals and the associated prayers and texts documented by ethnographers in indigenous communities in Mesoamerica over the last century and a half seem to emphasize the four cardinal directions, the center, and a basic three-tiered cosmos (e.g., Wisdom 1940; Ichon 1973; Villa Rojas 1988, 122–134; Sandstrom 1991; Tedlock 1992; Monaghan 1995; see also Klein 1982; Prechtel and Carlsen 1988). Rarely do we encounter examples of multilayered cosmologies in a context that is not otherwise relying heavily on Christian ideas and concepts. This is surprising if the multilayered cosmos was indeed part of a common and widespread precolumbian worldview, which would have further facilitated a merging with the Euro-Christian or Dantean cosmology.

PAINTING THE UNIVERSE—BUT WHOSE UNIVERSE?

Having reached the conclusion that precolumbian evidence for a multilayered universe does so far not exist, we must ask ourselves the question of how and why the vertical structure got adopted? To possibly answer this, we have to examine the early colonial context in which documents such as the *Codex Vaticanus A* are most likely to have been produced. First of all, it is important to recognize the prejudices and the cultural blindfolds of the Spaniards and other Europeans who recorded their impressions of the New World. Åke Hultkrantz rightly noted that "Landa's version of the Mayan belief in a life after death is incomplete and is colored by his own Catholic expectations" (Hultkrantz 1979, 238–239). In general, the friars worked from a preconceived model of the pagan "alternative" to Christianity, based primarily on Greco-Roman religion, myths, and cosmology. Tzvetan Todorov points to a series of equivalences between Aztec and Roman gods in Sahagún's description of the Aztec deities, and refers to one example in which the goddess Chicomecoatl is called another Ceres. In Todorov's view, "this kind of comparison is indeed very widespread in the writings of the period," and "Sahagún works according to a plan he has established following his first contacts with the Aztec culture but also in terms of his idea of what a civilization can be [and he] imposes his conceptual framework on the Aztec lore . . . Sahagún expects the Aztec gods to resemble the Roman gods" (Todorov 1984, 231–233; see also D'Olwer 1987, 111–112; Clendinnen 1991, 154–155).

However, images like the cosmogram of the *Codex Vaticanus A* should not only be seen as a result of Dominican and Franciscan attempts to conceptualize the Mesoamerican cosmography according to a Euro-pagan tradition.

Rather, we believe there are indications that such images are also a result of the appropriation of Euro-Christian ideas by well-educated colonial Nahuas. Many of the manuscripts, codices, and other documents produced in the early colonial period were not only the work of the friars themselves, since Indian scribes and informants who had been taught by the friars contributed considerably to the manuscripts, not least to the production of images.

This leads us to the early Franciscan schools such as the Colegio de Santa Cruz in Tlatelolco and the working methods of the friars (Ricard 1966, 207–235; Kobayashi 1985; Robertson 1994; Baudot 1995; Bremer 2003; Boone 2003). The school in Tlatelolco opened in 1536, and the sons of the Indian nobility, the so-called *colegiales*, were taught, among other things, grammar, geometry, astronomy, and elemental theology (Baudot 1995, 112–113; Escalante Gonzalbo 2008a). Some of the *colegiales* were also trained in the arts: "At the Mendicant primary schools, such as San José de los Naturales, and at the Colegio de Santa Cruz in Tlatelolco, Nahua youths learned the figural styles and iconographic system of late Gothic and Renaissance Europe" (Boone 2003, 145). Thomas Bremer describes the students of the Franciscan schools as hybrid *colegiales* with a bicultural education, being "neither natives nor Europeans" (Boone 2003, 20), and he points out how Sahagún's *Historia General* was the work not only of Sahagún himself, but also of "the hybrid *colegiales*, Indians by birth and quasi-Europeans by education and training," who according to Bremer, "did the greatest part of the work" (Bremer 2003, 16). Elizabeth Boone (2003) and Pablo Escalante Gonzalbo (2003) also note the strong European influence in the paintings of the *Historia General*. While conducting research, Escalante Gonzalbo had access to many of the books that were originally placed in the Franciscan monasteries and schools, and he provides abundant examples of how native artists copied images or altered images from the Bible or from books on natural history (Escalante Gonzalbo 2003, 174; see also Mathes 1982; Kobayashi 1985, 271–274; Peterson 1988, 278; Baird 1986; Bricker and Miram 2002, 34–36). For example, "representations of Indian gods and priests . . . have a distinctly Christian flavour, and one can occasionally detect the presence of specifically Christian themes behind certain images" (Escalante Gonzalbo 2003, 177). Jeanette Peterson makes a similar observation and notes that part of the *Florentine Codex* "displays a complex layering of native and Euro-Christian value systems," and further argues that the Nahua artists "sometimes neglected appropriate and available Aztec subject matter in preference for Christian models" (Peterson 1988, 290).

In 1579 the book *Rhetorica Christiana* was printed in Perugia, Italy. The author, Franciscan friar Diego de Valadés (1533–1582?), was most likely born in Extremadura, Spain, and emigrated with his family to the New World at an early age. Alternatively he would have been the son of a Spanish conquistador and a native Tlaxcalan woman (Lara 2008, 63). The volume was directed towards European literati who took interest in mnemonic techniques, and its full title begins as follows: "Christian rhetoric for the purpose of speaking and praying in public, with examples of both actions, chiefly taken from the histories of the Indies" (Lara 2008, 62). Thus, Valadés's work provides us with a unique insight in the manner of preaching and teaching the Christian beliefs.

One of the many illustrations in the *Rhetorica* (Valadés 1579) shows the mentor of Valadés, Friar Pedro de Gante (c. 1480–1572), preaching to the Indians by means of pictures (figure 1.5). A text accompanying the image states that "because the Indians lack letters, it was necessary to teach them by means of some illustrations" (cited after Edgerton 2001, 116). As noted by Samuel Edgerton: "While Valadés's main subject was to explain in detail the art of literary rhetoric—how language can most effectively serve the teaching of Christian doctrine—he . . . believed that the same techniques and intention were equally applicable to the visual arts" (Edgerton 2001, 114). According to Robert Ricard, Friar Luis Caldera was using identical visual means of persuasion: "Carrying large pictures representing the sacraments, the catechism, heaven, hell, and purgatory, he went from village to village" (Ricard 1966, 104). Some of the pictures used for teaching the catechism were also represented in the *Rhetorica*. Thus, three full-page images show heaven, earth, and hell as divided into layers. One of these illustrations shows God sitting on his throne inside a cloud at the uppermost layer as Jesus reclines in his lap. The devil, being worshipped by sinners, is placed at the bottom of the page (figure 1.6). The layout of this illustration and the utilization of it by the friars sum up one of the crucial points of this chapter.

From these analyses and studies of the schools and cultural backgrounds of the teachers and the native students we get the impression of an environment in which an almost inevitable blending of old and new was the norm (Escalante Gonzalba 2008b). It was in this bicultural setting that many of our primary sources on the nature of the Mesoamerican cosmic structure were formed both in text and in image.

DANTE'S UNIVERSE

A central argument of this chapter relies on the assumption that the vertical layout of the cosmos in the postcolumbian sources, with multiple layers

FIGURE 1.5. *The Franciscan Pedro de Gante preaching to Nahuas by using images showing episodes in the life of Christ. (From Diego de Valadés's* Rhetorica Christiana, *1579. Nettie Lee Benson Latin American Collection, University of Texas Libraries, The University of Texas at Austin.)*

FIGURE 1.6. *Diego de Valadés's image of "the great chain of being"* (scala naturae)—*heaven, earth, and hell—divided into layers.* (From Diego de Valadés, Rhetorica Christiana, *1579. Nettie Lee Benson Latin American Collection, University of Texas Libraries, The University of Texas at Austin.)*

of heaven and underworld, can reasonably be ascribed to the influence by the Spanish friars on the *colegiales* who authored the manuscripts. In other words, we need to address the following question: In what way did the Euro-Christian ideas about the rendering of cosmos influence the reports from the *colegios* in New Spain?

To answer this question, we need to consider the ideological background of the Franciscans and Dominicans. At the time of the conquest, these two monastic orders held cosmological views similar to what could be labeled a Dantean cosmology. Although Dante Alighieri (1265–1321) did not belong to the either the Franciscan or the Dominican Order himself, he was strongly influenced by the ideological views held by the two orders of mendicant friars. During his youth Dante attended philosophical and theological schools and became familiar with the central Franciscan and Dominican teachings (Toynbee 1900, 48, 66; Hawkins 2007, 126–127). In the *Divine Comedy* (written 1308–1321; see Alighieri 1939) he indirectly expresses a strong affiliation with these teachings through his appraisal of the values of poverty (see *Purgatorio* 12, 20; *Paradiso* 6, 11). In a central scene of his famous epic Dante meets the *sapienti*, the teachers of the Church (*Paradiso* 10). At this crucial theological moment, he lets Thomas Aquinas, as a representative of the Dominican Order, praise St. Francis, while the Franciscan Bonaventura in turn praises St. Dominic, showing us that Dante held these two orders in especially high regard (Havely 2004, 4–5; Jacoff 2007, 12–3; Mazotta 2007, 7). Although Dante wrote his *Divine Comedy* some 200 years before the Spanish invasion of Mexico, his description of the geography of heaven and hell was more widespread than ever in the late fifteenth and early sixteenth centuries. One of the most famous illustrations of the *Comedy*, by the Florentine artist Sandro Botticelli, dates to about 1480–1495 (Schulze 2000). The first translations into Latin were published in 1417, and one of these was prepared by a Franciscan bishop, Juan de Serravalle (Friedrich 1949). About a decade later, the first Spanish translation appeared, and throughout the fifteenth century and into the beginning of the sixteenth century the *Comedy* had tremendous influence on Spanish theological poetry (Friedrich 1949, 45–47).[3] At this point in time, the great epic had reached the status of a quasi-biblical text, based on its own claim to have been inspired by vision (Hollander 2007, 273).

In the *Comedy*, Dante travels as a pilgrim through the three realms of the dead (Inferno, Purgatory, and Heaven). The Roman poet Virgil guides him through Inferno and Purgatory, while his ideal woman, Beatrice, guides him through Heaven. Thus, the arrangement of the cosmos provides the compositional and narrative backbone of the *Comedy*. The Inferno consists of nine concentric circles, resembling an inverted pyramid. In each of the nine circles, the dead are suffering according to a gradual increase in wickedness, culminating at the icy center of the earth, where the three-headed Satan resides.

While this underworld is placed below the holy city of Jerusalem, Purgatory is found on the exact opposite side of the world. The earthly Purgatory consists of a mountain with seven terraces, each corresponding to one of the deadly sins. But as Toynbee notes, "the seven terraces, together with the Antepurgatory and the Terrestrial Paradise, form nine divisions, thus corresponding to the nine circles of Hell, and the nine spheres of Paradise" (Toynbee 1900, 202). After climbing the seven terraces, Dante reaches the Garden of Eden, from where he can resume the last part of the trip, which leads him to the heavenly spheres and the residency of God.

Dante's heaven represents the medieval equivalent of Aristotle's nine heavenly spheres, an account of heaven that was commonly accepted throughout Catholic Europe at the time. Aristotle arrived at the cosmological conception of nine heavenly spheres through astronomical observation and reasoning. He explained the seemingly incoherent movements of the heavenly bodies in terms of different spheres: each planet as well as the sun and the moon had its own separate sphere, around which it moved independently, thus causing the different bodies to move in different cycles. Apart from the five visible planets, plus the sun and the moon, the fixed stars also had their own separate sphere, the firmament, and behind the fixed stars was the sphere of the *primum mobile*, the "first moved" or "first mover," which caused the movement of the rest of the spheres, thus constituting nine heavenly spheres. Adopting the Aristotelian scheme of the heavenly spheres, Dante reaches a theological-astronomical geography of the cosmos, consisting of nine layers of the underworld, seven or nine layers on earth, and again nine layers of the heavens in a most symmetrical fashion, reflecting the ideals of the time (Aveni 1992, 168).

Although it seems likely that Spanish friars such as Sahagún were well acquainted with Dante, it is not known when the *Divine Comedy* was first brought to New Spain and whether it formed part of the Franciscan and Dominican monastery libraries during the earliest years of the postcolumbian era. The inventory lists of the library of the Colegio de Santa Cruz in Tlatelolco of 1572, 1574, and 1582 do not mention the *Comedy* (Mathes 1982). However, these lists are not necessarily complete and it is uncertain what the contents of the library were previous to these relatively late inventory lists. Some volumes were privately owned and at the disposal of the friars, some were transported from Europe, and some others were bought in Mexico (Kobayashi 1985, 271–272). There is little doubt, however, that Dante's work arrived in Mexico in this period. A sixteenth-century copy of the *Comedy* has been found in the National Library in Mexico City by Escalante Gonzalbo and may well originally have belonged to one of the convent libraries (Pablo Escalante Gonzalbo, personal communication 2008). From the year 1600 we have a detailed invoice from a shipment of books from Sevilla to one Martín de Ibarra in Veracruz, which refers to two copies of the *Comedy* in Italian among the almost 700 volumes listed (Green and Leonard 1941, 2, 12).

What is crucial to our argument, however, is not so much whether the Franciscans and Dominicans in Mexico had one or several copies of Dante's *Comedy* in their possession, but rather that they were familiar with the Dantean cosmology with its multiple layers in heaven and hell, and that this worldview would seem a natural point of reference in the interpretation of the local accounts and renderings of the geography of the cosmos.

It is an indisputable fact that, from the time of early Renaissance Italy until the conquest of Mexico, and throughout Catholic Europe, the general understanding of the cosmology of the heavens was closely akin to the Aristotelian version adopted by Dante. The Aristotelian scheme was commonly combined with the likewise originally Greek idea of the four elements—earth, water, air, and fire—all together constituting cosmograms of somewhere between the nine heavenly layers of Aristotle and 13 layers, including earth as one of the elements. This pattern of diversity in the representations of the heavens is curiously repeated in the two main sources for the theory of stratification in the Mesoamerican sky, the *Codex Vaticanus A*, which has 13 layers in the heavens, and the *Selden Roll*, which has nine (figure 1.7). Furthermore, the author of *Vaticanus A* made an explicit reference to the Euro-Christian cosmic layout, describing Omeyocan as: "el lugar donde está el Creador" and "la primera causa" (see Anders, Jansen, and Reyes 1996, 45).

Most often, Renaissance cosmograms were circular, geocentric representations, showing the earth in the center, encircled by the three other elements, followed by nine or 10 additional spheres (Heninger 1977). But now and then we encounter what could be described as extracted "slices" of the circular representations, schematically showing the layers placed horizontally on top of each other (figure 1.8), forming a square representation not unlike the stacked layers of the *Selden Roll*, to which we will shortly return for a more comprehensive description.

Interestingly, we also find examples of how the geocentric universe was translated into "mythological terms" in the sense that classical deities were depicted as presiding "over the numerous components of the Ptolemaic universe" (Heninger 1977, 174). In Natalis Comes's *Mythologiae* from 1616,

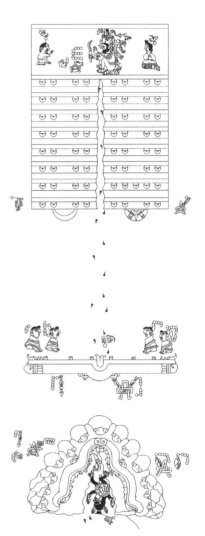

FIGURE 1.7. *The first two scenes of the sacred narrative of the* Selden Roll. *(MS. Arch. Selden A 72(3), Bodleian Library; drawing by Victor Medina.) In the first scene, to the left, are nine heavenly layers, and to the right, contrasting the horizontally layered heaven is the more traditional Mesoamerican representation of the underworld. Here seven caves can be reached through the open maw of the earth god.*

Roman celestial, terrestrial, and aquatic gods are represented in a layered universe (figure 1.9), and it is worth quoting Heninger's description of the scene at length:

> At the bottom it reveals the gods of the underworld . . . Immediately above this lowest region, there is a region filled with aquatic gods . . . Next there is a region populated by various terrestrial gods and a river god . . . The lowest of the celestial spheres is that of the Moon, divided into three ranks of nonde-

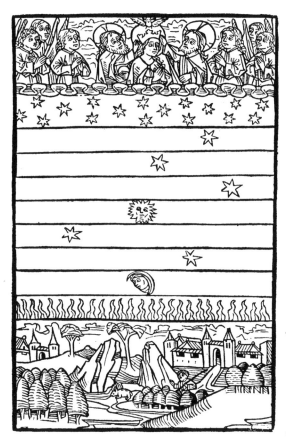

FIGURE 1.8. *Illustration from Konrad von Megenberg's* Das Buch de Natur *(1349–1350), showing the heavenly realm divided into layers. (After Heninger 1977; The Huntington Library, San Marino, California.)*

script deities, and followed by a sphere for each of the other planets . . . Outside the planetary spheres, presumably in pagan empyreum, the twelve major gods of the Roman pantheon float on clouds. (Heninger 1977, 174)

Thus, a geocentric, layered cosmogram that merges with a pagan pantheon was created. The resemblance to the *Vaticanus* image is striking. It is thus reasonable to assume that the Dantean multilayered worldview was taught or indirectly referred to at the schools of the Spanish friars in New Spain, and thus reached native artists who were either working for the friars or independently. Considering the prevalence of and fame attributed to Dante's work during late fifteenth- and early sixteenth-century Spain, as well as Dante's affiliation with the Franciscan and Dominican Orders, we believe this to be a highly plausible scenario.

FIGURE 1.9. *Roman gods associated with the celestial, terrestrial, and aquatic spheres are represented in a layered universe on the frontispiece of Natalis Comes's* Mythologiae, *1616. (After Heninger 1977; The Huntington Library, San Marino, California.)*

IMAGES OF EARLY COLONIAL HYBRID UNIVERSES

As already indicated, the worldview expressed in the *Vaticanus A* seems to fit the pattern of production of religious images in the early colonial period we have outlined above (see also Díaz 2009, 35–41). The cosmogram runs over two pages of the codex (1v–2v), with heavenly layers 13–4 represented on 1v and with the creator god Tonacatecuhtli prominently portrayed seated in the thirteenth and uppermost layer. Continuing on 2r we find layers 3–2 at the top of the page, the moon occupying the second layer. However, this layer actually constitutes the first layer of the celestial sphere since the next layer, Tlalticpac, refers to the earth. The crust of the earth also forms the first layer of the underworld (the earth layers' special double function appears to be indicated visually in that they are placed outside the otherwise continuous series of layers). Next follow layers 2–9 of the underworld, with Mictlan as the ultimate and final station of the dead.

Another fascinating representation of the heavenly layers can be found in the Mixtec manuscript known as the *Selden Roll*, or *Codex Selden II*, which is dated to the sixteenth century (Burland 1955, 12–13; Glass and Robertson 1975, 196; Boone 2000, 152–153; Jansen and Pérez 2007, 105–107) (see figure 1.7). The first part of the document shows nine layers, with the sun and moon occupying the first and lower level. A path, marked by footprints, moves across or penetrates the sequence of layers, which are all marked with stars. This heavenly layout is reminiscent of what we saw on the *Lápida de los Cielos* (see figure 1.4), but as we shall see there are also signs of possible Euro-Christian additions and reconfigurations. In the ninth and uppermost layer the culture hero Lord 9 Wind (wearing the characteristic mask and headdress of the wind god) is seated between a couple. Both the man and the woman are named 1 Deer, and they are equivalents of Ometecuhtli and Omecihuatl (or the closely related Tonacateuchtli and Tonacacihuatl), the Aztec origin couple and gods of creation and procreation (Nicholson 1971, 397–398, 410–411; see also Seler 1996, 16; Miller and Taube 1993, 154).[4] The divine ancestral pair was said to reside in Omeyocan, a mythical place of duality, birth, and descent (Nicholson 1971; Miller and Taube 1993, 40–41, 127–128, 154). Interestingly, in precolumbian times this couple, found in most of Mesoamerica, was associated with caves and the earth rather than the heavens (Nielsen and Brady 2006).

In contrast to the *Vaticanus A* scene, we do not find a corresponding multilayered representation of the underworld in the *Selden Roll*. Here a row of footprints enters the open maw of the earth monster and reaches *Chicomoztoc*, or "The Seven Caves," a well-known Mesoamerican mythological place. Thus, the *Selden Roll* does not present the underworld as a mirror image of the layers of the sky, but couples the latter with a traditional Mesoamerican mythological location associated with the number seven (Miller and Taube 1993, 60). The seven caves are situated within a mountain and are arranged at the same level, much like the houses or caves of the *Popol Vuh*. To return to the placement of the creator couple in the upper layer, this image seems to be heavily inspired by Euro-Christian and Dantean imagery. The ancestral couple thus constitutes a near perfect parallel to Adam and Eve as they are standing in the Garden of Eden at the summit of Mount Purgatory in the famous fresco *The Comedy Illuminating Florence* by Domenico di Michelino (1417–1491) in the cathedral of Santa Maria del Fiore in Florence, Italy (see Nielsen and Sellner Reunert 2015, fig. 7). The fresco dates to 1465 and shows Dante holding the *Divine Comedy*. On the left side of the painting sinners are being led to Hell, on the right side the walls and towers of Florence, and behind Dante the terraces of Purgatory and the heavenly layers. Another striking depiction of the Euro-Christian cosmos comes from Nicole Oresme's *Book of the Heaven and the World* from 1377, which shows God seated in the uppermost of nine layers of stars, with the moon occupying the lowest level (see Nielsen and Sellner Reunert 2009, 410, fig. 6). Both the *Vaticanus A* and the *Selden Roll* thus merge precolumbian elements and deities with European and Christian concepts, and obvious parallels seem to have been sought: Ometecuhtli and Omecihuatl take the place of either Adam and Eve or all-embracing God; Omeyocan, the mystical abode of the first two humans is likened to Paradise, and the original division of heaven and underworld in a number of horizontal regions are transformed into a vertical structure akin to the illustrations that native artists and *colegiales* would have seen and copied from volumes in the Franciscan and Dominican schools and workshops. In keeping with this, the lowest level of the underworld shown in the *Vaticanus A* correlates with Dante's vision of Satan residing in the pit of the ninth circle or layer of the underworld.

Neither of the two representations discussed here can be said to be exact copies, but rather reinterpretations, of the Dantean and Christian cosmology aligned with a Mesoamerican belief system. The cosmographies of the *Codex Vaticanus A* and the *Selden Roll* are "neither native nor European" to use Bremer's phrase (Bremer 2003, 20); they both reflect a colonial-period merging of a native central Mexican concept of the universe with the Euro-Christian Dantean cosmology. Most important, we suggest that the native artists used European models when they drew and reformulated their vision of the universe, and thereby the horizontality of the precolumbian era was rescheduled into a vertical structure. In cases like the *Selden Roll*, a preexisting, seemingly "layered" heaven like that of the *Lápida de los Cielos* was readily merged with the Dantean model.

CONCLUSIONS

In this chapter we have presented and discussed the evidence bearing on the Mesoamerican vision of the structure of the cosmos, and we have argued that on the present evidence the long-assumed, multilayered model cannot be maintained. Rather, our examination has revealed that a basic three-tiered model combined with a strong emphasis on the horizontal divisions of each layer is more likely to have been the dominating view before the Spanish invasion and the introduction of Euro-Christian ideas relating to the cosmos. We also showed how the two crucial images from the *Vaticanus A* and the

Selden Roll may be explained as the product of hybrid native artists. We wish to emphasize that the clash of Christianity and Mesoamerican religions was a collision of a religious literate doctrine on the one hand, and a group of closely related but localized and less precisely formulated religious traditions on the other. As Pascal Boyer points out, "doctrines are not necessarily the most essential or important aspect of religious concepts. Indeed, many people seem to feel no need for a general, theoretically consistent expression of the qualities and powers of supernatural beings" (Boyer 2001, 140). Thus, while both Dominicans and Franciscans probably expected and sought one common and coherent version of the Aztec cosmos, the Aztec religious elite saw no problem in having several slightly diverging versions existing at any one time, none of these excluding the others, as there was no inflexible doctrine on the matter. That multiple interpretations of the underworld and its configuration can be considered equally correct, coherent, and acceptable by members of the same population or ethnic group is also demonstrated by Knab (2009), and with this in mind, we can perhaps better understand the flexibility and the ease with which the Euro-Christian concepts were integrated into existing beliefs (Díaz, chapter 3, this volume).

It has not been our point to reject the possibility of a vertically structured universe (with layers beyond the three basic layers), and we cannot and do not wish to deny that there may have been local traditions with such perception of the cosmological architecture. What we argue is that no such evidence exists at present and that all evidence pointing to a belief in a multilayered universe is postcolumbian and produced in a context where Euro-Christian ideas and thoughts were deliberately imported, taught, and consciously imitated and adopted. An overview of the precolumbian and ethnographic sources seems to favor a strong horizontal and directional emphasis, coupled with the belief in a basic three-leveled cosmos. The multilayered Mesoamerican universe is thus almost certainly a hybrid of original precolumbian concepts of the cosmos and Euro-Christian visions. The conclusions reached in this essay thus serve to exemplify how at times "common knowledge" and generalized terms and concepts need to be reviewed. Many arguments, scientific narratives, and "truths" clearly and unsurprisingly consist of several layers of assumptions and generalizations.

ACKNOWLEDGMENTS

We would like to express our thanks and gratitude to a number of friends and colleagues who have been helpful during the process of researching and writing the present chapter and previous versions thereof: Berenice Alcántara Rojas, Una Canger, Ana Díaz, Hugo García Capistrán, Pablo Escalante Gonzalbo, Elizabeth Graham, Christophe Helmke, Casper Jacobsen, Stephen Houston, Bodil Liljefors Persson, Amos Megged, Diana Magaloni Kerpel, Katarzyna Mikulska, Jørgen Nybo Rasmussen, Karl Taube, Gabrielle Vail, and María Teresa Uriarte. In addition, thanks and lots of love to Mette, Freja, and Liv and to Laura and Rose—our wives and daughters. Finally, we must not forget to state clearly that any errors and misinterpretations remain our sole responsibility.

NOTES

1. Cosmologies including a multilayered universe are, however, found in regions bordering the Americas and Mesoamerica. As part of initiation rituals, central Asian and Siberian shamans thus climb trees or wooden posts notched with "seven or nine *tapty*, representing the seven or nine celestial levels" (Eliade 1964, 275). Considering the many fundamental mythological themes and cosmological concepts that Mesoamerica shares with this area, it could be argued that the multilayered universe in Mesoamerica was derived from such an ancient substratum of religious ideas and worldviews (Berezkin 2002, 2004; see also Nielsen and Sellner Reunert 2008). Likewise derived from that substratum may be the belief in a multilayered universe—consisting of four strata in the sky as well as in the underworld—recorded among peoples of the American Southeast (e.g., Lankford 2007, 18–20; Duncan 2011). However, these occurrences do not change the fact that we have no precolumbian evidence of such beliefs in Mesoamerica, and we find it to be a critical maneuver to transfer the multitiered model from other cultural regions to that of Mesoamerica.

2. The three examples of *ilhuicame* are to be found in the *Doctrina Christiana en lengua Española y Mexicana: hecha por los religiosos de la orden de sancto Domingo* (1548, fol. XVIr); Bernardino de Sahagún's *Psalmodia christiana y sermonario de los sanctos del año en lengua mexicana* (1583, fol. 234r); and in the *Cantares mexicanos* (fol. 2r) (Alcántara Rojas 2011).

3. Lazaro Iriarte shows that Franciscans were active in both translating and commenting on the *Comedy*. Thus, an Italian Franciscan, Accursius Bonfantini († c. 1338) was among the first to publish a commentary to Dante's work (Iriarte 1984, 131).

4. For a critical review of scholars' translations and subsequent potential misinterpretations of crucial Nahuatl terms such as *Ometeuchtli*, *Omecihuatl*, and *Omeyocan*, see Haly (1992). Haly suggests that *Omeyocan* should be translated as "the place of bones" rather than "the place of duality," since *ome* ("two") is in fact the stem *omi* ("bone") (Haly 1992, 280–281). If correct, this reinterpretation has profound consequences for long-held views on Aztec cosmology and religion.

REFERENCES

Alcántara Rojas, Berenice. 2011. "Ilhuicac, tlalticpac y mictlan: Cosmologías nahuas y cristianas en el discurso de evangelización del siglo XVI." Paper presented at the Ciclo de Conferencias Cosmologías indígenas, nuevas aproximaciones. Museo Nacional de Antropología, Mexico, January 27.

Alighieri, Dante. 1939. *The Divine Comedy*. 3 vols. Trans. John D. Sinclair. Oxford: Oxford University Press.

Anders, Ferdinand. 1963. *Das Pantheon der Maya*. Graz: Akademische Druck- und Verlagsanstalt.

Anders, Ferdinand, Maarten Jansen, and Luis Reyes García, eds. 1996. *Religión, costumbres e historia de los antiguos mexicanos: Libro explicativo del llamado* Códice Vaticano A. Mexico: Akademische Druck- und Verlagsanstalt, Sociedad Estatal Quinto Centenario, Fondo de Cultura Económica.

Anderson, Arthur J. O., and Charles E. Dibble, trans. 1978. *Florentine Codex. General History of the Things of New Spain by Fray Bernardino de Sahagún*. Book 3: *The Origins of the Gods*. Santa Fe, NM: University of Utah, School of American Research.

Andrews, J. Richard, and Ross Hassig, trans. 1984. *Treatise on the Heathen Superstitions That Today Live among the Indians Native to This New Spain, 1629*, by Hernando Ruiz de Alarcón. Norman: University of Oklahoma Press.

Ashmore, Wendy. 1992. "Deciphering Maya Architectural Plans." In *New Theories on the Ancient Maya*, ed. Elin C. Danien and Robert J. Sharer, 173–184. Philadelphia: University of Pennsylvania, University Museum.

Astor-Aguilera, Miguel Angel. 2010. *The Maya World of Communicating Objects: Quadripartite Crosses, Trees, and Stones*. Albuquerque: University of New Mexico Press.

Aveni, Anthony. 1992. *Conversing with the Planets: How Science and Myth Invented the Cosmos*. New York: Kodansha International.

Baird, Ellen T. 1986. "Sahagún's *Primeros Memoriales* and *Codex Florentino*: European Elements in the Illustrations." In *Smoke and Mist: Mesoamerican Studies in Memory of Thelma D. Sullivan*, ed. J. Kathryn Josserand, and Karen Dakin, 15–40. International Series 402. Oxford: British Archaeological Reports.

Baudot, Georges. 1995. *Utopia and History in Mexico: The First Chroniclers of Mexican Civilization, 1520–1569*. Niwot: University Press of Colorado.

Berezkin, Yuri E. 2002. "Some Results of Comparative Study of American and Siberian Mythologies: Applications for the Peopling of the New World." *Acta Americana* 10 (1): 5–28.

Berezkin, Yuri E. 2004. "Southern Siberian: North American Links in Mythology." *Acta Americana* 12 (1): 5–27.

Bierhorst, John, trans. 1992. *History and Mythology of the Aztecs: The* Codex Chimalpopoca. Tucson: University of Arizona Press.

Boone, Elizabeth H. 2000. *Stories in Red and Black: Pictorial Histories of the Aztecs and Mixtecs*. Austin: University of Texas Press.

Boone, Elizabeth H. 2003. "The Multilingual Bi-visual World of Sahagún's Mexico." In *Sahagún at 500: Essays on the Quincentenary of the Birth of Fr. Bernardino de Sahagún*, ed. John F. Schwaller, 137–166. Berkeley, CA: Academy of American Franciscan History.

Boone, Elizabeth H. 2007. *Cycles of Time and Meaning in the Mexican Books of Fate*. Austin: University of Texas Press.

Boyer, Pascal. 2001. *Religion Explained: The Evolutionary Origins of Religious Thought*. New York: Basic Books.

Bremer, Thomas S. 2003. "Reading the Sahagún Dialogues." In *Sahagún at 500: Essays on the Quincentenary of the Birth of Fr. Bernardino de Sahagún*, ed. John F. Schwaller, 11–29. Berkeley, CA: Academy of American Franciscan History.

Bricker, Victoria R., and Helga-Maria Miram, trans. and notes. 2002. *An Encounter of Two Worlds: The Book of Chilam Balam of Kaua*. Middle American Research Institute 68. New Orleans, LA: Tulane University.

Burland, Cottie A., comment. 1955. *The Selden Roll. An Ancient Mexican Picture Manuscript in the Bodleian Library at Oxford*. Berlin: Verlag Gebr. Mann.

Cantares Mexicanos. n.d. Manuscript 1628, Biblioteca Nacional de México.

Carlson, John B. 1981. "A Geomantic Model for the Interpretation of Mesoamerican Sites: An Essay in Cross-Cultural Comparison." In *Mesoamerican Sites and World-Views*, ed. Elizabeth P. Benson, 143–215. Washington, DC: Dumbarton Oaks Research Library and Collection.

Carlson, John B. 1997. *The Margarita Structure Panels and the Maya Cosmogonic Couplet of Ancestral Emergence*. Early Classic Copan Acropolis Program 13. Philadelphia: University of Pennsylvania Museum.

Christenson, Allen J., trans. 2003. *Popol Vuh: The Sacred Book of the Maya*. Winchester and New York: O Books.

Clendinnen, Inga. 1991. *Aztecs: An Interpretation*. Cambridge: Cambridge University Press.

Codex Fejérváry-Mayer. 1971. Graz: Akademische Druck- und Verlagsanstalt.

Codex Tro-Cortesianus (Codex Madrid). 1967. Graz: Akademische Druck- und Verlagsanstalt.

Codex Vaticanus 3738 (Codex Vaticanus A). 1979. Graz: Akademische Druck- und Verlagsanstalt.

Díaz, Ana. 2009. "La primera lámina del *Códice Vaticano A*: ¿Un modelo para justificar la topografía celestial de la antigüedad pagana indígena?" *Anales del Instituto de Investigaciones Estéticas* 31 (95): 5–44.

Díaz, Ana. 2011. "Las formas del tiempo: Tradiciones cosmográficas en los calendarios indígenas del México Central." PhD diss., Universidad Nacional Autónoma de México, Facultad de Filosofía y Letras.

Díaz, Ana, ed. 2015. *Cielos e inframundos: Una revisión de las cosmologías mesoamericanas*. Mexico City: National Autonomous University of Mexico.

Díaz, Ana, and Berenice Alcántara Rojas. 2011. "Las esferas celestes pintadas con palabras nahuas: Anotaciones marginales en un ejemplar de la *Psalmodia christiana* de Sahagún." *Estudios de Cultura Náhuatl* 42: 193–201.

Dibble, Charles E., and Arthur J. O. Anderson, trans. 1961. *Florentine Codex: General History of the Things of New Spain by Fray Bernardino de Sahagún*. Book 10: *The People*. Santa Fe, NM: University of Utah Press, School of American Research.

Dibble, Charles E., and Arthur J. O. Anderson, trans. 1969. *Florentine Codex: General History of the Things of New Spain by Fray Bernardino de Sahagún*. Book 6: *Rhetoric and Moral Philosophy*. Santa Fe, NM: University of Utah Press, School of American Research.

D'Olwer, Nicolau. 1987. *Fray Bernardino de Sahagún (1499–1590)*. Salt Lake City: University of Utah Press.

Doctrina Christiana en lengua Española y Mexicana: hecha por los religiosos de la orden de sancto Domingo (1548). Facsimile edition. 1944. Madrid: cultura Hispánica.

Duncan, James R. 2011. "The Cosmology of the Osage: The Star People and Their Universe." In *Visualizing the Sacred: Cosmic Visions, Regionalism, and the Art of the Mississippian World*, ed. George E. Lankford, F. Kent Reilly III, and James F. Garber, 18–33. Austin: University of Texas Press.

Edgerton, Samuel Y. 2001. *Theaters of Conversion: Religious Architecture and Indian Artisans in Colonial Mexico*. Albuquerque: University of New Mexico Press.

Eliade, Mircea. 1964. *Shamanism: Archaic Techniques of Ecstasy*. New York: Bollingen Foundation.

Escalante Gonzalbo, Pablo. 2003. "The Painters of Sahagún's Manuscripts: Mediators between Two Worlds." In *Sahagún at 500: Essays on the Quincentenary of the Birth of Fr. Bernardino de Sahagún*, ed. John F. Schwaller, 167–191. Berkeley, CA: Academy of American Franciscan History.

Escalante Gonzalbo, Pablo. 2008a. "El Colegio de la Santa Cruz de Tlatelolco." *Arqueología Mexicana* XV (89): 57–61.

Escalante Gonzalbo, Pablo, ed. 2008b. *El arte cristiano-indígena del siglo XVI novohispano y sus modelos europeos*. Cuernavaca: Centro de Investigación y Docencia en Humanidades del Estado de Morelos.

Estrada-Belli, Francisco. 2006. "Lightning Sky, Rain, and the Maize God: The Ideology of Preclassic Maya Rulers at Cival, Peten, Guatemala." *Ancient Mesoamerica* 17: 57–78.

Freidel, David, Linda Schele, and Joy Parker. 1993. *Maya Cosmos: Three Thousand Years on the Shaman's Path*. New York: Quill/William Morrow.

Friederich, W. P. 1949. "Dante through the Centuries." *Comparative Literature* 1 (1): 44–54.

Garibay, Ángel María. 1965. *Teogonía e historia de los mexicanos: Tres opúsculos del siglo XVI*. Mexico: Porrúa.

Glass, John B., and Donald Robertson. 1975. "A Census of Native Middle American Pictorial Manuscripts." In *Handbook of Middle American Indians*, vol. 14, part 3, edited by Howard F. Cline, 81–252. Austin: University of Texas Press.

Graham, Elizabeth. 2009. "Close Encounters." In *Maya Worldviews at Conquest*, ed. Leslie G. Cecil and Timothy W. Pugh, 17–38. Boulder: University Press of Colorado.

Green, Otis H., and Irving A. Leonard. 1941. "On the Mexican Book Trade in 1600: A Chapter in Cultural History." *Hispanic Review* 9 (1): 1–40.

Grube, Nikolai. 2003. "Appendix 2: Epigraphic Analysis of Altar 3 of Altar de los Reyes." In *Archaeological Reconnaissance in Southeastern Campeche Mexico: 2002 Field Season Report*, ed. Ivan Šprajc. http://www.famsi.org/reports/01014/01014 Sprajc01.pdf.

Haly, Richard. 1992. "Bare Bones: Rethinking Mesoamerican Divinity." *History of Religions* 31 (3): 269–304.

Havely, Nick. 2004. *Dante and the Franciscans: Poverty and the Papacy in the Commedia*. Cambridge: Cambridge University Press.

Hawkins, Peter S. 2007. "Dante and the Bible." In *The Cambridge Companion to Dante*, ed. Rachel Jacoff, 125–140. Cambridge: Cambridge University Press.

Headrick, Annabeth. 2007. *The Teotihuacan Trinity: The Sociopolitical Structure of an Ancient Mesoamerican City*. Austin: University of Texas Press.

Heninger, Simeon Kahn, Jr. 1977. *The Cosmographical Glass: Renaissance Diagrams of the Universe*. San Marino, CA: The Huntington Library.

Hollander, Robert. 2007. "Dante and His Commentators." In *The Cambridge Companion to Dante*, ed. Rachel Jacoff, 270–280. Cambridge: Cambridge University Press.

Hopkins, Nicholas A., and J. Kathryn Josserand. 2012. *Directions and Partitions in Maya World View*. Accessed January 30, 2012. http://www.famsi.org/research/hopkins/directions.html.

Hultkrantz, Åke. 1979. *The Religions of the American Indians*. Berkeley: University of California Press.

Ichon, Alain. 1973. *La religión de los totonacas de la sierra*. Mexico City: Instituto Nacional Indigenista.
Iriarte, Lazaro. 1984. *Der Franzikusorden: Handbuch de franziskanischen Ordensgeschichte*. Altötting: Verlag der Bauerischen Kapuziner.
Jacoff, Rachel. 2007. "Introduction to *Paradiso*." In *The Cambridge Companion to Dante*, ed. Rachel Jacoff, 107–124. Cambridge: Cambridge University Press.
Jansen, Maarten. 1988. "The Art of Writing in Ancient Mexico: An Ethno-Iconological Perspective." *Visible Religion: Annual for Religious Iconography* 6: 86–113.
Jansen, Maarten, and Gabina Aurora Pérez Jiménez. 2007. *Encounter with the Plumed Serpent: Drama and Power in the Heart of Mesoamerica*. Boulder: University Press of Colorado.
Karttunen, Frances. 1992. *An Analytical Dictionary of Nahuatl*. Norman: University of Oklahoma Press.
Klein, Cecelia. 1975. "Post-Classic Mexican Death Imagery as a Sign of Cyclic Completion." In *Death and the Afterlife in Pre-Columbian America*, ed. Elizabeth P. Benson, 69–85. Washington, DC: Dumbarton Oaks.
Klein, Cecelia. 1982. "Woven Heaven, Tangled Earth: A Weaver's Paradigm of the Mesoamerican Cosmos." In *Ethnoastronomy and Archaeoastronomy in the American Tropics*, ed. Anthony F. Aveni, and Gary Urton, 1–35. Annals of the New York Academy of Sciences 385. New York: New York Academy of Sciences.
Knab, Tim. 1991. "Geografía del inframundo." *Estudios de Cultura Náhuatl* 21: 31–58.
Knab, Tim. 1993. *A War of Witches: A Journey into the Underworld of the Contemporary Aztecs*. New York: HarperCollins.
Knab, Tim. 2009. *The Dialogue of Earth and Sky: Dreams, Souls, Curing, and the Modern Aztec Underworld*. Tucson: The University of Arizona Press.
Knowlton, Timothy W. 2009. "Composition and Artistry in a Classical Yucatec Maya Creation Myth: Prehispanic Ritual Narratives and Their Colonial Transmission." *Acta Mesoamericana* 20: *The Maya and Their Sacred Narratives: Text and Context in Maya Mythologies*: 111–121.
Knowlton, Timothy W. 2010. *Maya Creation Myths: Words and Worlds of the Chilam Balam*. Boulder: University Press of Colorado.
Kobayashi, José María. 1985. *La educación como conquista: Empresa franciscana en México*. Mexico City: El Colegio de México.
Krickeberg, Walter. 1971. *Altmexikanische Kulturen*. Berlin: Safari-Verlag.
Krickeberg, Walter, Hermann Trimborn, Werner Müller, and Otto Zerries. 1968 [1961]. *Pre-Columbian American Religions*. London: Weidenfeld and Nicolson.

Kubler, George. 1973. "Science and Humanism among Americanists." In *The Iconography of Middle American Sculpture*, ed. Ignacio Bernal, 163–167. New York: Metropolitan Museum of Art.

Lankford, George E. 2007. "Some Cosmological Motifs in the Southeastern in the Southeastern Ceremonial Complex." In *Ancient Objects and Sacred Realms: Interpretations of Mississippian Iconography*, ed. F. Kent Reilly, and James F. Garber, 8–38. Austin: University of Texas Press.

Lara, Jaime. 2008. *Christian Texts for Aztecs: Art and Liturgy in Colonial Mexico*. Notre Dame, IN: University of Notre Dame Press.

León-Portilla, Miguel. 1956. *La filosofía náhuatl*. Mexico City: Instituto Indigenista Interamericano.

León-Portilla, Miguel. 1988. *Time and Reality in the Thought of the Maya*. Appendix by Alfonso Villa Rojas. Norman: University of Oklahoma Press.

López Austin, Alfredo. 1988. *The Human Body and Ideology: Concepts of the Ancient Nahuas*, 2 vols. Salt Lake City: University of Utah Press.

López Austin, Alfredo. 1997. *Tamoanchan, Tlalocan: Places of Mist*. Niwot: University Press of Colorado.

Martin, Simon, and Nikolai Grube. 2008. *Chronicle of the Maya Kings and Queens: Deciphering the Dynasties of the Ancient Maya*. Revised edition. London: Thames and Hudson.

Mathes, Miguel. 1982. *Santa Cruz de Tlatelolco: La primera Biblioteca Académica de las Américas*. Mexico City: Secretaría de Relaciones Exteriores-Archivo Histórico Diplomático Mexicano.

Matos Moctezuma, Eduardo. 1987. "Symbolism of the Templo Mayor." In *The Aztec Templo Mayor*, ed. Elizabeth H. Boone, 185–209. Washington, DC: Dumbarton Oaks Research Library and Collection.

Mazotta, Guiseppe. 2007. "Life of Dante." In *The Cambridge Companion to Dante*, ed. Rachel Jacoff, 1–3. Cambridge: Cambridge University Press.

Mendoza, Rubén G. 1977. "World View and the Monolithic Temples of Malinalco, Mexico: Iconography and Analogy in Pre-Columbian Architecture." *Journal de la Société des Américanistes* LXIV: 63–80.

Mignolo, Walter D. 1995. *The Darker Side of the Renaissance: Literacy, Territoriality, and Colonization*. Ann Arbor: University of Michigan Press.

Mikulska, Katarzyna. 2008. "El concepto de *ilhuicatl* en la cosmovisión nahua y sus representaciones gráficas en códices." *Revista Española de Antropología Americana* 38 (2): 151–171.

Mikulska, Katarzyna. 2010. "¿Cuchillos de sacrificio? El papel del contexto en la expresión pictórica mesoamericana." *Itinerarios* 12: 125–154.

Miller, Mary, and Simon Martin, eds. 2004. *Courtly Art of the Ancient Maya*. San Francisco and London: Fine Arts Museum of San Francisco/Thames and Hudson.

Miller, Mary, and Karl Taube. 1993. *An Illustrated Dictionary of the Gods and Symbols of Ancient Mexico and the Maya*. London: Thames and Hudson.

Monaghan, John. 1995. *The Covenants with Earth and Rain: Exchange, Sacrifice, and Revelation in Mixtec Sociality*. Norman: University of Oklahoma Press.

Monaghan, John. 2000. "Theology and History in the Study of Mesoamerican Religions." In *Supplement to the Handbook of Middle American Indians*. Volume 6: *Ethnology*, ed. John D. Monaghan, and Barbara W. Edmonson, 24–49. Austin: University of Texas Press.

Nicholson, Henry B. 1971. "Religion in Pre-Hispanic Central Mexico." In *Handbook of Middle American Indians*, vol. 10, part 1, ed. Gordon F. Ekholm and Ignacio Bernal, 395–446. Austin: University of Texas Press.

Nicholson, Henry B. 1976. "Preclassic Mesoamerican Iconography from the Perspective of the Postclassic: Problems in Interpretational Analysis." In *Origins of Religious Art and Iconography in Preclassic Mesoamerica*, ed. Henry B. Nicholson, 157–175. Los Angeles: University of California Los Angeles, Latin American Center.

Nielsen, Jesper, and James E. Brady. 2006. "The Couple in the Cave: Origin Iconography on a Ceramic Vessel from Los Naranjos, Honduras." *Ancient Mesoamerica* 17: 203–217.

Nielsen, Jesper, and Toke Sellner Reunert. 2008. "Bringing Back the Dead: Shamanism and the Maya Hero Twins' Journey to the Underworld." *Acta Americana* 16 (1): 49–79.

Nielsen, Jesper, and Toke Sellner Reunert. 2009. "Dante's Heritage: Questioning the Multi-Layered Model of the Mesoamerican Universe." *Antiquity* 83 (320): 399–413.

Nielsen, Jesper, and Toke Sellner Reunert. 2015. "Estratos, regions e híbridos: Una reconsideración de la cosmología mesoamericana." In *Cielos e Inframundos: Una revision de las cosmologías mesoamericanas*, ed. Ana Díaz, 25–64. Mexico: Universidad Nacional Autónoma de México, Instituto de Investigaciones Históricas, Fideicomiso Felipe Teixidor y Montserrat Alfau de Teixidor.

Pasztory, Esther. 1983. *Aztec Art*. New York: Harry N. Abrams.

Peterson, Jeanette F. 1988. "The *Florentine Codex* Imagery and the Colonial '*Tlacuilo*.'" In *The Work of Bernardino de Sahagun: Pioneer Ethnographer of Sixteenth-Century Aztec Mexico*, ed. J. Jorge Klor de Alva, Henry B. Nicholson, and Eloise Quiñones Keber, 273–293. New York: State University of New York/Institute for Mesoamerican Studies/University at Albany.

Pohl, John M.D., Virginia M. Fields, and Victoria I. Lyall. 2012. "Children of the Plumed Serpent: The Legacy of Quetzalcoatl in Ancient Mexico." In *Children of the Plumed Serpent: The Legacy of Quetzalcoatl in Ancient Mexico*, ed. Virginia M.

Fields, John M.D. Pohl, and Victoria I. Lyall, 15–49. London and Los Angeles: Scala Publishers Limited and Los Angeles County Museum of Art.

Prechtel, Martin, and Robert S. Carlsen. 1988. "Weaving and Cosmos amongst the Tzutujil Maya." *Res: Anthropology and Aesthetics* 15: 122–132.

Quilter, Jeffrey. 1996. "Continuity and Disjunction in Pre-Columbian Art and Culture." *Res: Anthropology and Aesthetics* 29/30: 303–317.

Ricard, Robert. 1966 [1933]. *The Spiritual Conquest of Mexico*. Berkeley: University of California Press.

Robertson, Donald. 1994 [1959]. *Mexican Manuscript Painting of the Early Colonial Period: The Metropolitan Schools*. Norman: University of Oklahoma Press.

Roys, Ralph L., trans. 1933. *The Book of Chilam Balam of Chumayel*. Washington, DC: The Carnegie Institution of Washington.

Roys, Ralph L. 1943. *The Indian Background of Colonial Yucatan*. Washington, DC: The Carnegie Institution of Washington.

Sahagún, Bernardino de. 1583. *Psalmodia Christiana y Sermonario de los sanctos el año, en lengua mexicana*. Mexico: Pedro Ocharte.

Sahagún, Bernardino de. 1997. *Primeros memoriales*. Trans. and notes by Thelma D. Sullivan. Norman and Madrid: University of Oklahoma Press/Patrimonio Nacional/Real Academia de la Historia.

Sandstrom, Alan R. 1991. *Corn Is Our Blood: Culture and Ethnic Identity in A Contemporary Aztec Village*. Norman: University of Oklahoma Press.

Schulze Altcappenberg, Hein-Th., ed. 2000. *Sandro Botticelli: The Drawings for Dante's Divine Comedy*. London: Royal Academy of Arts.

Seler, Eduard. 1996. "The World View of the Ancient Mexicans." In *Collected Works in Mesoamerican Linguistics and Archaeology*, vol. 5, ed. Frank E. Comparato, 3–23, Culver City, CA: Labyrinthos.

Sharer, Robert J., and Loa P. Traxler. 2006. *The Ancient Maya*. Stanford, CA: Stanford University Press.

Stuart, David. 2011. *The Order of Days: The Maya World and the Truth about 2012*. New York: Harmony Books.

Sugiyama, Saburo. 1993. "Worldview Materialized in Teotihuacan, Mexico." *Latin American Antiquity* 4 (2): 103–129.

Tedlock, Barbara. 1992. *Time and the Highland Maya*. Revised edition. Albuquerque: University of New Mexico Press.

Thompson, J. Eric S. 1934. "Sky Bearers, Colors, and Directions in Maya and Mexican Religion." In *Contributions to American Archaeology* 10, vol. II. 209–242. Washington, DC: Carnegie Institution of Washington.

Thompson, J. Eric S. 1954. *The Rise and Fall of Maya Civilization*. Norman: University of Oklahoma Press.

Thompson, J. Eric S. 1970. *Maya History and Religion*. Norman: University of Oklahoma Press.
Toby Evans, Susan. 2004. *Ancient Mexico and Central America: Archaeology and Culture History*. London: Thames and Hudson.
Todorov, Tzvetan. 1984. *The Conquest of America: The Question of the Other*. New York: Harper and Row.
Toynbee, Paget. 1900. *Dante Alighieri: His Life and Works*. London: Methuen.
Tozzer, Alfred. 1941. *Landa's* Relación de las Cosas de Yucatán: *A Translation*. Papers of the Peabody Museum of American Archaeology and Ethnology. Cambridge, MA: Harvard University.
Valadés, Fray Diego. 1579. *Rhetorica Christiana*. Perugia, Italy.
Villa Rojas, Alfonso. 1988. "Appendix A: The Concepts of Space and Time among the Contemporary Maya." In *Time and Reality in the Thought of the Maya*, 2nd ed., enlarged, by Miguel León-Portilla, 113–159. Norman: University of Oklahoma Press.
Winning, Hasso von. 1969. *Pre-Columbian Art of Mexico and Central America*. London: Thames and Hudson.
Wagner, Elisabeth. 2006. "Maya Creation Myths and Cosmography." In *Maya: Divine Kings of the Rain Forest*, ed. Nikolai Grube, 280–293. Cologne: Tandem Verlag GmbH/Könemann.
Willey, Gordon R. 1973. "Mesoamerican Art and Iconography and the Integrity of the Mesoamerican Iconological System." In *The Iconography of Middle American Sculpture*, ed Ignacio Bernal, 153–162. New York: Metropolitan Museum of Art.
Wisdom, Charles. 1940. *The Chorti Indians of Guatemala*. Chicago, IL: University of Chicago Press.
Zantwijk, Rudolph van. 1981. "The Great Temple of Tenochtitlan: Model of Aztec Cosmovision." In *Mesoamerican Sites and World-Views*, ed. Elizabeth P. Benson, 71–86. Washington, DC: Dumbarton Oaks Research Library and Collection.

2

Incorporating Mesoamerican Cosmology within a Global History of Religion

Some Considerations on the Work of Lorenzo Pignoria

SERGIO BOTTA

In the interreligious confrontation between the indigenous cultures of Mesoamerica and Christianity, a key role was played by the "encounter" between different cosmologies.[1] In this perspective, the aim of this chapter is to investigate how the indigenous cosmology included in the renowned *Codex Vaticanus Latinus 3738* (also identified as *Codex Vaticanus A* or *Codex Ríos*)[2] was incorporated by the European antiquarian culture at the beginning of the seventeenth century. As would be noticed, the theological underpinnings of the *Codex Vaticanus A* allowed the Paduan antiquarian Lorenzo Pignoria to promote a Christian assimilation of the indigenous worldview and the manufacturing of a consequent global history of religion. Indeed, the worldviews that faced each other in colonial times generated a new "religious field" (Bourdieu 1971) within which practices and strategies were mobilized in order to gain a monopoly of symbolic capital (Bourdieu 1973). Such competitive processes could be observed in the construction of diverse colonial religious discourses,[3] but not all modern discursive products served the same function.[4]

At the risk of oversimplifying the complexity of colonial processes, it is necessary to distinguish different phases of the interreligious confrontation. In the Bourdesian field constituted by "local" encounters between Europeans and indigenous cultures, a first type of discursive production on cosmologies was intended to extirpate idolatry and colonize the indigenous

DOI: 10.5876/9781607329534.c002

imaginary (Gruzinski 2004). Several cases show how the reproduction of authoritative rhetorical models functioned as a tool for the Europeans to perform an alleged abdication of the indigenous perspective. The reproduction of Christian rhetorical dialogues within the famous *Colloquios de los Doce* by Bernardino de Sahagún (1997 [1564]) authorized the legitimacy of colonial rule and the European symbolic capital (Klor de Alva 1989). However, as the analysis of the Sahaguntine work shows, the literary performance of the ultimate "triumph" of Christianity was a fictional resource capable of generating an *illusio*, the belief that fiction constitutes reality (Bourdieu 1973). Therefore, "colonial *illusio*" conceals the agonistic tensions between social actors, relocating the social and political conflict to a "supernatural" level. In this perspective, it is possible to look at some colonial sources produced during the first phase of interreligious confrontation in New Spain. For instance, the renowned *Codex Vaticanus A* could be seen as a performative space of direct and indirect dialogue able to produce hybrid cosmologies (Knowlton and Vail 2010). As noted by Díaz, new cosmologies represent "more than mere syncretism or hybridization, understood as a collision between two cultures that interact and absorb each other, we should perhaps conceive of forms of knowledge that are flexible, active, and procedural" (Díaz, introduction to this volume). On the one hand, such negotiations show that European observers were committed to deconstruct and reconstruct indigenous cosmologies, mediating their symbolic capital and incorporating them within a Euro-Christian cosmological pattern;[5] on the other hand, these dynamics reveal that indigenous social actors were symmetrically engaged in a work of mimetic defense of their own hermeneutical instruments. However, their reinterpretation of Euro-Christian ideas indicates that Novohispanic cosmologies were not only forms of hybridization but also the result of a "struggle for recognition" (Honneth 1995). Regarding the religious field, different representations of the cosmos express the competition in order to gain control over supernatural resources and "naturalize" the colonial social order.

Such an interpretative framework, which needs to be verified through an investigation into the multiple forms of negotiation produced in specific contexts, enables us to see how different the construction of these first hybrid cosmologies is compared to that of cultural objects shaped in different historical and geographical contexts. In a later phase of the relationship between the Old and New Worlds—for example, since the second half of the sixteenth century—Europeans became fully aware of the impact these events exerted on their history (Elliot 1970), and started to reflect on the very meaning of the notion of "discovery"—or "invention" (O'Gorman 1958)—of the Americas as a "global" process (Gruzinski 1988).⁶ European reception of hybrid cosmologies—locally produced for colonial control—reshaped in different contexts, converted into dispositifs through which indigenous religious histories could be conceived of as particular paths within a unique and global history of religion dominated by Christianity (among others, Sheehan 2006).

As we see below, during the second phase of interreligious confrontation, Europeans tended to consider cultures in terms of genealogies, as a series of complicated derivations and transmissions from the book of Genesis to the present. As noted by Ryan, genealogical interest is essentially antihistorical: "Genealogy was powerful medicine against the confusions introduced by novelties, and it served as an effective prophylaxis against the impact of the new worlds. If anything, the discovery of exotic peoples intensified the genealogical preoccupations of the seventeenth century by giving antiquarians a whole new fund of data to incorporate into traditional genealogical schemes" (Ryan 1981, 532).

The present chapter aims to analyze the appendices that in 1615 the Paduan antiquarian Lorenzo Pignoria included within a mythographic work by Vincenzo Cartari, published for the first time in Venice in 1556 under the title *Le immagini con la spositione de i dei degli antichi* (Cartari 1556). This work, originally composed to disseminate knowledge about images of ancient Greek and Roman gods, contained information about Japan, India, and the Americas, which Pignoria added with the aim of producing a "comparative mythology"⁷ that would take into account an altered "geopolitical" background.

Concerning the Mesoamerican gods, Pignoria collected images mainly from the *Codex Vaticanus A* and reconsidered its hybrid cosmology, taking a step towards a definite appropriation of indigenous religion. The new cosmology proposed by Pignoria not only reproduced the hybrid one contained in the *Codex Vaticanus A*—which combined indigenous and Christian features—but also proposed a global representation of idolatry: through a distinctive comparative reflection, the antiquarian outlined a historical development of religion.

In this perspective, Pignoria's work perfectly fits within the cultural movement aimed to reconstruct the modern concept of religion (Nongbri 2013) that emerged during this period "in response to a *crisis of comparison* caused by the increasingly overwhelming evidence for diversity in human belief and modes of worship" (Hanegraaff 2016, 588). The study of religion changed decisively in the second half of the sixteenth century, as the amount of information about Europe's past and the present of America and Asia increased (Miller 2001, 182). Religion in colonial contexts was not only "a Christian theological concept arbitrarily imposed on the rest of the world. Additionally, and more crucially, it was a necessary technical requirement of the comparative enterprise" (Hanegraff 2016, 589).

In this perspective, modern discourses on "other" religions (Botta 2013)—built through a groundbreaking use of comparison—played various roles in different historical and cultural contexts. Hybrid cosmologies produced for colonial reasons—for example, in the *Codex Vaticanus A*—were interpreted and employed diversely in changed circumstances. According to Domenici, which use Fabian to think the social life of Mesoamerican objects, any cultural object can be contemporary for different viewers in different times (Domenici 2017). A cultural object, going through a variety of historical contexts during its cultural biography, "elicited and continues to elicit different discourses on cultural otherness" thanks to its "enduring coevalness" (Domenici 2016b, 80, 91). Therefore, the "dialogue" between the hybrid cosmology produced within the *Codex Vaticanus A* and the Christian cosmology proposed by Pignoria—as a product of an antiquarian culture that tries to incorporate exotic worlds into traditional genealogical schemes—shows that "collecting" Indian objects[8] as a form of expropriation represents the darker side of European exploration (Quiñones Keber 1995a). Therefore, according to Quiñones Keber, it is possible to look at Mesoamerican cosmologies, assembled and reproduced in colonial sources, as consciously constructed cultural objects, mainly altered for European consumption. In this sense, the transformation of Mesoamerican cosmology by Pignoria can be interpreted as an effective way to incorporate religious otherness into an expanding Western culture.

FROM CARTARI'S MYTHOGRAPHY TO PIGNORIA'S ANTIQUARIANISM

In the early seventeenth century, Pignoria published an addition to the work by Vincenzo Cartari, *Le immagini con la spositione de i dei degli antichi*. The latter was one of the most important iconographic and mythographic treatises devoted to pagan gods, which contributed to the dissemination of a body of knowledge, uncertain and dispersed until that moment. The *Immagini* represented the culmination of a Renaissance interest in classical antiquity that developed from works such as the *Genealogia deorum gentilium* by Giovanni Boccaccio (Seznec 1939). Since the second half of the sixteenth century, Cartari's work gained popularity throughout Europe, for it was published in a vernacular language and immediately translated into Latin, English, French, and German (Lein 2002, 226). Aiming to investigate the image of ancient gods, it paradoxically represented the affirmation of a textual culture: even though

Cartari's intent was to provide artists with a complete mythological and iconographic repertoire, the 1556 edition did not contain any image and relied only on literary sources. Starting with the Venetian edition of 1571—*Le Imagini de gli Dei de gli Antichi*—98 images were created by painter Bolognino Zaltieri, who was directly inspired by Cartari's texts (Cartari 1571). By the end of the sixteenth century, and throughout the seventeenth century, *Imagini* became one of the most important "manuals" for visual artists in western Europe.

Cartari died in 1581 at the beginning of a renewed concern for pagan gods, due to recently discovered lands and cultures. While still successful in Europe, at the beginning of the seventeenth century Cartari's treatise needed a new discursive organization and a broader repertoire of images (Maffei 2013, 61). Accordingly, a revised edition was published by Cesare Malfatti in 1608, containing a brief addition to the text that points to a novel interest in the surprising appearance of "modern polytheisms" (Maffei 2013, 61–62).[9] After a few years, editor Pietro Paolo Tozzi commissioned Pignoria to assemble a new collection of images and produced an appendix to Cartari's work in order to compare ancient gods with the idols from Asia and America.

Lorenzo Pignoria was born in Padua in 1571. He was educated by Jesuits and continued his studies at the university in Padua, moving within the circle of humanist Giovanvincenzo Pinelli. In 1602, he was ordained a priest and entered the service of Bishop Marco Corner as a librarian. Eventually, he became canon of Treviso (Lein 2002, 226). In Padua, Pignoria belonged to a line of scholars "who were trained as Aristotelians and distanced themselves from the methods and views of the Platonists" (Mulsow 2005, 189). In Rome from 1605 to 1607, he gained access to a circle of scholars and antiquarians devoted to archaeological studies and engaged in a lively exchange of ideas, images, and material objects (von Wyss-Giacosa, in press).

While researching ancient and modern gods, Pignoria built a complex network of correspondents (Volpi 1992b), among whom were Galileo Galilei, Nicolas Fabri de Peiresc, Hans Georg von Hohenburg Herwart—who would help Pignoria collect materials and sources—and Girolamo Aleandro, secretary of Cardinal Francesco Barberini and a Vatican librarian, who would allow him access to the *Codex Vaticanus A* (Mason 1997, 19). From the correspondence with Galileo, it appears that at that time Pignoria was looking for any novel information related to some "Indian idols," not only Egyptian but also Indian, Chinese, Japanese, Burmese, from the East Indies, Peru, Mexico, and the rest of New Spain as well (Miller 2001, 202; Volpi 1992a; 1992b, 103; 1996).[10]

Pignoria contributed to three new editions of Cartari's treatise, published respectively in 1615, 1626, and 1647. He added renovated illustrations made by the engraver Filippo Ferroverde, which replaced those by Zaltieri (Seznec 1931, 268). The new edition of Pignoria was infused with an interest in visual sources and in collecting antiquities, an essential expression of the antiquarian culture emerging at the beginning of the seventeenth century (Burke 2003; Momigliano 1950; Mulsow 2005). In 1615, Pignoria expanded Cartari's treatise by adding a few appendices, further extended in 1626 (Cartari and Pignoria 1626). The 1647 edition remained unchanged, since it was published after Pignoria's death from plague in 1631 (Cartari and Pignoria 1647). The first two appendices are known respectively as *Annotationi di Lorenzo Pignoria al libro delle Imagini del Cartari* and *Aggionta all'Imagini del Cartari del sig. Lorenzo Pignoria*. In this section, Pignoria inserted a large number of *vere et reali immagini* dedicated to classical antiquity through the description of ancient artifacts—statues, figurines, coins, cameos, exotic objects—selected from his collection and from those he had been able to observe in Padua, Rome and France from correspondents and friends such as Nicolas-Claude Fabri de Peiresc or Paolo Gualdo. This section embodies a sort of "visual turn" (Burke 2003, 273) that expresses an unprecedented use of material objects and images. Besides, the appendices represent a new way of writing a *historia* of religion, which, combining textual evidence with numismatics and eyewitness descriptions, completely change the view of the past (Mulsow 2005). The novel importance of empiricism, "of seeing with one's own eyes and of studying specific objects in scholarly investigation—a *cognitio singularium*—is a thoroughly characteristic and essential feature of the mindset of an antiquarian" (von Wyss-Giacosa, in press). Scholars revealed an unusual awareness about the hypothetical nature of their own historical theses, in "a space of uncertainty and sparked insoluble discussions" (Mulsow 2005, 190). Thus, this visual turn "not only extended the subject matter of history but also redefined historical method" (Burke 2003, 296). In adopting this perspective, Pignoria introduced strong elements of discontinuity with respect to the original structure of the work by Cartari, also from the point of view of the methodology used in the analysis of the images of ancient gods. While the latter was formed by a mythographic *corpus*, based on the descriptive power of words to represent the iconography of the gods, Pignoria paid more attention to visual aspects and established an analytical methodology founded on the constant relationship between text and illustration (Maffei 2013, 63–64), a form of "investigating religion visually" (von Wyss-Giacosa 2012).

Pignoria's main contribution to the study of Mesoamerican cosmology is the appendix devoted to the description of Mexican, Indian, and Japanese gods: the *Seconda parte delle Imagini de gli dei indiani*. As noted, Pignoria's perspective is radically different from the literary approach chosen by Cartari. He repeatedly uses the illustrations of archaeological objects as hermeneutical instruments. Such a methodology clearly appeared in Ferroverde's drawings that, while referring to the images of Zaltieri, revealed a specific attention to antiquarian sources and archaeological discoveries.

This new section—dedicated to what, according to Peter Burke, might be called "alternative antiquities" (2003, 282)—is of great importance, as it extends the focus to the gods of the whole world, reflecting a renewed comparative urgency. As for the images from India and Japan, Pignoria based his analysis on objects directly observed, thanks to European collectors, as well as on data from Jesuit accounts.

However, the sources of the appendix that Pignoria devoted to the gods of Mesoamerica had a different origin. The "American section" of the 1615 edition is composed of 12 illustrations. Among these, seven drawings are copies of images contained in the aforementioned *Codex Vaticanus A*, which Ferroverde could directly examine. In the 1626 edition, Pignoria added one more illustration taken from the same codex: a figure of the god Quetzalcoatl, which corresponds to folio 31r and was not depicted after a direct consultation of the codex—at that time Pignoria was back in Padua—but thanks to an engraving by Philips van Winghe. Among other engravings of the "American section," two idols were copied from images sent to Pignoria by the scholar of Egyptian art and antiquities Hans Georg Herwarth, who collected them for the Kunstkammer of the Duke of Bavaria in Munich. The first of these images has been considered somewhat problematic by commentators. Some of them believed that it represents a wooden mask of Kiwasa, a god worshiped in Florida and Virginia (Maffei 2013, 72), but its precise origin has been argued (Feest 1992, 96). In addition, Christian Feest (1986) identified the second image as a Taino *zemi*. Finally, Pignoria inserted in this "American section"—for specific discursive needs—three Egyptian and classical images that did not constitute a direct source for "ethnographic" understanding of Mesoamerican religions but that still play a key role in the assimilation of the cosmology contained in the *Codex Vaticanus A*.

Therefore, the central part of the Mexican *corpus* is composed of seven images that Pignoria copied from the *Codex Vaticanus A*, together with the corresponding texts that the antiquarian extensively used to build his own discourse on the nature of the Mexican idols. At page V of the 1615 edition, Pignoria depicts an image of the Mexican god called Homoyoca, while at page VII the same god is recognized as Hometeutle. Illustrations and glosses came from folio iv of the *Codex Vaticanus A*, a page of the codex that contains some of the most valuable and controversial data on Mesoamerican cosmology. In the following section, after adding an image of the Egyptian gods Isis and Osiris at page VII, Pignoria draws four double images from folio 2v of the codex, which represent four pairs of "underworld" gods (Anders, Jansen, and Reyes García 1996, 42–43). At page XIV, Pignoria copies the image of the "messenger of the god Citlallatonac," which deeply transforms the character depicted in folio 7r of the codex. Finally, at page XVII we find an image of the god Quetzalcoatl, taken from folio 7v. After two classical illustrations—at pages XX and XXII—Pignoria introduces the "American idols" (pages XXIV and XXVI), copied for him by Herwarth. Scholars repeatedly noticed the casual order and the subjective selection of the images Pignoria acquired from the codex. However, it should not be underestimated that—if an exception could be made for the Quetzalcoatl inserted in the edition of 1626—all the images selected by Pignoria came from the "cosmological section" (formed by the drawings contained in pages IV–IOV) of the *Codex Vaticanus A* (Anders, Jansen, and Reyes García 1996, 39–75).

Before shifting back to the interreligious confrontation between Pignoria and the cosmology contained in the codex, it is useful to focus briefly on how the antiquarian came to be acquainted with the document preserved in the Vatican Library (Domenici 2016a). Some have considered that the *Codex Vaticanus A* was probably produced in Mexico on Italian commission (Jansen 1984); other scholars have suggested that the codex was painted in Italy (Glass and Robertson 1975). Certainly, it appeared in Italy in the period between 1566 and 1589. The "Item 3738 in the Rainaldi inventory of the Vatican collection, compiled between 1596 and 1600" (Ehrle 1900, 9–10; Quiñones Keber 1995a, 240) shows that the codex was certainly kept in the Vatican Library when Pignoria started his research on Indian antiquities. He explicitly quoted the Dominican Pedro de los Ríos (Lein 2002, 228), the last commentarist of the *Codex Telleriano-Remensis*—painted by an indigenous painter in Mexico (Quiñones Keber 1995b)—of which the *Codex Vaticanus A* was a partial copy with glosses translated into Italian (Anders, Jansen, and Reyes García 1996, 23–33). The codex represents a "revealing early example of the encounter of Renaissance Italy with a non-western cultural artifact" (Quiñones Keber 1995a, 231). It is the only colonial Mexican manuscript to feature a lengthy, systematically composed Italian text to explicate its images. As the commentary in the codex was written in Italian, it is clear that the intention was to produce a document directed to an Italian audience (Mason 1997, 19). Pignoria informs us that he obtained the images of Mexican idols through the mediation of Senator Malipiero Ottaviano, who received them from Cardinal Marco Antonio

Amulio.¹¹ The latter had been prefect of the Vatican Library between 1565 and 1566 (Anders, Jansen, and Reyes García 1996, 15–19; Domenici 2016b, 351–352).

PIGNORIA AND THE ASSIMILATION OF MESOAMERICAN COSMOLOGY

The order assigned by Pignoria to the images of Mexican idols has been recently studied by the art historian Margit Kern. She laments that most scholars have not noticed that "the description of the Mesoamerican gods follows a particular chronology based on the model of the Christian narrative of salvation with no explicit explanation of any such narrative strategy in the text" (Kern 2016, 18). According to her, Pignoria inscribed Mesoamerican cosmology into a chronology comprising various key events from the Bible, extending from the God of creation, the original progenitors of mankind, and the Flood to the Annunciation and all the way to the introduction of Christianity (Kern 2016, 22). This "narrative of salvation" would correspond to a "Neoplatonic body of thought and mark out Pignoria as an adherent of *philosophia perennis*" (Kern 2016, 22). According to Kern, Pignoria supported a historical and genealogical argument according to which it was possible to consider "Egypt as the land of origin of New Spain's cults" (Kern 2016, 18).

However, the hypothesis advanced by Kern collides with the Aristotelian theological and philosophical formation of Pignoria, which distances him from the supposed Neoplatonic body of thought. According to Mulsow, Italian scholars such as Pirro Ligorio, Girolamo Aleandro, Lorenzo Pignoria, or Giacomo Filippo Tommasini indeed received Neoplatonism and Hermeticism in their antiquarian studies. It is true that mythographers of the first renaissance culture did not "hesitate to repeat over and over—though with an increasing apparatus of knowledge about antiquity—late antique allegorical interpretations of the myths about the gods." However, "around 1600, philological and antiquarian knowledge tended more and more to destroy these old allegories" (Mulsow 2005, 187). Moreover, there is no sign in Pignoria's appendices of a *philosophia perennis* that—tracing a genealogical connection between Egyptian and Mexican idolatry—would have incorporated Mesoamerican gods into a sort of *prisca theologia*, with paradoxical effects on colonial rhetoric. Actually, Pignoria did not positively assess Mesoamerican cosmology. He interpreted both Egyptian and Mexican idolatry as a form of "barbarism" that had produced a sort of *pseudo Theologia* (Cartari and Pignoria 1615, XIII).

The deconstruction of Mesoamerican cosmology is performed relentlessly by means of rhetorical arguments relating it to the imitative action of the devil, which delegitimizes any possible positive analogy with Christianity (Seznec 1931, 274). However, Pignoria does not seem aware of the fact that the hybrid cosmology of the *Codex Vaticanus A* was already an attempt to assimilate indigenous religion. Therefore, the antiquarian amplified the diabolic interpretation of those potentially comparable features, without understanding that comparability itself was manufactured by the hybridization shaped by and in the glosses of the codex.

In this perspective, Pignoria's discourse on Mesoamerican gods opens with an Egypto-genetic argument that aims to outline a comprehensive history of religion (Miller 2001, 202) in which Egypt is conceived of as the place of origin of superstition (MacCormack 1995, 88–93).[12] Pignoria asserts a diffusionist theory of idolatry (Rubiés 2006),[13] based on the authority of Herodotus, who had suggested the prominence of Egyptian culture in antiquity. In this comprehensive and innovative history of religion, Pignoria proposes a negative and "exclusive" evaluation of Egypt, drawing a clear line between true and false religion: "His diffusionism results from his efforts to keep the history of religion within the framework of *historia sacra*: the distinction between true and false religion preestablishes the conditions according to which he develops his views. This then leads almost inevitably to a diffusionist model, both for Hebrew culture with regards to true religion and for Egyptian culture with regards to false religion" (Mulsow 2005, 195). Furthermore, Pignoria elaborated a comparative and visual methodology, based on an innovative use of images and objects. As an expert on Egyptian objects and iconography, he found similarities with Mexican idols. The correspondences he found between the images of American and Egyptian "idols" constituted the starting point for a *vera narratio*, a process in which the concept of *conformità* became a multilayered heuristic instrument (von Wyss-Giacosa, in press).[14] As a result, starting from the hypothesis of the existence of "conformity" between Egypt and Mexico, Pignoria traced a *historia sacra* and incorporated Mesoamerican cosmology within a global history of religion.

Emphasizing the resemblance between Mexican and Egyptian gods,[15] Pignoria finally opens his description with a portrait of the god Homoyoca (figure 2.1), who is compared to the Roman Jupiter (Cartari and Pignoria 1615, IV). This attempt to compare the Mexican and the Roman gods is not present in the *Codex Vaticanus A*. However, the innovation does not entail that Pignoria encouraged a positive evaluation of Mesoamerican religion. On the contrary, as evident for example, by his use of the concept of *misera gentilità*, Pignoria exposes an entirely negative theological judgment (Lein 2002, 229).

FIGURE 2.1. *The god Homoyoca, implicitly compared to the Roman god Jupiter via an image redrawn from the* Codex Vaticanus A, *IV. Compare figure 2.2. (Cartari and Pignoria, 1615, V.)*

The image of Homoyoca provides a clear expression of the antiquarian methodology of Pignoria. To demonstrate the "conformity" of Mesoamerican gods, the antiquarian applies a visual comparison that—despite the fact that the illustration was clearly deduced from the *Codex Vaticanus A* (figure 2.2)—produced an autonomous meaning. First, Pignoria isolates the god from its original context, eradicating the representation of the cosmos that, as noted before, is a basic feature of the representation contained in folio IV of the codex. In that contest, the sequence organized into the first three pages of the codex shows a complex vertical representation of the cosmos divided into three levels. It depicts "the Aztec cosmos with its thirteen celestial layers and nine levels of the underworld" (Quiñones Keber 1995a, 235). Contrary to what occurs with the Homeyoca image offered by Pignoria, in folio IV of the *Codex Vaticanus A* the god is placed in the upper part of the representation, as the highest of the nine upper skies was considered its dwelling (otherwise, in folio 2r the middle region and the nine layers of the underworld are depicted).

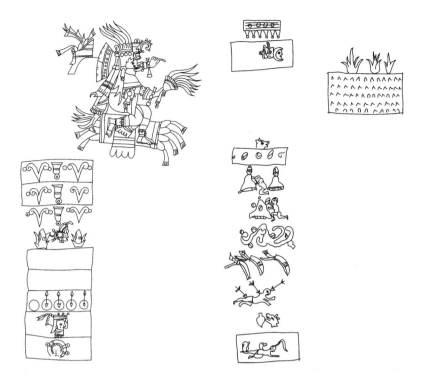

FIGURE 2.2. *The representation of the upper nine layers in the* Codex Vaticanus A. *(Folio 1v, sixteenth century. Drawing by Victor Medina.)*

Below the pictorial representation of the god, celestial layers are represented by nine horizontal strips painted in different colors, to which alphabetic glosses with the respective Nahuatl names were added. However, Pignoria produced a definitive westernization of the indigenous cosmology that had already been displayed in the *Codex Vaticanus A*, both on a pictographic and a textual ground. Actually, in the Italian glosses of folio 1v, the highest of the nine layers of the sky became the dwelling of a creative god, considered as a sort of Mesoamerican "first cause" and consequently comparable with the hegemonic god of the Romans.

The aim of this chapter is not trying to understand whether the representation of the cosmos can be considered "authentically" Mesoamerican or rather a product of the hybridization that took place during the colonial era (Nielsen and Sellner 2009). More interestingly, Pignoria obstinately pursued the path of a Christianization of Mesoamerican cosmology. First, he identically reproduced the data contained in the explanatory glosses of the codex,[16] which proposed a theological discussion around the cosmological content of the pictorial images, so that "the text, written by two scribes, represents a European intrusion" (Quiñones Keber 1995a, 236). Consequently, the original vertical representation of the cosmos contained in the *Codex Vaticanus A* is, paradoxically, distributed by Pignoria on a chronological level: Homoyoca is transformed into a creator god who opens the road, by analogy with the biblical account, of a *historia sacra*. The transformation of the Mexican god in a First Cause represents the theological instrument for establishing its potential comparability. Indeed, Pignoria hopes that—despite their barbarism—the indigenous people could have a glimpse of "natural light" that allows them to know the mystery of the Holy Trinity.[17] Consequently, the Mexican god can be conveyed, through the application of the visual methodology, within a global history of religion. Pignoria applies again his Egyptian interpretation by inserting two medallions of Osiris and Isis (which are not originally included in the *Codex Vaticanus A*) within the illustration of the god Homoyoca.

These two figures were taken from the Mensa Isiaca (also known as the Tabula Bembina) (Pignoria 1605), an Egyptianizinging Roman table, probably dating from the reign of the emperor Claudius and unearthed around 1520 (Curl 2005, 110; Mason 2001, 133–134). Pignoria studied the Mensa Isiaca in 1605 and took from it material that would later appear in the appendix to Cartari's work (Maffei 2013, 70; Seznec 1931, 278). This unique visual empiricism allowed him to prove his hypothesis. The images taken from the altar of Isis were juxtaposed to Homoyoca "in such a way that on the one hand all the idols resemble one another in their kneeling posture, and on the other the overall composition bears a formal similarity to the Egyptian images of gods with their framed cartouches" (Kern 2016, 18).[18] As a result, the interpretation of this image produces a step forward towards the "colonial semiosis" (Mignolo 1995) begun by the commentators of the *Codex Vaticanus A*. In the latter, the theological discussion was the product of a competitive negotiation between text and images. In the former, instead, Pignoria proposed an even more effective "theological intrusion." In order to confirm the hypotheses concerning the Egyptian origin of the gods, Pignoria adds an image of Osiris and Isis that he personally observed in the Mensa Isiaca (figure 2.3), proposing a comparison based on the analysis of the gods' gestures and bodies (Lein 2002, 231; Mulsow 2005, 187).

Later, Pignoria expounds a brief genealogy of Mexican gods, recovering the texts of the *Codex Vaticanus A*. He enhances the potential analogy with biblical accounts and shows that Homoyoco generated—before the Flood—Cipatoual and his wife Xumoco, interpreted as if they were a sort of Adam and Eve

FIGURE 2.3. *Osiris and Isis. Pignoria sought to buttress his theory of the Egyptian origin of Mesoamerican gods by employing similar poses and gestures in his cross-cultural images of deities. (Cartari and Pignoria, 1615, VII.)*

(Lein 2002, 231). Then, at pages IX–XII, four pairs of gods are depicted, which came from folio 2v of the *Codex Vaticanus A* and are described in detail in the glosses of folio 3r. The illustrations of the four couples follow a completely distinct order compared to that of the codex. Besides, Pignoria mentions only the names of male gods, associating them with classical underworld deities (Lein 2002, 232).[19]

The next two images represent a step forward towards a theological interpretation of the Mesoamerican gods. The first one—at page XIV—is a representation of a divine messenger announcing the birth of Quetzalcoatl (figure 2.4), represented at page XVII (Mason 1997, 20). The first of these two images reveals a profound transformation, both visually and textually, compared to that included in folio 7r of the *Codex Vaticanus A*.

Again, Pignoria offers a decontextualisation of the image of the "messenger of the god Citlallatonac" included in the codex. The isolation of the figure from

FIGURE 2.4. *The "messenger" of Citlallatonac. Pignoria redrew an image from the Codex Vaticanus A (see figure 2.5) to evoke the Christian archangel Gabriel and his announcement to the Virgin Mary of the coming birth of Christ. (Cartari and Pignoria, 1615, XIV.)*

its original pictographic context produces the eradication of the mythological significance of the episode, described in the glosses of the codex as a ceremony of flowers that would take place during the fourth age, that of the goddess Xochiquetzal (Anders, Jansen, and Reyes García 1996, 63–69) (figure 2.5).

This separation from the mythological context allows Pignoria to redefine completely the image of the codex. On the one hand, the representation loses any connection to mythical events, as shown by the disappearance of the flower from the hand of the messenger of the god Citlallatonac. His body has been rotated by ninety degrees by Pignoria, thus producing a figure rushing with outstretched arms from left to right, which would immediately remind any viewer familiar to some degree with Christian iconography of the Annunciation (Mason 1997, 20). Hence, "the birth of the God Quetzalcoatl had been announced to his virgin mother by a figure who reminded Pignoria of the angel Gabriel" (MacCormack 1995, 93).

FIGURE 2.5. *Pignoria redrew the central image of this page (7r) of the* Codex Vaticanus A *to evoke Christian images of angels (see figure 2.4), and thus visually to compare the announcement of the miraculous birth of Quetzalcoatl with the archangel Gabriel's annunciation to the Virgin Mary of her impending miraculous conception and the birth of Jesus. (After A. Aglio; Kingsborough 1830, vol. 4.)*

It is interesting to note that a symmetrical change also takes place on the textual level. Once again, Pignoria follows the text of the *Codex Vaticanus A*, summarizing it and eliminating every mythological subject.[20] He extends his interpretation of the miraculous birth of Quetzalcoatl by explicitly identifying the messenger with the archangel Gabriel. But Pignoria also places a strong emphasis on diabolical imitation (as was not done by the commentator of the codex), by citing a passage from the Second Epistle of Paul to the Corinthians (2 Cor 11:14): *Satanas transfiguratur in Angelum lucis*, "Satan himself is transformed into an angel of light."[21] Through the usage of the polemical Pauline arguments against the "pseudo-apostles," Pignoria rejects any possible Neoplatonic interpretation of the birth of Quetzalcoatl.[22] Therefore, the following image of Quetzalcoatl (figure 2.6)—in his manifestation as god of the wind—becomes the cornerstone of Pignoria's *historia sacra*. As noted before, the image and the text reproduced the corresponding representation contained in folio 7v of the *Codex Vaticanus A*.

FIGURE 2.6. *Quetzalcoatl, adapted by Pignoria from the* Codex Vaticanus A *(see figure 2.7), with subtle Christianizing features, such as the cross on the object to the right. (Cartari and Pignoria, 1615, XVII.)*

Compared to the text of the codex,[23] Pignoria makes only some minor changes in the discourse with his selection of information to convey and, of course, his lack of knowledge of the Nahuatl language.[24] However, significant changes are produced by Pignoria on a visual ground (figure 2.7). On the one hand, his figure is once more isolated from its original pictographic context; on the other, some subtle differences in the depiction of the god that Pignoria introduces are extremely important. For instance, Jansen has pointed out that, if we look at the incensory to the right of the god, it can be noticed that Ferroverde added on it a cross that was absent in the codex (Jansen 1984, 78n104). This curious novelty carries Pignoria's discourse on extremely challenging grounds. Far from being a simple reproduction from the codex, Pignoria's image hides a complex epistemological operation, which acts through subtle interventions in order to incorporate within his "Mexican" images a significantly Christian view. Pignoria could unlock the decisive transition of its *historia*

FIGURE 2.7. *Quetzalcoatl in the* Codex Vaticanus A *(folio 7v), which Pignoria adapted in figure 2.6. Note the crosses on his cape, perhaps representing to some a prehispanic "preparation for the Gospel." (After A. Aglio [Kingsborough 1830, vol. 4].)*

sacra describing the four reasons—once again of a visual kind—that justify his comparative exercise. In the first place, the antiquarian noted the presence of a headdress—which he considers to be a diabolical object—that could also be found in classical ceremonies (Cartari and Pignoria 1615, XV–XIX). The second and third elements of the depiction are the *lituus* that Quetzalcoatl carries in his right hand—that resembles those of the Roman augurs—and the *cornucopia* that the Mexicans used as an incensory and that was mentioned in classical legends. Finally, the fourth element—the cross—is obviously crucial to propose an original discourse compared to the description of Quetzalcoatl contained in the *Codex Vaticanus A*.[25] Apparently, some chroniclers interpreted the presence of the crosses as a sign of a "preparation for the Gospel." In this direction, Pignoria was only exposing the existence of such hypotheses proposed by chroniclers—among which Botero stands out—who considered the precolonial presence of crosses in the New World as a way to pave the way for the Gospel (Cartari and Pignoria 1615, XVIII).

Nevertheless, Pignoria did not argue in favor of or against these genealogical hypotheses, nor did he try to support the assumption of a prehispanic *preparatio evangelica*. Actually, he was only advocating his own antiquarian methodology. In this direction, he warned the readers about the fact that the presence of prehispanic crosses visually reminded him of a rare and similar one he had seen on a "Medaglia di Constantino il Grande" (Medal of Constantine the Great) (Kern 2016, 21) (figure 2.8). From his point of view, similarities between these material objects, whose presence is confirmed thanks to a careful use of antiquarian empiricism, is once again sufficient to produce a visual sort of intercultural epistemology.[26]

It is obvious that the comparability with the Constantinian cross acts also as a rhetorical instrument allowing Pignoria to resume his *historia sacra*. In conclusion, the antiquarian could add an image of Constantine (figure 2.9), crowned by Victory. Atop his staff is the chi-rho symbol, visually alluding to his motto, *Hoc signo victor eris* ("In this sign will you triumph"), which gives the definitive meaning to the assimilation of Mesoamerican cosmology into a Christian genealogy.

Contrary to what was claimed by Kern, the Constantinian reference does not legitimize a Neoplatonic "narrative of salvation," or proximity with a *philosophia perennis* (Kern 2016, 22), but rather a sort of theological-political recapitulation of the history of inclusion of every form of paganism. In the age of Reformation, the appropriation of the image of the Roman emperor played a key role in the competition among kings and the papacy, at least until the Peace of Westphalia in 1648. The Constantine symbology provided the rhetorical tools to legitimize opposing propagandistic discourses. While European nations proposed an image of Constantine as a precursor to a "spiritual empire"—marking, for example, the imperial triumphs of the Habsburgs—the papacy continued to promote the image of the emperor as protector of a Church that intended to play an active role within European politics; the papacy still hoped to exercise a *potestas indirecta in temporalibus* (Biasiori 2013; Freiberg 1995). In the years when the work by Pignoria was completed, the commemoration of Constantine was a significant element in

FIGURE 2.8. *The image of the cross represented in a Constantine medal. (Cartari and Pignoria, 1615, XX.)*

the reconstruction and decoration of the Lateran Palace by Pope Sixtus V (Quednau 2013). At the same time, Paul V wanted to paint the new Vatican archive room with scenes of gifts, privileges, and duties offered to the church by great rulers, among which Constantine stood out (Friberg 1995). Therefore, the glorification of the Roman emperor represented an instrument of political self-representation for the papacy: Constantine epitomized the founding image of a dominion over the entire world, of the final conversion of the pagans and the destruction of all idols. If Constantine was *primus imperatorum christianae fidei propagator*, Pignoria uses his image to incorporate definitively Mesoamerican idolatry into a Christian theological-political cosmology.

CONCLUSION

Documentary sources have their own independent cultural life. In the change of historical and cultural contexts, they become part of different orders

FIGURE 2.9. Hoc signo victor eris. *Pignoria employed the image of the emperor Constantine, a convert from paganism to Christianity, as an emblem of the spreading spiritual and political power of the Church into Mesoamerica. (Cartari and Pignoria, 1615, XXII.)*

of discourse. As noted, the hybrid cosmology produced in the "local" confrontation between Christianity and indigenous religions initially served its colonial function, conveying the competition for the monopoly on supernatural resources in New Spain. Nevertheless, the *Codex Vaticanus A* was intended since its very creation to carry out part of its cultural life in another context. Arriving in Italy, its hybrid cosmology was resignified in order to exercise a new function. In the environment in which was inserted by Pignoria, it no longer served to legitimize a local colonial order, but to contribute to the naturalization of a renewed concept of religion. The discourse on alternative antiquities was mobilized to promote the global triumph of Christianity over religious otherness. The ethnographic information on the confusion and novelties of "alien" religions should not remain without explanation. They generated a genealogical preoccupation that led to the questioning of traditional interpretative models. The antiquarian culture, of which Pignoria is a pivotal exponent, showed its paradoxical character. The crisis of comparison produced the need for a new empiricism, capable of molding a revolutionary historical method. This revolution, however, also continued to play its antihistorical function. The *vera narratio* regarding indigenous cultures is nothing but a particular form of a universal *historia sacra*. Renewed cosmology is not a *philosophia perennis*. It should be included and subordinated to a global order, within which a universalized notion of religion pursues the legitimization of European culture and Christianity.

ACKNOWLEDGMENTS

I would like to thank Ana Díaz for her invitation to contribute to the present volume, as well as for her careful editing. Davide Domenici provided me with useful information on Italian collections and suggestions that significantly improved the quality of this chapter. Paola von Wyss-Giacosa allowed me to read an unpublished text on Pignoria. Finally, Chiara Ghidini carefully read an earlier draft of this text and provided useful suggestions. However, the responsibility for the chapter's contents and for any errors or misunderstandings is solely mine.

NOTES

1. On the "religious encounter" between Europeans and indigenous cultures, see Burkhart (1989) and Pardo (2004).

2. On the *Codex Vaticanus A*, see, among many others, Anders, Jansen, and Reyes García (1996) and Domenici (2016b).

3. For a discussion on the concept of "colonial discourse" in Latin American studies, see, among others, Adorno (1988); Klor de Alva (1992); Mignolo (1995); Seed (1991).

4. As introduction to a Foucaldian discursive perspective in the study of religion, see von Stuckrad (2003).

5. On the destruction and reconstruction of Mesoamerican cosmology in the *Codex Vaticanus A*, see Díaz (2009); 2015; Mikulska (2008); Nielsen and Sellner Reunert (2009, 2015); Schwaller (2006).

6. "After conquest had given way, more or less, to the needs of settled government, making sense of the Americas became a matter of state, rather than simply an expression of curiosity. The establishment of an administrative apparatus of empire underpinned a need for more secure information about the land and its inhabitants" (Miller 2001, 186).

7. "Pourtant, si fragile qu'il soit, le petit essai de Pignoria ébauchait les travaux futurs sur l'origine des religions et leur diffusion dans le monde. Il donnait l'un des

premiers exemples de la méthode comparative appliquée à ces problèmes, presque vierges encore. Il enseignait aussi qu'en ces matières il faut savoir être hardi, au point d'admettre ce qui parait invraisemblable tout d'abord. Pignoria, nous l'avons vu, blâme comme d'une faiblesse les géographes anciens d'avoir tenu pour de pures légendes toutes les relations de voyage et d'expédition aux pays très lointains" (Seznec 1931, 280).

8. For an introduction to the collection of American object in Italian modern history, see Feest (1995) and Heikamp (1976).

9. "Nell'isole scoperte gl'anni passati da Spagnoli, che ora si addimandano Mondo nuovo, perché agli antichi furono incognite, si è trovato che quei popoli venerano alcuni idoli fatti qual di creta, qual di legno e qual di pietra" (Cartari 1608, 17).

10. "In Galleria io stimavo che ci fosse qualche Idolo Indiano, perchè nella Vigna di S. A. in Roma io viddi pitture di que' paesi; et dalle gallerie degl'altri Prencipi io ho pure havuto qualche curiosità di questa sorte. Et noti V. S. ch'io non domando cose Egittie, ma Indiane, come della China, del Giapone, del Pegù, et parti simili dell'Indie Orientali; dell'Occidentali ancora, come Perù, Mexico, Nova Spagna etc" (Favaro 1929–1939, XI, 414).

11. "Tutte le sopra registrate immagini con le notitie principali di esse, accresciute però da me con qualche raffronto Historico, & co' Paralleli delle atiche superstitioni d'altri popoli, io le ho hauute dall'Illustriss. Sig. OTTAVIANO MALIPIERO Senatore grauissimo & d'amabilissima placidità di natura. Furono per quanto ho inteso del Cardinale AMVLIO gloriosa memoria, & io le stimo assai più, che alcune altre narrationi d'huomini poco versati, che vanno in volta, & si leggono tutto'l di" (Cartari and Pignoria 1615, XXIII).

12. "HERODOTO sensato scrittore, & non così bugiardo, come volgarmente è tenuto, parlando dell'Egitto, scrive, che ha cose piu marauigliose, che qual si voglia altro paese; & che sopra ogn'altra parte del mondo, si vedono in questa opere, alle quali la penna de' Scrittori non arriua. E veramente quella d'Herodoto non si può chiarore hiperbole, vedendoli piene le carte e sacre e profane, della grandezza, delle forze, delle ricchezze di quel grandissimo, e nobilissimo Regno" (Cartari and Pignoria 1615, I).

13. "Hora se gl'Egittij havessero cognitione dell'Indie Occidentali ò nò, molto c'è che dubitare; tuttauia Benedetto Aria Montano nel suo Apparato alla Biblia Reggia, tiene, che la terra Ophir nominata ne' Libri de Re, & nei Paralipomeni fosse il Perù & la Noua Spagna" (Pignoria 1615, IV). On a diffusionist theory by Pignoria, see Miller (2001, 203); Mulsow (2005, 194); Seznec (1931, 278–279). On the geographical controversy about Ophir and the New World, see Gliozzi (1977).

14. "Et in somma per tutto questo, che chiamano nouo mondo, tanto nell'Occidente, quanto nell'Oriente, io ho auertito tanta la conformità fra le superstitioni Egittiane, & quelle del Paese, che ho hauuto a marauigliarmi alcune volte" (Cartari and Pignoria 1615, XXVII).

15. "Ma lasciando da parte le auttorità, io mi voglio valere in questo proposito, d'vna congettura non punto debole, & è, che i popoli di questa parte di mondo si sono conformati in maniera della fabbrica de gl'Idoli loro con le imagini delle Deità Egittie, che niente più" (Cartari and Pignoria 1615, IV).

16. "Voleva dire questo in quell'Idioma tanto quanto il Creatore del tutto, ovvero la prima causa, e lo chiamavano ancora Hometeutle, quanto signore di Tre dignità, o signore tre . . . E li otomies. Chiamauano la stanza di questo loro Dio Narihnepamuhca, che volea dire sopra le none compositioni, o per altro nome Homeiocam, cioè luoco del signor trino" (Cartari and Pignoria 1615, VI). "Questo vuol tanto dire come il luogo dov'è il Creatore del tutto, ò la prima causa. Chiamanlo per un altro nome Hometeule, che vuol tanto dire come sig[nor]e di tre dignità, ò signore tre, e li Otomies chiamavano questo luogo dove lui è hiue narichnepaniucha, che vuol'dire sopra le viiii composture ò compos[izio]ni; et per altro nome homeiocan .i. [id est] luogo del S[ign]ore Trino" (*Codex Vaticanus A*, folio 1v).

17. "E cli qua si vede apertamenre quanto sia vero quello, che scriue S. Paolo, che le cose inuisibili di Dio, dall'homo si comprendono bene spesso per mezo di quelle, che si vedono; poiche in mezo a questa barbarie riluceua pure vn poco di lume di noue cause superiori, che noi chiamiamo Cieli, & della prima causa, nella quale adombrauano cosi à modo loro l'ineffabile misterio della Santissima Trinità" (Cartari and Pignoria 1615, VI).

18. "Hora questo Homoyoca, & nelli abbigliamenti, & nella positura io direi, che fosse tolto poco meno che di peso da gl'Egittij, appresso a quali Osiride in tale maniera si figuraua, come si vede, & io notai già nell'antichissima mensa Isiaca del Serenissimo Signor Duca di Mantoua, nell'orlo della quale dicisette volte si vede una simile Imagine, variata però in quanto à gli ornamenti" (Cartari and Pignoria 1615, VI).

19. "Haueuano oltre questo i Mexicani il Dio Miquitlantecatle, che voleua dire il Signore dell'Inferno, per altro nome Tzitzimitl, il medesimo che Lucifero; & quello con alcuni altri della medesima classe, haueua la gamba dritta ranicchiata" (Cartari and Pignoria 1615, VIII).

20. "Qui fingono li miserabili certi sogni della loro cecità, dicendo che un Dio, che se diceva Citlallatonac, ch'è quello segno, che si vede in cielo detto strada di S[an]to Iacobo ò via latea, manda un amb[asciato]re dal cielo con u[n]a ambasciata à una vergi[n]e che era in tulan, che si diceva chimalman, che vuole dire rotella, la quale haveva due sorelle, l'una hochitlique [xochitlicue], e l'altra conatlicue [coatlicue]" (*Codex Vaticanus A*, folio 7r).

21. "Rappresentauano in pittura questi vn'Ambasciatore del Dio Citlallatonac (così chiamauano la via Lattea) mandato ad vna Vergine, che habitaua in Tulan detta per nome Chimalma, cioé Rotella, alla quale disse l'Ambasciatore, che Dio voleua, che essa concepisse vn figliolo; il quale fu conceputo senza congiontione d'huomo, & fu chiamato Quetzalcoatle; sì che questo Ambasciatore fu'l Gabriele (se così e lecito

à dire) di questi miseri; & cosi *Satanas transfiguratur in Angelum lucis*" (Cartari and Pignoria 1615, XIII).

22. "Qvesto è il ritratto dell'Ambasciatore sopradetto, nel quale io ho con qualche marauiglia fatto riflessione sopra l'ornamento del capo, che è molto simile a que cartocci, che gl'Egitij piantauano in capo al loro Harpocrate, come si può vedere nella statua, ch'io ho appresso di me" (Cartari and Pignoria 1615, XV).

23. "Q[uest]o quetzalcoatla[co] topiltzin, che vuol dire n[ost]ro molto caro figl[iol]o vedendo che non cessavano li peccati, et travagli del mondo, dicono che così come è stato il p[rim]o, che principiò ad invocare li Dei et fare loro sacrificij, così è stato il p[rim]o che hà fatto penitenza à fine di placare li Dei, accio che perdonassero al suo populo. Dicono, che sacrificò se medesimo cavando il p[ro]p[ri]o sangue con spine et altre sorti di penitenza" (*Codex Vaticanus A*, folio 7v).

24. "Hora questo Quetzalcoatl fu chiamato ancora Topilczin, cioe mio molto amato figliolo, e dicono, che nascesse con l'vso di ragione, & che fosse'l primo, che cominciasse, ad inuocar li Dei, e far loro sacrificij, co'l suo sangue medesimo, che si cauaua dalla persona con spine, & in altre maniere" (Cartari and Pignoria 1615, XV).

25. "La quarta più notabile & più riguardeuole dell'altre è la figura della Croce, che si vede in tre luochi, due nel mantello, & vna nel corpo dell'Incensiere, che cosi chiamauano i paesani quello, che noi habbiamo nominato Cornucopia. E veramente che questa non sia Croce io non dubito punto, & questo tanto più, quanto si vede, che nostro Signore Iddio, per sua misericordia, fece strada grande alla preparatione dell'Euangelio in alcuno di quelli paesi" (Cartari and Pignoria 1615, XVIII).

26. "Et in proposito mi souuiene di notare, come vna similisiima se ne vede in vna rarissima Medaglia dì Costantino il Grande, non publicata ne auuertita da alcuno, ch'io sappia, a quest'hora, della quale ho posto il dissegno per hauerla io appresso di me. Io stimo non poco questa Medaglia, poiche pare, che molti si siano accordati a credere, che a Costantino apparisse il legno de la Croce in aria, (come scrivono tutti gli Historici Ecclesiastici di que' tempi) nelle due prime lettere del nome di CHRISTO scritto in greco, come portarono poi nell'Insegna maggiore dell'essercito gli Imperatori seguenti. Et veramente la congettura non è irragioneuole, si per la rarità delle Medaglie di Costantino con la croce, si per la testimonianza di Costanzo suo figliuolo, che fece battere monete, come qui sotto co'l motto HOC SIGNO VICTOR ERIS" (Cartari and Pignoria 1615, XIX–XXI).

REFERENCES

Adorno, Rolena. 1988. "El sujeto colonial y la construccion cultural de la alteridad." *Revista de Crítica Literaria Latinoamericana* 14 (28): 55–68.

Anders, Ferdinand, Maarten Jansen, and Luis Reyes García. 1996. *Religión, costumbres e historia de los antiguos mexicanos: Libro explicativo del llamado* Códice Vaticano A. Mexico: Akademische Druck und Verlagsanstalt/Sociedad Estatal Quinto Centenario/Fondo de Cultura Económica.

Biasiori, Lucio. 2013. "Costantino e i re della prima Età moderna (1493–1705): Imperatore cristiano o re sacerdote?" In *Costantino I. Enciclopedia costantiniana sulla figura e l'immagine del cosiddetto Editto di Milano*, ed. Alberto Melloni: III, 18–30. Roma: Treccani.

Botta, Sergio. ed. 2013. *Manufacturing Otherness: Missions and Indigenous Cultures in Latin America*. Newcastle upon Tyne: Cambridge Scholars Publishing.

Bourdieu, Pierre. 1971. "Genèse et structure du champ religieux." *Revue française de sociologie* 12 (3): 295–334.

Bourdieu, Pierre. 1973. *Esquisse d'une theorie de la pratique, précédé de trois essais d'ethnologie Kabyle*. Genève: Droz.

Burke, Peter. 2003. "Images as Evidence in Seventeenth-Century Europe." *Journal of the History of Ideas* 64 (2): 273–296.

Burkhart, Louise. 1989. *The Slippery Earth: Nahua-Christian Moral Dialogue in Sixteenth-Century Mexico*. Tucson: University of Arizona Press.

Cartari, Vincenzo. 1556. *Le immagini con la spositione de i dei degli antichi: Raccolte per Vincenzo Cartari*. Venezia: Francesco Marcolini.

Cartari, Vincenzo. 1571. *Le Imagini de gli Dei de gli Antichi, nelle quali si contengono gli Idoli, Riti, ceremonie, and alter cose appartenenti alla Religione de gli Antichi*. Venezia: Giordano Ziletti.

Cartari, Vincenzo. 1608. *Le Imagini de gli Dei de gli Antichi, del Signor Vincenzo Cartari Reggiano, novamente ristampate et ricorrette [...]*. Padova: Pietro Paolo Tozzi.

Cartari, Vincenzo, and Lorenzo Pignoria. 1615. *Le Vere e Nove Imagini de gli Dei de gli Antichi di Vicenzo Cartari Reggiano [...]*. Padova: Pietro Paolo Tozzi.

Cartari, Vincenzo, and Lorenzo Pignoria. 1626. *Seconda Novissima editione delle Imagini de gli dei delli antichi di Vincenzo Cartari Reggiano [...]*. Padova: Pietro Paolo Tozzi.

Cartari, Vincenzo, and Lorenzo Pignoria. 1647. *Imagini de gli dei delli antichi di Vincenzo Cartari Reggiano [...]*. Venezia: Tomasini.

Curl, John Stevens. 2005. *The Egyptian Revival: Ancient Egypt as the Inspiration for Design Motifs in the West*. London and New York: Routledge.

Díaz, Ana. 2009. "La primera lámina del *Códice Vaticano A*. ¿Un modelo para justificar la topografía celestial de la antigüedad pagana indígena?" *Anales del Instituto de Investigaciones Estéticas* 31 (95): 5–44.

Díaz, Ana. 2015. "La pirámide, la falda y una jicarita llena de maíz tostado: Una crítica a la teoría de los niveles del cielo mesoamericano." In *Cielos e inframundos:*

Una revisión a la cosmología mesoamericana, ed. Ana Díaz, 65–107. México: Universidad Nacional Autónoma de México-Instituto de Investigaciones Históricas, Fideicomiso Felipe Teixidor y Montserrat Alfau de Teixidor.

Domenici, Davide. 2016a. "Nuovi dati per una storia dei codici messicani della Biblioteca Apostolica Vaticana." *Miscellanea Bibliothecae Apostolicae Vaticanae* XXI: 341–362.

Domenici, Davide. 2016b. "The wandering 'Leg of an Indian King': The Cultural Biography of a Friction Idiophone Now in the Pigorini Museum in Rome, Italy." *Journal de la société des américanistes* 102 (1): 79–104.

Domenici, Davide. 2017. "Missionary Gift Records of Mexican Objects in Early Modern Italy." In *The New World in Early Modern Italy, 1492–1750*, ed. Elizabeth Horodowich and Lia Markey, 86-102. Cambridge: Cambridge University Press.

Ehrle, Franz. 1900. *Il manoscritto messicano Vaticano 3788 detto il Codice Rios*. Rome: Danesi.

Elliot, John H. 1970. *The Old World and the New (1492–1650)*. Cambridge: Cambridge University Press.

Favaro, Antonio. 1929–1939. *Le opere complete di Galileo Galilei*. Firenze: G. Barbera.

Feest, Christian. 1986. "Zemes idolum diabolicum. Surprise and Success in Ethnographic Kunstkammer Research." *Archiv für Völkerkunde* 40: 181–198.

Feest, Christian. 1992. "North America in the European Wunderkammer before 1750, with a Preliminary Checklist." *Archiv für Völkerkunde* 46: 61–109.

Feest, Christian. 1995. "The Collecting of American Indian Artifacts in Europe, 1493–1750." In *America in European Consciousness, 1493–1750*, ed. Karen Ordahl Kuppermann, 324–360. Chapel Hill: University of North Carolina Press.

Freiberg, Jack. 1995. "In the Sign of the Cross: The Image of Constantine in the Art of Counter-Reformation Rome." In *Piero della Francesca and His Legacy*, ed. M. Aronberg Lavin, 67–87. Hannover and London: Studies in History of Art.

Glass, John B., and Donald Robertson. 1975. "A Census of Native Middle American Pictorial Manuscripts." In *Handbook of Middle American Indians*, 14: *Guide to Ethnohistorical Sources*, ed. Howard F. Cline, 81–252. Austin: University of Texas Press.

Gliozzi, Giuliano. 1977. *Adamo e il nuovo mondo: La nascita dell'antropologia come ideologia coloniale: dalle genealogie bibliche alle teorie razziali (1500–1700)*. Milan: Franco Angeli Editore.

Gruzinski, Serge. 1988. *La colonisation de l'imaginaire: Sociétés indigènes et occidentalisation dans le Mexique espagnol, XVIe–XVIIIe siècle*. Paris: Gallimard.

Gruzinski, Serge. 2004. *Les quatre parties du monde: Histoire d'une mondialisation*. Paris: La Martinière.

Hanegraaff, Wouter. 2016. "Reconstructing 'Religion' from the Bottom Up." *Numen* 63: 577–606.

Heikamp, Detlef. 1976. "American Objects in Italian Collections of the Renaissance and Baroque: A Survey." In *First Images of America: The Impact of the New World on the Old*, ed. Fredi Chiapelli, Michale J.B. Allen, and Robert L. Benson, 1: 464–467. Berkeley: University of California Press.

Honneth, Axel. 1995. *The Struggle for Recognition: The Moral Grammar of Social Conflicts*. Cambridge: Polity Press.

Jansen, Maarten. 1984 "El Codice Rios y Fray Pedro de los Rios." *Boletin de Estudios Latinoamericanos y del Caribe* 36: 69–81.

Kern, Margit. 2016. "Pictorial Theories by Missionaries in Sixteenth-Century New Spain: The Capacities of Hieroglyphs as Media in Transcultural Negotiation." *Art in Translation* 8 (3): 1–31.

Kingsborough, Edward. 1830. *Antiquities of Mexico: Facsimiles of Ancient Mexican Paintings and Hieroglyphs*. vol. 4. Londres: A. Aglio.

Klor de Alva, Jorge J. 1989. "La historicidad de los coloquios de Sahagún." *Estudios de Cultura Náhuatl* 15: 147–184.

Klor de Alva, Jorge J. 1992. "Nahua Colonial Discourse and the Appropriation of the (European) Other / Le Discours colonial des Nahuas et l'appropriation de l'Autre (européen)." *Archives des sciences sociales des religions* 77: 15–35.

Knowlton, Timothy, and Gabrielle Vail. 2010. "Hybrid Cosmologies in Mesoamerica: A Reevaluation of the *Yax Cheel Cab*, a Maya World Tree." *Ethnohistory* 57 (4): 710–739.

Lein, Edgar. 2002. "Imagini Degli Dei Indiani. La representación de las divinidades indianas por Vincenzo Cartari." In *Herencias indígenas, tradiciones europeas y la mirada europea*, ed. Helga von Kügelgen, 225–258. Madrid and Frankfurt am Main: Iberoamericana-Vervuert.

MacCormack, Sabine. 1995. "Limits of Understanding: Perception of Greco-Roman and Amerindian Paganism in Early Modern Europe." In *America in European Consciousness. 1493–1750*, ed. Karen Ordahl Kuppermann, 73–129. Chapel Hill: University of North Carolina Press.

Maffei, Sonia. 2013. "Cartari e gli dèi del Nuovo Mondo: Il trattatello sulle *Imagini de gli dei indiani* di Lorenzo Pignoria." In *Vincenzo Cartari e le direzioni del mito nel Cinquecento*, ed. Sonia Maffei, 61–119. Roma: Ginevra Bentivoglio Editoria.

Mason, Peter. 1997. "The Purloined Codex." *Journal of the History of Collections* 9 (1): 1–30.

Mason, Peter. 2001. *The Lives of Images*. London: Reaktion.

Mignolo, Walter. 1995. *The Darker Side of the Renaissance: Literacy, Territoriality, and Colonization*. Ann Arbor: University of Michigan Press.

Mikulska, Katarzyna. 2008. "El concepto de *ilhuicatl* en la cosmovisión nahua y sus representaciones gráficas en códices." *Revista Española de Antropología Americana* 38 (2): 151–171.

Miller, Peter. 2001. "Taking Paganism Seriously: Anthropology and Antiquarianism in Early Seventeenth-Century Histories of Religion." *Archiv für Religionsgeschichte* 3: 183–209.

Momigliano, Arnaldo. 1950. "Ancient History and the Antiquarian." *Journal of the Warburg and Courtauld Institutes* 13 (3/4): 285–315.

Mulsow, Martin. 2005. "Antiquarianism and Idolatry: The Historia of Religions in the Seventeenth Century." In *Historia, Empiricism and Erudition in Early Modern Europe*, ed. Gianna Pomata and Nancy Siraisi, 181–209. Cambridge: MIT Press.

Nielsen, Jesper, and Toke Sellner Reunert. 2009. "Dante's Heritage: Questioning the Multi-Layered Model of the Mesoamerican Universe." *Antiquity* 83 (320): 399–413.

Nielsen, Jesper, and Toke Sellner Reunert. 2015. "Estratos, regions e híbridos: Una reconsideración de la cosmología mesoamericana." In *Cielos e Inframundos: Una revision de las cosmologías mesoamericanas*, ed. Ana Díaz, 25–64. Mexico: Universidad Nacional Autónoma de México, Instituto de Investigaciones Históricas, Fideicomiso Felipe Teixidor y Montserrat Alfau de Teixidor.

Nongbri, Brent. 2013. *Before Religion: A History of a Modern Concept*. New Haven, CT: Yale University Press.

O'Gorman, Edmundo. 1958. *La invención de América; el universalismo de la cultura de Occidente*. México: Fondo de Cultura Económica.

Pardo, Osvaldo F. 2004. *The Origins of Mexican Catholicism: Nahua Rituals and Christian Sacraments in Sixteenth-Century Mexico*. Ann Arbor: University of Michigan Press.

Pignoria, Lorenzo. 1605. *Vetustissimae tabulae Aeneae sacris Aegyptiorum simulachris coelatae explicatio*. Venezia: Rampazzetto and Franco.

Quednau, Rolf. 2013. "Architettura e iconografia costantiniana a Romafra Rinascimento e modern." In *Costantino I. Enciclopedia costantiniana sulla figura e l'immagine del cosiddetto Editto di Milano*, ed. Alberto Melloni, II: 737–758. Roma: Treccani.

Quiñones Keber, Eloise. 1995a. "Collecting Cultures. A Mexican Manuscript in the Vatican Library." In *Reframing the Renaissance: Visual Culture in Europe and Latin America, 1450–1650*, ed. Claire J., Farago, 229–242. New Haven, CT: Yale University Press.

Quiñones Keber, Eloise. 1995b. *Codex Telleriano-Remensis: Ritual, Divination, and History in a Pictorial Aztec Manuscript*. Austin: University of Texas Press.

Rubiés, Joan-Pau. 2006. "Theology, Ethnography, and the Historicization of Idolatry." *Journal of the History of Ideas* 67 (4): 571–596.

Ryan, Michael T. 1981. "Assimilating New Worlds in the Sixteenth and Seventeenth Centuries." *Comparative Studies in Society and History* 23 (4): 519–538.

Sahagún, Bernardino de. 1997 [1564]. *Primeros memoriales*. Trans. and notes by Thelma D. Sullivan. Norman and Madrid: University of Oklahoma Press/Patrimonio Nacional/Real Academia de la Historia.

Schwaller, John F. 2006. "The *Ilhuica* of the Nahua: Is Heaven Just a Place?" *The Americas* 62 (3): 391–412.

Seed, Patricia. 1991. "Colonial and Postcolonial Discourse." *Latin American Research Review* 6 (3): 181–200.

Seznec, Jean. 1931. "Un essay de mythologie comparée au début du XVIIe siècle." *Mélanges d'archéologie et d'histoire* 48: 268–281.

Seznec, Jean. 1939. *La Survivance des dieux antiques*. London: The Warburg Institute.

Sheehan, Jonathan. 2006. "Thinking about Idols in Early Modern Europe." *Journal of the History of Ideas* 67 (4): 561–570.

Volpi, Caterina. 1992a. "Le vecchie e nuove illustrazioni delle Immagini degli Dei degli antichi di Vincenzo Cartari (1571 e 1615)." *Storia dell'Arte* 74: 48–80.

Volpi, Caterina. 1992b. "Lorenzo Pignoria e i suoi corrispondenti." *Nouvelles de la République des Lettres* 2: 71–123.

Volpi, Caterina. 1996. *Le immagini degli dèi di Vincenzo Cartari*. Rome: Edizione De Luca.

von Stuckrad, Kocku. 2003. "Discursive Study of Religion: From States of the Mind to Communication and Action." *Method and Theory in the Study of Religion* 15: 255–271.

von Wyss-Giacosa, Paola. 2012. "Investigating Religion Visually: On the Role and Significance of Engravings in Athanasius Kircher's Discourse on Idolatry." *Asdiwal* 7: 119–150.

von Wyss-Giacosa, Paola. In press. "Through the Eyes of Idolatry: Lorenzo Pignoria's Argument on the *Conformité* of Ethnographic Objects from the West and East Indies with Egyptian Idols." In *Through Your Eyes: Religions and Beliefs as Intercultural Mirror (16th–18th Centuries)*, ed. Giovanni Tarantino and Paola von Wyss-Giacosa, with the collaboration of Giuseppe Marcocci. Leiden: Brill.

3

Dissecting the Sky

Discursive Translations in Mexican Colonial Cosmographies

ANA DÍAZ

The first two folios of the colonial codex *Vaticanus A* (1v, 2r) include a figure that shows a complete image of the Mexican cosmos (figure 3.1). The structure of the cosmogram is shown as fixed and divided into 11 upper levels, the surface of the earth, and the eight layers of Mictlan, the underworld. This image served as the primary visual source for Eduard Seler's (1961) reconstruction of the ancient Mexican cosmography, which the author later complemented with a list of 13 gods mentioned in the *Histoire du Mechique* (2002, 155) and depicted in the codices *Borgia* (9–13), *Borbonicus* (1–18), and *Tonalamatl Aubin*. Seler named these characters "lords of the days" (see Köhler 2000) and located them on every level of the sky, associating one deity with every vertical layer of the sky and with every hour of the day (1961, 29, 34–36). This assimilation of astronomical principles led Seler to reconfigure his cosmographical structure, rearranging the upper and lower vertical layers of the cosmos as a double-pyramidal structure in which the sky was occupied by the 13 lords of the day and the underworld by the Nine Lords of the Night (see figure 3.2).[1]

Although these two models (the structure of 13/9 vertical layers and the double-pyramidal structure) seem to differ greatly, they are part of the same discursive strategy because both imply a fixed structure of levels, each one inhabited by a god/element, and because both Mexican cosmographical models are based on astronomical principles. Indeed, astronomy seems to be the

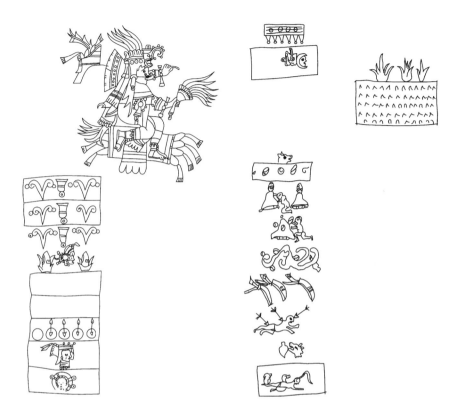

FIGURE 3.1. *The Mexican cosmos, as represented in the* Codex Vaticanus A *(fol. 1v–2r, sixteenth century). The figure is divided in two pages. Folio 1v shows the nine upper levels of the sky (the three upper layers are named* teotl, *and the other six are named* ilhuicatl*), and the composition is reigned by Tonacatecuhtli, the god who is sitting above the structure. In the next folio (2r) we see the rest of the cosmogram, beginning with the remaining two sky levels, the earth (depicted as a field crowded with plants in the right corner of the page), and under it the eight levels of the underworld. (Drawing by Victor Medina.)*

main conceptual reference articulating the different reconstructions of the Mexican cosmography in colonial and modern sources.

Despite the fact that the image in the codex *Vaticanus A* on which the "13/9 levels model" is based is the only known cosmographical picture that includes a vision of the vertical realms of the cosmos, this general cosmic configuration has been widely accepted by scholars, who have projected its different versions onto other Mesoamerican regions, aiming to identify a panregional common worldview (Tozzer 1907; Thompson 1934, 1954, 1970; Holland 1961,

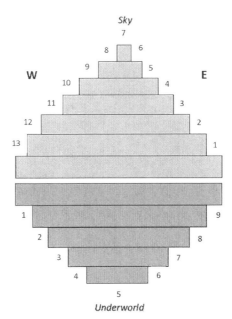

FIGURE 3.2. *Cosmographical reconstruction following the model of the double-pyramid structure of Eduard Seler (1961, 31–33). Each number stands for a god. (Reconstruction by Ana Díaz.)*

1963; León Portilla 1966, 1994; Matos Moctezuma 1987, 2008; Tavárez, chapter 5, this volume).[2]

However, the Mexican cosmography depicted in the codex *Vaticanus A* contrasts with the descriptions written in other contemporaneous sources, such as the *Histoire du Mechique* (2002), also used by Seler to build his model, and the *Historia de los mexicanos por sus pinturas* (2002). As shall be seen, at least five different versions of the Mexican cosmographical composition can be found in these manuscripts.

The information used in these colonial sources was supposedly derived from traditional images, drawn in precolumbian codices (*Historia de los mexicanos* 2002, 24–25), suggesting that the compilers of the *Historia* and the *Histoire*, and the painters of the codex *Vaticanus A*, may have all used the same cosmographical picture (or visual repertoire) as their primary source. If that is the case, the differences in the descriptions of the Mexican cosmos may be due to the authors' rearticulation of the model to respond to their specific objectives or to regional/cultural variations. Nevertheless, it is possible that no original

Mexican cosmogram model existed and that the authors and painters of the colonial sources invented it in each case in order to illustrate their cosmographical descriptions.

To understand the differences between these sources, it is crucial to analyze them in their context of use and production, identifying the historical modifications of the pictorial and literary indigenous colonial discourses. This approach has shown to be fruitful in understanding the reconstruction strategies used by the authors of the colonial sources (Nielsen and Sellner Reunert 2009; chapter 1, this volume; Díaz 2009, 2011; Botta, chapter 2, this volume; Vail, chapter 4, this volume). In addition, to better comprehend the Mexican cosmological principles, it is vital to complement and compare this information with that derived from the analysis of pre-Columbian sources. Much headway has been made in this respect by authors who have studied the visual and/or linguistic codes employed by the precolumbian societies of central Mexico to refer to specific cosmological topics (Klein 1982; Mikulska 2008, 2010, and chapter 8, this volume). Therefore, if we follow the transformations of the visual codes employed in prehispanic and postcontact sources, we may be able to obtain a broader understanding of the main topics and codes of the Mexican cosmographical discourse.

The aim of this chapter is to contrast the visual discourses of the Mexican cosmos in colonial and precolonial sources. To do so, the essay is divided into two sections. The first explores the contradictions between the descriptions of the cosmos in colonial sources, focusing on the representations of the sky, and it considers the possibility of identifying the original images that may have served as their model. Specifically, it examines the content, context, and production process of the early colonial sources that describe the Mexican cosmography, focusing on the reconstruction strategies used by the authors, including the ways in which the introduction of the Christian cosmographical discourse affected the reconfiguration of indigenous cosmological conceptions in the colonial written sources.[3]

Following from this analysis, the second section of the chapter argues for the conception of a cosmos as a living entity that is multiple and dynamic. It examines prehispanic visual sources, using the sky, or Citlalinicue, as an entry point. Citlalinicue is a main character in the foundational narratives or cosmogonies, who in colonial sources was disarticulated in order to build a cosmographical fixed structure with her iconographical elements. Thus, the sky becomes an entry point to begin the discussion and review of Nahua cosmological and cosmographical principles.[4]

The second analysis shows that while the examination of the colonial sources allows for a division between the sky and the other realms of the cosmos, such as the underworld. Excluding other spaces, which seem to be part of the world, too, for example: the place of dreams or the alternative time-spaces of previous cosmogonic eras. Through this analysis the sky emerges as a body and, at the same time, as a threshold. It seems to be an entity who assumes different aspects or qualities, revealing a multiple and dynamic constitution. Therefore, this arrangement reveals an emphasis on regional-historical cosmogonical and cosmographical principles, in contrast to our universal (theological or scientific) worldview.

1. NAHUA COSMOGRAPHIES AS DESCRIBED IN EARLY COLONIAL WRITTEN SOURCES

In this section I review the information found in early colonial sources about the composition of the Mexican cosmos, in particular the sky, and explore two of the main strategies that the authors of the manuscripts used in their reconstructions of the cosmos: the creation of a cosmographical arrangement that would be recognized by a Christian readership, and the extraction of elements from the native narratives.

Historia de los Mexicanos por sus pinturas (*History of the Mexicans as Told by Their Paintings*)

This manuscript is essential to the reconstruction of the Mexican cosmological discourse because it includes original narratives that describe the origin of the world, the gods, natural phenomena, humankind, and the calendar, as well as the beginning of Mexican history. The author of the text is unknown, but some hypotheses suggest Andrés de Olmos as the compiler.

Historia de los Mexicanos is a small booklet, written in Spanish, with no images. Dyer estimates its production date between 1541 and 1547, based on the study of watermarks and paleography (Dyer 1996, 109–110). It was bound in a volume called *Libro de Oro*, which is a compilation of texts that includes the *Memoriales* of Motolinía. The physical analysis of the booklet shows that it includes two texts written by different scribes at different times (*Historia de los mexicanos por sus pinturas*, n.d.).

The main text of the manuscript was written by the first scribe, and it occupies the entire booklet (fol. 150r–155r), except the covers, which were originally left blank. The composition of this text reveals that it is the final draft of a complex editorial and compilatory process. It was written with careful handwriting and fitted perfectly between the margins of each page.

The information is organized in chronological order throughout 20 brief but impeccably structured chapters, which must have been previously composed in order to obtain a coherent historical exposition. The orthographical mistakes found in Nahua words reveal that the scribe did not speak the native language, indicating that this manuscript was indeed the final version of a collective effort.

This document begins with the narration of the mythical origin of the world and ends with the moment the Spaniards first viewed the temple of Huitzilopochtli. This section includes no images, but a note in the upper margin of the first folio indicates that the narrative was drawn from a painting brought by Bishop Ramírez: "Esta relación se sacó de pintura que trajo don Sebastián Ramírez, Obispo de Cuenca" (fol. 150r; *Historia de los mexicanos* 2002, 25).

In the first part of this document is a brief reference to the Mexican cosmological structure. It introduces the first couple, Tonacatecli [sic] and his partner, Tonacaçiguatl [sic] or Suchequecatl [sic], who created the other gods and always inhabited the thirteenth sky (fol. 150r; *Historia de los mexicanos* 2002, 25).

The final version of the Mexican history written in the *Historia de los mexicanos* includes a text written by a second scribe, who had to use a tighter calligraphy in order to fit the text into the small space left in the verso of the last folio and the cover of the original booklet. This author had to add pages, with watermarks dated 1566 (Dyer 1996, 101–103) in order to complete the narrative. This second text offers another version of the Mexican history, emphasizing the Mexican royal genealogy, and adds complementary information about native cosmology and calendars. In this text we find a cosmographical description that gives specific information about the composition of the sky:[5]

> (1) These Mexican Indians thought that in the first sky were Citlalinicue and Citlalatonac; Tonacatecuhtli made them guardians of the sky; (2) in the second sky were the women with no flesh, the *tetzauhcihua* or *tzitztimime*; (3) the 400 men of five colors created by Tezcatlipoca; (4) the birds of all species; (5) snakes of fire, which became comets and lights; (6) all the winds; (7) the dust; (8) all the gods met there, and none could ascend to the place of Tonacatecuhtli and his wife, who were then supposed to inhabit the ninth sky (9). And they did not know the content of the remaining skies. (*Historia de los mexicanos por sus pinturas* 2002, 80; translation by Díaz)

In addition, the scribe describes how, after asking where the sun was located, the informants explained that "it was in the wind, and it walked during the day, not at night, because when it reached midday, it came back to the east, while its brightness went to the west; and [they say] that the moon walked behind the sun, and it [the moon] never reaches it [the sun]" (*Historia de los mexicanos* 2002, 80; translation by Díaz).

As can be seen, the first scribe only reports the existence of the thirteenth sky, inhabited by the original couple. Indeed, it seems that this place was the only one relevant to the narrative, as the structure of the sky is not mentioned. In contrast, the second scribe describes the sky as nine spaces (probably, but not clearly, as superimposed layers), which were occupied by different characters. In addition, the existence of other skies is reported: unnumbered and of unknown content. In this text the sun and moon seem to be secondary elements that move in the "sky of the winds."

Histoire du Mechique

Like *Historia de los mexicanos*, this document has served as a primary source for the study of Mexican cosmology and cosmography, although it also has no images. The text was written by a single scribe who did not speak Nahua, as revealed by the misspellings. It was bound in a volume that integrates different sources and was translated into French around 1548–1553 (Tena 2002, 118). The authorship of this text remains unknown, but the signature of André Thevet appears in the manuscript.[6]

The authors of the various documents assembled in the *Histoire du Mechique* collected information from Mexico-Tenochtitlan, Tetzcoco, and Chalco. Therefore, we can find different variations in the descriptions of the cosmos.

The Mexican Cosmos

In the *Histoire du Mechique*, the Mexicas are said to have believed in 13 skies. (1) In the first sky was the god of the years, Xiuhtecuhtli; followed successively by (2) Coatlicue, the earth goddess; (3) Chalchiuhtlicue, "House of a Goddess" [*sic*]; (4) Tonatiuh, the sun; (5) five gods of different color, called *Tonaleque*; (6) Mictlantecuhtli, god of the underworld; (7) Tonacatecuhtli and Tonacacihuatl; (8) Talocatecuhtli, god of earth; (9) Quetzalcoatzin, a main "idol"; (10) Tezcatlipoca, another great idol; (11) Yohualtecuhtli, god of the night or darkness; (12) Tlahuizcalpantecuhtli, god of the sunrise; and (13) in the thirteenth sky and higher were the god Ometeotl, "Two Gods," and a goddess called Omecihuatl, "Two Goddesses" (*Histoire du Mechique* 2002, 142–145; reconstruction of the Nahua names by Tena 2002).

The Tetzcocan Cosmos

While the cosmography of the people of Tetzcoco is not described as such in this source, it is referred to in a narrative. The chapter dedicated to the people of Tetzcoco tells the story of the creation of the sun.

> These Indians say there was a goddess, Citlaline [sic] [Citlalinicue], who sent her 1,600 sons to the city of Teotihuacan, next to Tetzcoco; they perished after their arrival to the city. Then, 26 years after the creation of the world, which had remained in darkness because of the lack of the sun, three gods met: Tezcatlipuca [sic], Ehecatl, and the goddess Citlalecue [sic] [Citlalinicue]. (*Histoire du Mechique* 2002, 152; translation by Díaz)

The Chalcan Cosmos

> In another province, called Chalco, they say that the water was the first thing in the world, but they do not know who created it. And that from the sky descended the gods, called Céacatl, Tezcatlipoca, and Chiconauhécatl, all of them sons of Citlainicue, goddess of the stars, who is said to have created the stars, the sun, and the moon . . . They also say that there are nine skies, but they do not know where the sun, the moon, the stars, and the gods are located (*Histoire du Mechique* 2002, 154–155; translation by Díaz).

In sum, the *Histoire du Mechique* provides three cosmological models. The Mexican cosmography depicts a sky divided into 13 levels, each of them inhabited by a god—as mentioned above, Seler used this source in his reconstruction of the cosmographical model in order to identify the gods of each layer of the sky. The Tetzcocan version does not define a cosmographical structure but tells the story of the creation of the sun by Citlalinicue, the mother of gods, with Ehecatl and Tezcatlipoca. Finally, upon being asked to explain their cosmography, the Chalcans answered that they knew that the sky had nine divisions, but they did not know their content or order. Indeed, instead of describing a structured composition of the sky, the Chalcans told the story of Citlalinicue, the mother of the gods and the celestial bodies. In fact, Citlalinicue, whose name can be translated as "Stars [Are] Her Skirt," was an embodiment of the starry sky among the Nahua of central Mexico, and her personification of the firmament is also reported in the *Codex Vaticanus A*.

Codex Vaticanus A

This manuscript, written in Italian in folio format and richly illustrated, includes information from different regions of New Spain, especially from central Mexico. It was compiled by Pedro de los Ríos (ca. 1562–1566), who also produced the codex *Telleriano Remensis* for ecclesiastical authorities. Its original title is *Indorum cultus, idolatria, et mores* (*Worship, Idolatry, and Customs of the Indians*), but it is known as the *Codex Ríos* or *Codex Vaticanus A*.

Batalla (2006) identifies two different sources in this document: the text, which he calls the European book, and the pictures, which he calls the indigenous book. I agree with his identification of two discursive forms in dialogue in the volume, but as I have argued elsewhere (Díaz 2009, 2011), the images depicted in this codex express indigenous concepts through new, European strategies of visualization. That is, they do not coincide with a traditional precolumbian visual culture. The first folios of the manuscript, which include the description of the cosmos, exemplify this operation. Here, the text and the image are complementary but autonomous sources that together create a clear and coherent cosmological exposition for a Christian reader.

The Glosses

The first folio of the codex displays two texts. The first gloss (folio 1v) introduces the cosmos of the natives of New Spain. The second text (folios 1v–2r) reproduces the names of the levels of the sky of the Mexican cosmography. There are connections but also differences between the two registers.

> TEXT 1. DESCRIPTION OF THE COSMOS: HOMOYOCA [OMEYOCAN]
>
> This means "the place where the Creator of everything is located," or "the First Cause." They call him by another name Hometeule [Ometeotl], which means "Lord of Three Dignities," or "Lord Three"; and the Otomies called this place, where he is *Hive narichnepaniucha* [this is a misspelled Nahua phrase], which means "above the nine compositions," and by another name Homoyocan [*sic*], "the place of Lord Three," who, according to the opinion of many elders, "created with his word Cipactonal and his wife, who was called Oxomoco; they are the ones who were before the flood, who created Tocatiutle [*sic*] as we shall say in what follows" (Anders et al. 1996, 44; translation by Díaz).
>
> TEXT 2. THE NAMES OF THE LEVELS OF THE SKY
> 11. Teotl. tlatlauhca quasi dicat cielo rosso . . . (red "god")
> 10. Teotl. cocauhca (cozauhca). q[uasi] d[icat] cielo giallo a . . . (yellow "god")
> 9. Teotl. yztaca. q[usai] d[icat] cielo bianco . . . (white "god")

8. Yztapal nanazcaya. q[uasi] d[icat] cielo delle rose. . . . (where the *iztapalli* stones grind)[7]
7. Ylhuicatl. xoxo uhca. Q[uasi] d[icat] cielo verde . . . (sky [of] green)
6. Ylhuicatl. yayauhca. q[uasi] d[icat] cielo verde e negro . . . (sky [of] dark)
5. Ylhuicatl. mamaluacoca . . . (sky [of] *mamaluaztli* [astronomical constellation])
4. Ylhuicatl. huixtutla . . . (sky where the *huixtutla* abounds)[8]
3. Ylhuicatl. tunatiuh . . . (sky [of] Sun)
2. Ylhuicatl. tz[i]tlalicoe . . . (Sky [of] Citlalincue)
1. Ylhuicatl tlalocaypanmeztli . . . (Sky [of] Moon over the Tlalocan)

This cosmological exposition is complemented and contradicted by information in folio 2v, where the glossist writes that "they say that these four gods or demons [the ones depicted in folio 2r] had their own women, and the same is said about the gods of the sky, who each had his own [wife]" (*Vaticano A*, 2v; translation by Díaz). These marriages are not shown in the cosmography of folios 1v–2r, but do appear in the image depicted on the next folio (2v), where the *tlacuilo* introduced the lords of the underworld seated in front of their wives. In contrast, neither women nor couples of gods are depicted in the interior of the sky in this codex, but only a major male god, Tonacatecuhtli.

The Images

In folios 1v and 2r, the sky is depicted as composed of 11 stratified spaces (see Díaz 2009, 2019, for a detailed analysis of this image). However, the *tlacuilo* painted nine layers on folio 1v and the remaining two on the next folio. Thus, the painter respected the nine divisions described in the gloss of the *Homoyoca*—which are described as part of the Otomi cosmography—but disrupted the continuity of the sky. This division left the starry sky, Citlalinicue, in the higher of the two bands depicted on folio 2r, located atop the earth. Citlalinicue is a main character of the Mexican cosmologies. In the written sources she is the creator and/or mother of the heavenly bodies and the gods who descended to earth. In this image she *is* the band located at the upper space depicted above the earth (fol. 2r).

The fragmentation of the 11 layers of the sky into 9 + 2 can be understood as an adaptation of native cosmological principles to a Christian cosmographical discourse. While the number 11 does not appear in Mexican cosmologies, in Christian cosmographies the upper realm of the cosmos was divided into

11 spaces: two elements located above the earth (air and fire), seven planets, the firmament (with the stars), and the First Cause (the heaven of God), located outside the composition.[9] These 11 divisions of the sky were conceived as absolutely autonomous spheres, each occupied by a character or astronomical phenomenon.

Except for the highest level, Citlalinicue, the content of the other levels of the sky in this image is unclear, as the objects and characters depicted (except for the sun and the moon) are not mentioned in other sources. Nevertheless, the philological analysis of the names of the layers helps clarify the painter's intention. The three upper strata of the structure are named *teotl*, equivalent to "lord" or "god," and the rest are called *ilhuicatl*, meaning "sky" (see the glosses transcribed above). This distinction is consistent with the descriptive gloss, which erroneously translates Ometeotl (2-*teotl*) as "Lord Three," who inhabits the Omeyocan, the place of the First Cause. Thus, a theologian might easily identify the Christian Trinity and the First Cause of the Aristotelian cosmography, and assume an analogy between Christian and Mexican cosmological discourses.

This analogy is reinforced by the character depicted above the layers of the sky. He is identified in the gloss as Ometeotl, but the maize cobs attached to his seat (*petlatl icpalli*) and his crown of turquoise (*xiuhitzolli*) are associated with Tonacatecuhtli, the Lord of Maize, the material substance of man. Quiñones Keber (2003) has discussed the connections between Christ, as embodied in the Eucharist, and Tonacatecuhtli, the Lord of Our Substance. This connection between Mexican and Christian concepts is also confirmed by images such as the one reproduced in figure 3.3, where the three levels of creation are shown in a scheme that condenses the three persons of the Trinity into the physical image of Christ.

The complexity of the cosmogram of codex *Vaticanus A* cannot be exhaustively explained here, but it is clear that the figure combines Mexican iconographical elements into a visual composition that reproduces a Christian cosmographical discursive formula. The *tlacuiloque* tried to merge two different cosmological arrangements into one universal model: the nine divisions of the Mexican/Otomi sky, and the 11 spheres of the Christian metaphysics. Hence, the context of production of the image becomes essential to understanding the operative principles that guided the painters (and glossists) in the processes of translation and transmutation of codes and concepts.

This example illustrates one of the strategies of the authors and compilers of the colonial written sources to create new indigenous cosmographical models—the systematization of native knowledge into a cosmological-theological model

FIGURE 3.3. *The three levels of the creation as a* continuum *(sensibilis, intellectual and spiritual worlds). (Charles de Bouelles,* Physicorum elementorum . . . libri decem. *Paris, 1512, fol. 73. Drawing by Victor Medina.) In the lower level, we see the creatures of the "sensible world," including terrestrial animals; in the air we see the birds. In the second sphere we identify the planets and stars, the astronomical world. The third segmentum shows the angels and saints (spiritual world), and in the upper level, God (represented by Christ), the Primum Movile of the Cosmos.*

that integrated the vision of the people of New Spain. A second example can be found in the seventh book of the *Codex Florentine*, dedicated to natural philosophy, where the images do not seem to follow the traditional strategies of the Nahua visual culture, but rather a mixture of different European pictorial techniques and designs (Magaloni 2011). The main image of this book shows what seems to be a large calendar wheel that explains the function of the 52-year *xiuhpohualli*. In it, Spitler (2005) recognizes the influence of the Christian tradition, noting that the 13 circles in which the years are located are quite similar to the heavenly spheres of medieval Christian cosmographies.

A second, closely related strategy for the creation of these cosmographical and cosmological reconstructions was the identification of what the authors considered to be the distinctive "elements" or clusters of native cosmographies (mainly gods and astronomical bodies, such as the sun and the moon) and their subsequent arrangement in new formats. To do so, they dismembered them from their original discursive configurations: the indigenous narratives.

For example, the main characters of most descriptions seem to be the couples of gods who inhabited the sky, concentrated in the upper level, or the last numbered sky—Citlalinicue being one of the most cited. These couples can be found in Mexican, Chalcan, or Tetzcocan foundational stories, which suggests that each group had its own original couple and creational narrative (*Historia de los mexicanos* 2002, 25, 80; *Histoire du Mechique* 2002, 154–155; *Codex Vaticanus A* fol. 2v). However, the colonial written sources focus on the structure of the cosmos rather than on these characters. Indeed, their roles and relationships are secondary to their roles as content of each level.

Similarly, other main characters of the sagas, the sun and moon, were identified in the colonial manuscripts as secondary, who were located in specific levels of the sky (*Codex Vaticanus A* fol. 1v; *Historia de los mexicanos* 2002, 80; *Histoire du Mechique* 2002, 142–145). However, some references in these sources suggest that they did not have a precise and constant location. First, the informant of the *Historia du Mechique* does not mention the sun and the moon among the list of gods and elements located in the different divisions of the sky. Indeed, he indicates that they moved in the sky of the winds (*Historia de los mexicanos* 2002, 80). Second, in the *Histoire du Mechique*, the Chalcans answered that they did not know where these elements were located. This confusion cannot be attributed to ignorance on the part of the Chalcan informants, but rather offers a clue about the impossibility of applying the Christian operating principles to understanding the Chalcan world; Chalcans did not consider the cosmos to be structured as an immobile system of levels or spheres.

Finally, while Sahagún did include the narrative of the creation of the sun and the moon in the *Florentine Codex*, he considers it a simple fictitious story and justifies its inclusion by explaining that it was not possible to reconstruct the cosmography of the Mexicans because their knowledge of the matter was lacking (Sahagún 2002, 2, 689–690). The narrative is then followed by brief reports about meteorology, astronomy, and calendars—topics commonly included in European cosmographies.

The authors' selection of certain elements from native narratives and their subsequent reconstruction of the cosmos according to Christian principles can be considered two of the main strategies employed in these sources to reconfigure the native cosmos in terms comprehensible for the readers of New Spain. This reconfiguration also implied the unification of different traditions (Mexica, Tetzcocan, Chalcan, Otomi) into a single cosmological model that would be comprehensible for the Christian readers of the manuscripts. Bernardino de Sahagún had a similar intention, but he and his collaborators were not able to identify a cosmological configuration equivalent to the Christian model in their inquiries. Their failure to achieve this objective was reported in the prologue of the seventh book of the *Florentine Codex* (Sahagún 2002, 2, 689–690), dedicated to the presentation of the indigenous natural history.

This effort to create a unified universal model in the colonial sources can be said to respond to the necessity of translating cosmological information to the visual/discursive codes recognizable for a Christian reader. Indeed, there are many similarities between the cosmographies analyzed in this chapter. However, despite this unifying effort, important differences remain in their structure and their content. We can therefore assume that it is not possible to identify the "original" Mexican cosmographical model. This, in turn, has implications for the efforts of current research to identify a pan-Mesoamerican cosmological model.

If there was an ancient and pan-Mesoamerican model, as stable as established in the "13/9 levels model," we would expect it to be clearly reflected in the written sources examined here and in the large number of additional colonial reports and archaeological materials.[10] Indeed, if sources merely presented variations on the same content, such as a different list of gods, it could be argued that each group had its own pantheon. Likewise, if the only differences were among the names of the places within the same structure of the sky, we could again argue that these are regional variations on a same topic. However, the main differences between the sources discussed relate to the structural composition of the cosmos and the nature of the elements located in each level, that is, to the cosmographical discourse itself.

Therefore, it can be stated that the pan-Mesoamerican model proposed in recent decades is not supported by the colonial sources, despite the unifying effort of the authors of these manuscripts. In addition, understanding the strategies used by these authors to create an image of the cosmos—extracting elements from the native narratives and creating a cosmological image that would be recognized by a Christian readership—allows us to understand the principles that guided their translation and transmutation of codes and concepts.

This understanding of the context of production of the cosmological images sheds light on the creation of the cosmos as a structure of fixed divisions, emphasizing its divisions and displaying its content as it appears in astronomical Euro-Asiatic classical imagery. However, as we already suspect from the example of the sun and the moon, new insights can be gleaned by shifting our focus to sources that were not produced for a Christian readership—that is, to the characters of the literary narratives, who in colonial sources was disarticulated in order to build a cosmogrpahical fixed structure. This shift can help us understand the Nahua cosmological and cosmographical principles more fully. As we shall see, the fact that there is no pan-Mesoamerican model makes way for the possibility of a dynamic Mexican cosmology with multiple aspects.

2. THE COSMOS IN PRECOLUMBIAN VISUAL CULTURE FROM CENTRAL MEXICO

Cosmic Spaces: Configuration and Transit

As shown in the Mixtec historical codices, the different realms of the world were inhabited by a diversity of characters, and the connections between these realms allowed them to transit between them. The images in these sources also show that the content of these realms mirrored each other. In the interior of the sky, for example, we find palaces, temples, and other architectural structures, as well as mountains and rivers or lakes, reproducing the same arrangement of earthly towns (*Codex Vindobonensis*, 35, 47, 48, 52; *Codex Nuttall*, 16r–19r, 21r–22r; *Codex Colombinus*). A clear example of these two characteristics can be found in the *Codex Nuttall* (18r), where four male characters, commanded by 12-Wind, descend from the sky towards earth, while Lord 4-House and Lady 5-Snake, seated on their architectural platforms, watch their departure. The sun (identified by the calendrical name ?-Movement), located in the zenith, also witnesses the scene. The group of men carries important ritual objects, including a sacred bundle, intending to deposit it in the temple of a feathered serpent, on earth. Before reaching the temple, 5-Vulture, who is part of this committee of the sky, visits Lady 1-Eagle, who receives him in an aquatic cave located in Apoala (figure 3.4a).[11] Thus, this episode shows the similarity between the content of the different spaces, which were inhabited by different characters who could travel from one realm to another.[12] The same act seems to occur at codex *Vindobonesis* (figure 3.4b), where Lord 9-Wind descends to earth accompanied by two characters. As Klein (1982) observes, ropes are one of the main elements of the composition of precolumbian objects and served as guides in the descent and transit of persons and objects.

The similarity between the geographic-political composition of towns and the realms that they are deeply connected with (their sky, underworld, and place of dreams) is supported by ethnographic evidence (Knab 1991; see also Hull, chapter 6, this volume). To exemplify this operation, see figure 3.5, where

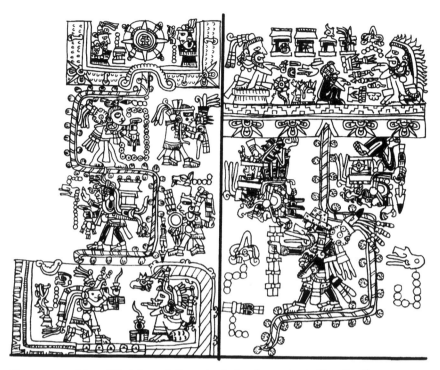

FIGURE 3.4. *(a) 12-Wind descends from the sky band accompanied by other three characters. The four men are walking on the rope that descends from the sky in the left scene. They carry ritual objects in their path to earth.* (Codex Nuttall, *18r. Mixtec, Postclassic; drawing by Ana Díaz) (b) 9-Wind descends from the sky accompanied by two characters. The main character can be recognized by his mask of Wind and the date that is depicted behind him (9-Wind). He descends to earth guided by a rope. Behind him come two characters, carrying houses (or temples) on their back. In the top of the image we see two lords speaking to 9-Wind (nude), giving him instructions before his departure to earth.* (Codex Vindobonensis, *48, Mixtec, Postclassic; drawing by Ana Díaz)*

Knab designed a reconstruction of the underworld of the Nahuas of San Miguel Tzinacapan. The image shows a place that reproduces the sociopolitical geography of an earthly town. Therefore, examples of Mixtec, Nahua, and Maya cosmology collected at different times reinforce the notion of a regional-historical cosmological constitution, in contrast to a universal (theological or scientific) cosmological worldview.

While the different spaces of the cosmos, as reproduced in images, share similar sociogeographical compositions—that is, they include houses, palaces,

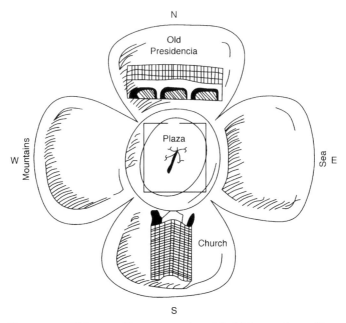

FIGURE 3.5. *Ethnographic reconstruction of the Nahua underworld, resembling a town on earth. (After Knab 2004; drawing by Victor Medina.)*

mountains, and rivers—despite their specific nature (aquatic, solid, dry, light, or dark), these spaces are not fixed and static but rather are characterized by a dynamic process of mutation, as discussed by Mikulska and Neurath (chapters 8 and 9, respectively, in this volume). This dynamic is also present in the narratives of the cosmological eras, in each of which the physical characteristics of the world, including the earth and its creatures, changed drastically. For example, in the sixteenth-century narratives, the sun was a different character with specific characteristics in each era (4-Ehecatl, 4-Atl, 4-Ocelotl). The last sun, Nanahuatl, became 4-Ollin after throwing himself into the fire. This act transformed his body, allowing it to shine with more intensity than that of his counterpart Tecuzistecatl and that of the suns of previous eras. Indeed, this cosmogonical dynamic is one of the main differences between Christian *perennial* metaphysics and Nahua/Mixtec *generative* cosmology, because it implies the configuration of history. The complexity of the dynamic cosmos shown in this section allows us to understand the sky as a space that is both a container and a threshold, dynamic and multiple.

FIGURE 3.6. *Two figures stand on a starry band, representing the sky, in the lower-right corner of the* Codex Vindobonensis, *13. (Drawing by Ana Díaz.)*

THE SKY AS BODY AND THRESHOLD: VISUAL CODES IN PRECOLUMBIAN SOURCES

In precolumbian sources from central Mexico, the sky can be identified by a starry border, or band, which surrounds the scenes that take place in its interior (figure 3.6). The same visual resource is used in Classic Maya iconography, where the clusters of sky, water, earth, and so on are added in bands which usually appear in binary compositions (Stuart and Houston 1994, 57–68; Schele and Miller 1986, 47; Tokovinine, chapter 7, this volume).

The band of the sky incorporates a large number of iconographical elements, including stars of different kinds (single eyes, white circles, bright stars, or "Venus" glyphs), flints, clouds, sunrays, dark mist, and so on (see Mikulska 2010, 2015). In codices. the band itself is marked by two or three lines of different colors, which are usually red and yellow, but occasionally other bands are added in gray, green, or black, depending on the characteristics of the sky.[13] These two, three, or four different-colored lines are also found in representations of

FIGURE 3.7. *The so-called* Altar of Venus, *a squared altar with starry iconography in its top band. (Museo Nacional de Antropología, Mexico City. Digital Archive of the Collections of the National Museum of Anthropology. Reproduction Authorized by the Instituto Nacional de Antropología e Historia. Secretaría de Cultura-INAH-MTM-MNA-CANON.)*

geological elements, such as earth's geological formations (mountains, fields) and bodies of water (rivers, lakes, swamps, sea), and they are also the colors of the higher skies depicted in the cosmogram of *Codex Vaticanus A*.

These visual codes for the representation of the sky through clusters of specific iconographical elements are found in Mixtec codices and in archaeological objects produced in central Mexico. For example, the Mexica Hall of the National Museum of Anthropology in Mexico City houses two objects—one square altar and one round *Cuauhxicalli* (figures 3.7 and 3.8)—which display starry white eyes in the upper band, divided by a double line from the lower band, where two different kind of starry eyes—a motif known as the Venus star (Seler 1963)—are carved. These iconographical elements thus suggest an identification between these sculptures and the sky.

Similar elements are found in the pictorial representation in the Mixtec *Codex Vindobonensis*, where three characters (3-Crocodile, 3-Rain, and

FIGURE 3.8. *A round altar with sky and solar iconography (*cuauhxicalli*). (Museo Nacional de Antropología, Mexico City. Digital Archive of the Collections of the National Museum of Anthropology. Reproduction Authorized by the Instituto Nacional de Antropología e Historia. Secretaría de Cultura-INAH-MTM-MNA-CANON.)*

3-House) are depicted conducting a ritual offering of a beheaded bird, a knot, and what seems to be a mat of green feathers, accompanied by the music of a sea-snail trumpet (figure 3.6). The scene takes place in the interior of the sky, represented as a blue band with bright starry eyes. The band bends at right angles in a bottom corner and is delimited at its inner edges by two thin lines, one red and one yellow. The offering of the three characters depicted in the codex replicates the same action that the users of the Mexican lithic objects would have carried out. In this case, the lithic altars embody the sky, upon which the offerings were to be deposited in order to be sent to the upper realm, where they may be received (Klein 2000; Taube 2009; Díaz 2012).

In these images, we can observe the aspect of the sky as container—as a place full of life, afterlife, and movement, where the lords, gods, or ancestors ruled and had impact on earth when they decided to interfere in human history. This is clearly shown in Mixtec codices, which describe how the lords of the sky sent ambassadors, heroes, and priests, such as 9-Wind, to earth (see figures 3.3 and 3.8 for examples). We also find the yellow/red lines that tend to characterize the clusters associated with the sky.

More than mere conceptual abstractions of layers of the sky, these yellow/red lines seem to indicate the organic material from which the world, and in this case the sky, was composed. The association between this red/yellow line and organic material can be observed in the representation of mutilated human and animal bodies (severed heads, arms, legs, feet), in which a thin red/yellow band surrounds the wounds, sometimes together with a flow of blood and the epiphyses of bones. In this sense, the story of the disembodiment of Tlalteotl that produced the upper and lower parts of the world stands as a paradigmatic example of the world's being composed of organic material (*Histoire du Mechique* 2002, 146). Thus, it can be argued that in certain contexts, the sky in precolumbian sources emerges as a body.

In addition, the examination of the clusters of iconographical elements related to the sky suggests that the sky was not only a sphere separate from other cosmic spaces, a container, but functioned at the same time as a threshold. An image depicted in the *Selden Roll* seems to confirm this. There the sun and the moon hang from the lower part of the sky band, composed in this case of eight starry bands. This element (the multiple-band sky) stands as the physical border between earth and the scene depicted in the interior of heaven, where Lord 1-Deer and Lady 1-Deer send 9-Wind to accomplish a mission on earth (figure 3.9). The image shows different moments condensed in a story. First, on the day 2-Deer, of year 13-Rabbit, the lord 1-Deer instruct 9-Wind to accomplish his mission on earth. On day 1-Lizard, year 1-Reed, the hero descends from the sky breaking through the cortex of eight layers of stars, accompanied by the Sun and the Moon in his journey. At 7-Reed, 9-Wind is received by four lords and descends to the interior of the earth (1-Rabbit). He returns to the sky after accomplishing his mission (see his footprints coming in two directions).

The representation of the sky in the *Selden Roll* as a huge modular complex with eight starry bands (Citlalinicue) one above the other is similar to the Citlalinicue band of the cosmogram depicted in *Codex Vaticanus A*, except that in this case each division has the same content. Thus, the painter of the *Selden Roll* reproduced a sky with nine partitions, but in his image the starry bands seem to condense and imply the conceptual definition of the sky, understood as the place where the astronomical bodies and meteorological phenomenon reside. Gods and ancestors, who send 9-Wind to earth, are depicted above this complex. Thus, I argue that in the *Selden Roll* the nine-layered band operates as a physical border that can be traversed by individuals.

This same principle can be seen to operate in the central section of the *Codex Borgia* (1993, 29–46),[14] where different characters with elongated bodies, identified as deities, function as organic borders that permit the transit of characters (figure 3.10).[15] These elongated characters have "sphincters" in the middles of their bodies, which permit the flux of characters, who transit from a narrative episode (the interior of a body) to the following scene (the next band-body). The sphincter is an iconographic element of precolumbian sources that allows the transit of creatures between different realms and that usually appears on the surface of the earth, water, and sky, represented with the same graphic cluster of red/yellow lines, indicating their organic nature in addition to their function as thresholds (figures 3.4 and 3.10). In the *Codex Borgia*, many of these characters have been associated with the sky, as they are adorned with stars, which makes them ideal in understanding the aspects of the sky as both body and threshold.

A detailed analysis of these elongated characters shows that the structure of their bodies allows them to serve as physical borders or frames, like the walls of a house, implying their spatial role. Eduard Seler

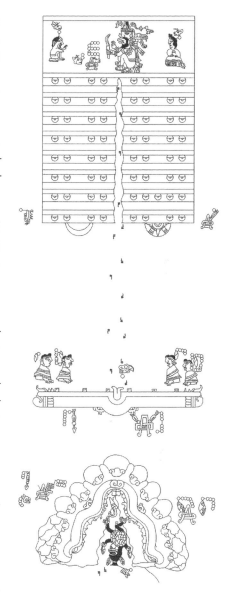

FIGURE 3.9. *Descent of 9-Wind to earth.* (Selden Roll, *first section. Mixtec codex, early colonial period. Drawing by Victor Medina.)*

DISSECTING THE SKY 121

FIGURE 3.10. *Band-goddesses dressed with embroided skirts. Their elongated bodies serve as frames in visual narratives, and their midbody "sphincters" permit characters to move between scenes of the narrative. (*Codex Borgia, 29–40. Reconstruction by Victor Medina.*)*

identified these creatures as "*borders* in the form of death goddesses" (Seler 1963, 10; emphasis added). Elizabeth Boone also states that the iconographic elements of the quadrilateral goddesses are indicators of death gods (skull heads, wild night hair, and conical headdress). However, Boone interprets this section of the codex as a cosmological narrative, and thus these quadrilateral deities are not interpreted as personifying Mictlan gods, but as primordial stages that preceded the creation of time and earth: that is, "they function less as specific gods than as major structural elements" (2007, 176). These structural elements organize units of action, generating narratives divided into episodes. Thus, Boone recognizes in these scenes elements of a Mexican cosmogony that show the first organization of space, and visions of emergence, birth, and sacrifice (2007, 175–176).

Similarly, rather than attempting to identify these creatures based on the colonial Nahua written sources, Nowotny highlights their function as frames and boundaries, that is, as spatial components: "the most frequent female figure [in the section] is the goddess who forms the frames and the separating strips" (2005, 265). According to him, the design of these bodies is reminiscent of the Rainbow, who surrounds the scenes in the Navajo sand paintings, creating subsections that the author identifies as rituals (2005, 266). In sum, these creatures can be understood as personified frames who structure the narratives (myths and/or rites) by serving as solid frontiers, but also as organic membranes that allow the flux of characters between the scenes.

The association of these deities with the sky thus allows us to further consider the multiple nature of the cosmic space, which is composed of organic matter and which operates as container and threshold simultaneously. However, the analysis of the band-shaped goddesses takes us further: they cannot be identified merely with one cosmic space—the sky[16]—and thus the examination of the sky leads us to other spaces. Indeed, as will be seen, the sky serves as an entry point into a cosmos that is not fixed.

While both Boone and Nowotny state that the iconographic identification of these band-shaped characters cannot be precise, they both draw attention to elements that may serve as potential indicators: (1) the skull heads and attributes of death, (2) the claws of earthly monsters, and (3) the stars added to the bodies and skirts of the deities.

Based on their skull heads (figure 3.9), adorned with a *cuexcoxhtechimalli*, which is presumed to be the main indicator of the identity of the character, Seler identified them as impersonators of the underground regions, which were crossed by Venus during his epic saga of rebirth. The identification of these skull-head characters as the regents of Mictlan is shared by other authors (e.g., Batalla 2008). However, the argument that these creatures might be impersonators of the Mictlan (Mictlantecuhtli or Mictlalcihuatl) because of the presence of attributes of death can be questioned, as many deities in the Mesoamerican pantheon, and not only lords of the underworld, are represented with a skull head (e.g., the *tzitzimime*, the *centzonhuitznahua*, the *cihuateteo*, Tlahuizcalpantecuhtli, Tlalteotl, Tlaltecuhtli, Mixtec Lady 9-Grass, Citlalinicue, and Coatlicue). Instead, these attributes may serve as markers of state: dead, ancient, primordial. These states do not imply a lifeless status; on the contrary, in some contexts they claim generative agency (López Austin 1973, 55; Olivier 2004, 31–36).

Similarly, the claws of earthly creatures do not necessarily suggest that these are earth deities, as many creatures, in different contexts, can share the attributes of the deities of the earth, or have claws. Nevertheless, what is striking is that it is impossible to link these images to one specific space—because they share attributes of earth, sky, and underworld—further destabilizing the notion of a cosmos with fixed spaces and specific contents or characters.

Taking this into account, the following section examines the third aspect of these elongated characters—the stars added to their bodies and skirts—in more detail and in relation to Citlalinicue, in order to consider the possibility of understanding objects as generative of cosmological discourses and configurations and thus of a cosmos that is constantly being constructed and produced under the same principles of production of objects (woven, depicted, or carved) and creatures (born or transformed throughout the different cosmogonic eras).

CITLALINICUE AND THE GENERATIVE ASPECTS OF OBJECTS

Nowotny observes that many of the band-shaped goddesses are adorned with stars, and hence "they could also be personifications of the heavenly bands" (2005, 267). For Boone, the quadrilateral deities are equivalent to the dark heavens that surrounded the creation of time before the arrival of the sun (2007, 179). In addition, Boone notes that one of the elongated deities could be the Milky Way, but she does not support a singular and absolute iconographical identification of these band-shaped characters (2007, 175–185) as does Milbrath, who identifies the band-shaped goddesses as the Milky Way (2013, 88), arguing that the images painted in this section refer to real astronomical events situated between 1495 and 1496.[17] Both Boone's and Nowotny's interpretations of the elongated characters underline the presence of astral bodies, either added to the bodies of the quadrilateral deities, or embroidered in the skirts of the band-shaped goddesses. But as noticed before, the identification of the elongated characters with specific astronomical (or cosmological) phenomena cannot be strongly supported.

The starry band might be identified as the Milky Way, but the relation is not obvious in Nahua and Mixtec sources—especially in the sources aforementioned in this text. Even so, however, there is a female character who appears in all the sources analyzed in the previous pages, including textual references and images: Citlalinicue, "Stars [Are] Her Skirt." She is one of the main characters of the Mexican cosmologies, represented as a lady who wears a skirt embroidered with stars, and in her nonhuman aspect, as a band/skirt of hanging eyes/stars (*Codex Vaticanus A*, fol. 2r).

Although all the band-shaped goddesses depicted in the *Codex Borgia* (29–46) are wearing skirts that include clusters associated with the sky, such as flints, bones, and different kinds of stars, I do not identify them all as Citalinicue, because the iconography embroidered in each skirt of these characters differs (figure 3.9). This suggests that these goddesses are not all impersonators of the same concept or force, Citlalinicue, but rather refer to different temporal/spatial dimensions, which can refer to different kind of skies, underworlds, moments, eras, or a mixture of these, each one identified by the distinctive attributes embroidered on the skirts. Nevertheless, understanding the generative force of Cilalinicue allows us to gain insight about the skirts and bodies of the elongated goddesses.

In the narratives of the *Histoire du Mechique*, Citlalinicue "Stars [Are] Her Skirt" emerges as one of the main characters of the Tetzcocan and Chalcan cosmogonical sagas. Both narratives emphasize her role as the mother of the gods who were sent to the earth—in another version to Teotihuacan—by her command; she also brought forth the sun, the moon, and the stars (*Histoire du Mechique* 2002, 152, 154–155). Mendieta writes that Citlalinicue gave birth to a flint, provoking the anger of her other children, who felt betrayed by their mother—like when Coatlicue got pregnant with Huitzilopochtli. Therefore, the gods threw a knife to the earth, which fell in Chicomoztoc, creating 1,600 gods upon contact (Mendieta 2002, 181–182). These references highlight the importance of Citlalinicue as a generative agent, whose actions take place in the sky, but had consequences in human history.

Similarly, Klein identified Citlalinicue as the first archetypal *tzitzimime*, the mother and grandmother of gods and stellar bodies, most of them identified as *tzitzimime*, too (2000, 38). She also states that certain monuments (altars, sculptures, platforms) reproduced the distinctive iconographical motifs of the mantles and clothes or skirts used by the *tzitzimime* that seemed to be produced in order to replicate and embody the generative powers of its original owners (Klein 2000, 45, 51; see figures 3.9 and 3.10).

Thus, the skirts worn by the band-shaped goddesses are not simply adornments, but essential components of the cosmological reconstruction. First, because certain Nahua gods and ancestors were identified based on the presence of an object or design: skirts (for Coatlicue, Citlalinicue, Ilancueitl, and Chalchiuhtlicue), smoking mirrors (for Tezcatlipoca), a shield (for Chimalman), and so on. This may also be the case with the band-shaped creatures in question, even if we cannot determine their names. Second, because the skirts include iconographic elements that may have incorporated or embodied the original source of energy of the characters depicted, a phenomenon similar to that of the production of *ixiptla* and sacred bundles or *tlaquimilolli*, which condensed the essence of the deities (Olivier 2006). While the nature of these

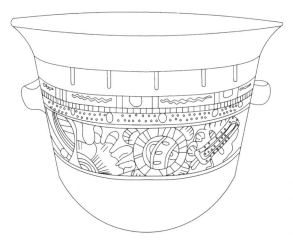

FIGURE 3.11. *Vessel with iconographic motifs that resemble an embroidered skirt. (Postclassic. Regional Museum of Palmillas. Yanga, Veracruz. Drawing by Victor Medina.)*

object-beings is complex (see Jansen 1982, 319–325; Olivier 2006, 199–225; Hermann 2008b; Ruíz Medrano 2011, 133–136; for *ixiptla* see Clendinnen 2010; Bassett 2015), their analysis is suggestive for the understanding of the agentive role of the referred skirts.

This discussion allows us to recognize new cosmological discourses and configurations in objects that we were previously not able to identify as such, such as the ceramic and ritual objects that incorporate reproductions of skirts and mantles (figure 3.11).

These objects are not properly cosmograms, as is the image on the first page of the *Codex Fejérvary-Mayer*, but instead are objects that reproduce cosmogonic concepts and that were used in ritual practices (see Neurath, chapter 9, this volume). Thus, in central Mexico the cosmological principles were drastically different from the cosmographical models based on Euro-Asiatic imaginaries.

Similarly, native cosmological principles went beyond objects, comprising their material production. Klein proposes a weavers' paradigm (1982), arguing the importance of focusing on productive practices, such as weaving, as special technologies that sustained the construction and reconstruction of the world (Klein 2000, 2008). The emphasis on weaving practices has also been emphasized by Prechtel and Carlsen (1988).

Klein's argument thus allows us to consider the weaving of skirts as productive of the cosmos. Indeed, her argument can be extended to the production of objects, beings, places, eras, and worlds that are constantly made and transformed through the inclusion of new qualities manifested in clusters. Thus, the act of making things (houses, mats, clothes, spaces of the world, and creatures) using a technology developed by man, as weaving, or painting, established human agency as the generative motor of the cosmos, including its spaces, creatures, and objects. The same seems to be the case with ceramics and ritual speaking; these activities combined specialized knowledge, creativity, and aesthetic choices to give shape, body, and matter to new things and beings.[18]

This principle of iconographical/metaphorical clusters can also be extended to names of characters such as Citlalinicue "Stars [Are] Her Skirt" and Citlalatonac "Shine of Stars," which condense a series of concepts that include the specific attributes associated to a character or aspect (of the sky). Thus, Citlalinicue and her husband Citlalatonac can be both identified as different generative aspects of the sky or *ilhuicatl*.[19] In this sense, these names produce visual and verbal metaphors that reproduce the same principles of organization as do clusters of iconographical elements when incorporated in skirts and objects. Hence, the articulation of word and image produces different kinds of embodiments (places, natural sources, deities, persons), which show the specific characteristics that are manifested in their current name/form—in this case the presence of different aspects of the sky—giving shape to a complex cosmos with a wide taxonomy of beings and places. This cosmogonic dynamic has no counterpart in Christian metaphysics.

In sum, Nahua visual, verbal, and material records thus become extremely useful for the cosmological reconstruction, because instead of suggesting a stratified, unalterable, and universal structured heaven, the graphic clusters show a wide sample of states of being that could be adopted by the sky, be it starry, clouded, dark, or bright (see Mikulska 2010, 2015, and chapter 8, this volume); be it divided into nine or 13 spaces, inhabited and transited by different entities, including time (see Tavárez, chapter 5, this volume); or be it container or threshold, body or space. These characteristics are mutable, historic and specific (see Vail and Hull, chapters 4 and 6, respectively, in this volume).

To end, these multiple and dynamic characteristics of the cosmos can be recognized in the colonial *Codex Florentine*. Here, the painters who worked with Sahagún offered a detailed description of the constitution of heaven, accompanied by an illustration in the appendix of the third book, which explains the destination of the dead. As in other parts of the manuscript, in this figure, which is the only one intended to reproduce the image of the sky in the *Codex Florentine*, the *tlacuilo* created a new code in order to represent this concept. The image (figure 3.12) shows a group of lords who seem to be floating above a group of women. These lords are warriors who died in combat, and are shown seated in the prestigious *petlatl icpalli*, or throne. Instead of wearing their warrior clothes, they are outfitted with the insignia of the *tecuhtli*, the *xiuhitzolli*.

FIGURE 3.12. *The solar sky, the destiny of warriors who died in combat.* (Codex Florentine, *book 3, appendix. Drawing by Victor Medina.)*

The ladies depicted seem to be closely related to the men, and their sad gesture reveals their role as widows, daughters, or mothers of these fallen men. Thus, this figure does not depict the physical characteristics of heaven as described in the written text that accompanies the illustration. Citlalinicue was finally displaced and in her place the *tlacuilo* presented a new image that highlighted two main messages: First is the personal relationship between the lords and ladies, which in the written text is emphasized by noting the strong link that extended after death through the execution of the proper rituals by the family of the departed. Second is the transformation of status experienced by the men, who died as warriors but who appear as lords in the sky, their new place.

The image thus condenses the ruling principles of the cosmologies of central Mexico presented in this chapter: they were dynamic and centered in the local history. They responded to generative action, present in ritual practice and did not imply a rigid universal structure. The sky, as the other realms of the world, was a place under constant construction. This was executed by ritual action and by the maintenance of strong social and familiar links with the ancestors and deceased, who were now living in faraway, but always connected, extending coexistence to alternative spaces.

CONCLUSION

This chapter traces the ways in which the visual configuration of the sky, as conceived in the visual culture of Nahuas and Mixtecs, was an object of transformation and semiotic translation in the early colonial sources. In sum, the postconquest documents tried to reconstruct a unique, universal, and static Mexican cosmographical model, which seem to be absent from both native visual culture and cosmogonic discourse. Instead of that unique prototype, Nahua and Mixtec cosmologies emphasized the local narratives, which explained the shape of the world and the history of its inhabitants, who were able to produce meaningful social links (familiar, political) in life and after death. The stories depicted and narrated in the written sources emphasized the connections between the current earth-town and alternative dimensions, including the sky, the underworld, the interior of water sources and mountains, and possibly the place of dreams. These alternative spaces also included other temporal realms, such as the previous cosmogonic eras, which remained vivid in memory/history.

Therefore, through the analysis of the operative and formal codes for depicting the image of the sky in central Mexico, it was possible to identify the use of clusters (stars of different kinds, rays, clouds, colors) that were integrated into sky bands. These bands were embodiments of different characters who displayed their distinctive attributes; for example, the generative qualities of Citlalinicue, a protagonist in Nahua cosmogonies, were expressed in the designs of her skirt. The written stories of the sixteenth century explained her role as the mother of gods and lineages, emphasizing the generative paradigm of a cosmos that is thought to be under constant construction and reconstruction. This dynamic process highlights the "production" of the cosmos, which operated under the same principles of production of objects (woven, depicted, or carved) and people (born or transformed through sacrifice). It is important to notice that the sky was conceived of as a place that was far away, in the upper realm, but it could be circulated in different directions by specific characters in special moments (like 9-Wind, 8-Deer, or the ritual specialists who acquired that knowledge).

Hence, cosmological knowledge of central Mexico encompassed aspects of history, sociopolitics, cosmogony, astronomy, meteorology, geography, and ritual technologies. In this sense, these cosmologies and cosmogonies contrast with classic Euro-Asian cosmographical and metaphysical configurations, as expressed in the colonial sources that sought to reconstruct the cosmological view of the inhabitants of New Spain.

NOTES

1. According to this model, the sky was a pyramid of seven bodies, where the Lords of the Day occupied the structure's 13 layers (6 ascending steps + 1 top + 6 descending steps = 13 upper positions). The underworld was made of nine positions, located on the five bodies of an upside-down pyramid (4 descending steps + 1 bottom + 4 ascending steps = nine positions) (Seler 1961, 31–33). See figure 3.1.

2. Some authors have criticized the prototype, but only to discuss the use of the stepped model instead of the vertical layered model, or to suggest minor changes, either in content (Krickeberg 1950; Caso 1967, 20; Nowotny 2005, 217–219; Becquelin 1995; Köhler 1995) or in structure (López Austin 1984).

3. I limit the discussion in this chaper to the analysis of the sky; nevertheless, the discussion can expand to include the underworld and other spheres of the cosmos. Unfortunately the complexity of the theme doesn´t allow me to expand the current discussion, which could be enriched with new data derived from the physical inspection of the *Codex Vaticanus A* in a missing page that was carefully cut in the early colonial period. It is highly probable that the page had depicted a figure of the underworld (Diaz, visual inspection in 2017). This theme will be discussed in a paper in preparation, on a new study of the Codex Vaticanus A, forthcoming.

4. In this chapter, we use the term *cosmology* to refer to the conceptual configuration of the world as conceived of by a specific society, and the term *cosmography* to indicate the graphical composition that shapes that cosmological discourse in visual terms. The cosmography was also a literary genre in European literature that included the explanation of the dynamic of the world and its visual representations. The figures and the written explanations cannot be disarticulated.

5. For the complete original text, see *Historia de los mexicanos* (2002, 80–83).

6. It has been suggested that Thevet was not the compiler or translator of the document, but one of its owners. He used part of the material in this manuscript to write his *Cosmographie universelle* (Jonghe 1906, 223; Tena 2002, 118).

7. *Iztapalli* or *iztapaltetl* are described as stones used to pave or glaze architectural structures. Tezozomoc identified them as blue stones (Gran Diccionario Náhuatl 2012); translation assisted by Alfonso Vite (2017).

8. *Huixtohtli* cannot be easily translated. It is used as an ethnonym, linked with the historic Olmecs, but it also appears as a root in the name of a goddess of the salt:

Huixtocihuatl. Nevertheless, it is not clear if the concept means "salt" (Gran Diccionario Náhuatl 2012); translation assisted by Alfonso Vite (2017).

9. The divisions of the sky into nine and 13 places is reported in different colonial sources (see *Codex Florentino* book X, 169, on Sahagún 1979; *Anales de Cuauhtitlán* 1975, 8; Ruiz de Alarcón 1984, 21, 22, 79, 166, 205, 223–225, 324–325, 362). The problem of the variation of repertoires is analyzed by Mikulska (2015; chapter 8, this volume); Nielsen and Sellner Reunert (2009; chapter 1, this volume).

10. Mesoamerican archaeological platforms of nine or 13 steps have often been interpreted as the materialization of the 13/9 levels of the cosmos. Nevertheless, most of these associations are not supported by local evidence but instead apply analytical models to explain the gaps that cannot be filled with local data.

11. See Anders et al. (1992b) and Hermann (2008a) for an analysis of this scene.

12. The transit between different realms is still present in contemporary ritual divinatory and medical practices, in which specialists have to travel to specific places in the sky in order to bring back the soul of a patient (Köhler 1995,100).

13. The red line appears in the *Selden Roll* (1) and *Codex Becker* II (1, 2, 4), and the red and yellow lines can be found in the codices *Vindobonensis and Nuttall*. Examples of the red–yellow band, with an additional different-colored or textured line, include a black band with white stars depicted in *Codex Borbónico* (16) and reptile skin depicted in *Codex Borgia* (3, 4, 5, 6, 20, 27, 28, 37, 49–52). The lithic objects of the National Museum of Anthropology described in this text also include this double band, complemented by other starry elements (see figures 3.6 and 3.7).

14. This section has been interpreted in different ways. It has been conceived of as an astronomical device that follows the visible transit of Venus-Quetzalcoatl. While Batalla Rosado has interpreted the scene as the mythic descent of Quetzalcoatl into the underworld (2008), Seler identified this section as a mythic explanation that condensed general astronomical knowledge (Seler 1963, 9–61). In a similar vein, Milbrath (2013) proposed that the codex registered specific astronomical events that took place between 1495 and 1496. As part of her methodology, she identified the scenes of this section as the 18 months of the year (Milbrath 1989, 1999, 2013). In contrast, Nowotny recognized in this narrative the representation of a ritual cycle (2005, 265–269), an idea supported by Anders and Jansen (1993). Finally, Boone identified in this narrative a Mexican cosmogonical saga (2007, 175–176). In my opinion, the reconstructions of Boone and Nowotny complement each other. For additional interpretations of this section, see Byland (1993), Byland and Pohl (1994), and Brotherston (1999).

15. This principle is also reproduced in the *Lápida de los Cielos* (Díaz 2012).

16. Katarzyna Mikulska analyzes the graphic representation of the sky and underworld, and the Nahuatl words used to refer to them. Her work reveals a clear

association between sky and underworld, conceived of as the place of ancestors and the dead (Mikulska, chapter 8, this volume).

17. While Milbrath's contributions to the discipline are fundamental, I do not agree with her interpretation of this section of the *Codex Borgia*. First, she identified the 18 months or *veintenas* based on isolated iconographic elements. After analyzing the figurative repertoires of the *xihuitl* and the *cecempoalapohualli* included in more than 20 codices and archaeological objects, trying to identify the original representation of the cycle of veintenas (Díaz 2011, 232–423), I arrived at conclusions similar to those of Kubler and Gibson (1951) and Brown (1977), who do not recognize the cycle of the year in this section. Second, the dates written in the *Codex Borgia* (29–46) are not complete, because none of them includes a numeral to complement the signs of the *tonalpohualli*; therefore, Milbrath's identification of the dates is not grounded on internal evidence. Finally, the identification of the quadrilateral and band-shaped deities as the Milky Way is supported by the projection of astronomical phenomena that fit her reconstruction but not by internal evidence: why would all these characters be the same entity, when the nominal attributes of their clothes are different?

18. The generation and reconstruction of the cosmos and its entities by the use of artifacts that become parts of the body or creatures of new generations, is clearly shown in the works compiled by Santos-Granero (2009) on the constructional views of Amazonian societies.

19. Both Citlalinicue and Citlalatonac are identified in some sources as the Milky Way (*Historia de los mexicanos* 2002, 80–81). While the authors do not explain the differences between these deities as impersonators of the Milky Way, I would argue that the association of Citlalinicue as the Milky Way is not supported by an absolute identification between the astronomical phenomenon and its impersonator, but by the concept that is shared by the terms Citlalinicue ("Skirt of Stars") and *Citlalatonac* ("Shine of Stars"). Indeed, the translation of concepts between European and Nahua astronomical knowledge should be carried out with caution. For example, there is no translation for "planet" in Nahuatl; these astronomical bodies seemed to be identified as *citlalli* ("stars"). As Schwaller (2006) shows in his analysis of the word *ilhuicatl*, Nahuatl vocabulary suffered derivations and modifications after the contact with Christian theology.

REFERENCES

Anales de Cuauhtitlán. 1975. In *Códice Chimalpopoca.*, trans. Primo Feliciano Velázquez, 3–118. México: Universidad Nacional Autónoma de México, Instituto de Investigaciones Históricas.

Anders, Ferdinand, Maarten Jansen, and Aurora Pérez Jiménez. 1992a. *Libro explicativo del llamado códice Vindobonensis*. Mexico: Akademische Druck-Verlagsanstalt, Fondo de Cultura Económica, Sociedad Estatal Quinto Centenario.

Anders, Ferdinand, Maarten Jansen, and Gabina Aurora Pérez Jiménez. 1992b. *Crónica Mixteca Libro explicativo del Códice Zouche-Nuttall*. México: Fondo de Cultura Económica, Akademische Druck- und Verlagsanstalt.

Anders, Ferdinand, and Marteen Jansen. 1993. *Los templos del cielo y de la obscuridad; oráculos y liturgia: Libro explicativo del llamado* Códice Borgia. Mexico: Sociedad Estatal Quinto Centenario/Akademische Druck- und Verlagsanstalt, Fondo de Cultura Económica.

Anders, Ferdinand, Maarten Jansen, and Luis Reyes García, eds. 1996. *Religión, costumbres e historia de los antiguos mexicanos. Libro explicativo del llamado* Códice Vaticano A. Mexico: Akademische Druck- und Verlagsanstalt, Sociedad Estatal Quinto Centenario, Fondo de Cultura Económica.

Bassett, Molly. 2015. *The Fate of Earthly Things: Aztec Gods and Bodies*. Austin: University of Texas.

Batalla Rosado, Juan José. 2006. "Estudio codicológico del *xiuhpohualli* del códice *Telleriano Remensis*." *Revista Española de Antropología Americana* 36 (2): 69–87.

Batalla Rosado, Juan José. 2008. *El Códice Borgia: Una guía para un viaje alucinante por el Inframundo*. Madrid: Biblioteca Apostólica Vaticana/Testimonio.

Becquelin, Pierre. 1995. "L´axe vertical dans la cosmologie maya." *Trace* 28: 53–59.

Boone, Elizabeth H. 2007. *Cycles of Time and Meaning in the Mexican Books of Fate*. Austin: University of Texas Press.

Brotherston, Gordon. 1999. "The Yearly Seasons and Skies in the *Borgia* and Related Codices." *Arara: Art and Architecture of the Americas* 2, paper 6. http://www.essex.ac.uk/arthistory/arara/issue_two/paper6.html.

Brown, Betty Ann. 1977. "European Influences in Early Colonial Descriptions and Illustrations of the Mexican Monthly Calendar." PhD diss., University of New México, Albuquerque.

Burland, Cottie Arthur. 1955. *The Selden Roll: An Ancient Mexican Picture Manuscript in the Bodleian Library of Oxford*. Berlín: Gebr. Mann.

Byland, Bruce. 1993. "Introduction and Commentary." In *The Codex Borgia: A Full-Color Restoration of the Ancient Mexican Manuscript*, ed. Giselle Díaz and Alan Rodgers, xiii–xxii. New York: Dover.

Byland, Bruce, and John Pohl. 1994. *In the Realm of 8-Deer: The Archaeology of the Mixtec Codices*. Norman: University of Oklahoma Press.

Caso, Alfonso. 1967. *Los calendarios prehispánicos*. México: Instituto de Investigaciones Históricas, UNAM.

Clendinen, Inga. 2010. "Reconstructing Religion in Sixteenth Century Mexico." *The Cost of Courage in Aztec Society*. Cambridge: Cambridge University Press.

Códice Borgia. 1993. Facsimile edition. Mexico: Sociedad Estatal Quinto Centenario/ Akademische Druck- und Verlagsanstalt, Fondo de Cultura Económica.

Díaz, Ana. 2019. *El cuerpo del tiempo: Códices, Cosmologías y tradiciones cronográficas del centro de México*. México: Instituto de Investigaciones Estéticas, UNAM, Bonilla Artigas editores.

Díaz, Ana. 2009. "La primera lámina del *Códice Vaticano A*: ¿Un modelo para justificar la topografía celestial de la antigüedad pagana indígena?" *Anales del Instituto de Investigaciones Estéticas* 31 (95): 5–44.

Díaz, Ana. 2011. "Las formas del tiempo: Tradiciones cosmográficas en los documentos calendáricos indígenas del México central." PhD diss., Universidad Nacional Autónoma de México.

Díaz, Ana. 2012. "La *Lápida de los cielos* del Museo Nacional de Antropología (México): Un recorrido por el imaginario celeste nahua a través de las piezas de la sala Mexica." *Anales del Museo de América* 20: 121–143.

Dyer, Nancy Joe. 1996. "Libro de oro, el manuscrito." In *Memoriales*, by Fr. Toribio de Benavente Motolinía, 97–118. México: El Colegio de México.

Gran Diccionario Náhuatl (GDN). 2012. Universidad Nacional Autónoma de México. http://www.gdn.unam.mx. Accessed February 15, 2017.

Heninger, S. K. 2004. *The Cosmographical Glass*. San Marino, CA: The Huntington Library and Art Gallery.

Hermann, Manuel. 2008a. *Códice Nuttall*; Segunda Parte: *Arqueología Mexicana*, Edición especial códices, 29.

Hermann, Manuel. 2008b. "Religiosidad y bultos sagrados en la mixteca prehispánica." *Desacatos* 27: 75–94.

Histoire du Mechique. 2002. In *Mitos e historias de los antiguos nahuas*, trans. Rafael Tena, 115–167. Mexico: Consejo Nacional para la Cultura y las Artes (Cien de México).

Historia de los mexicanos por sus pinturas. 2002. In *Mitos e historias de los antiguos nahuas*, trans. Rafael Tena, 15–114. Mexico: Consejo Nacional para la Cultura y las Artes, Cien de México.

Historia de los Mexicanos por sus pinturas. n.d. Manuscript JGI 31, *Manuscrito del Libro de Oro y Tesoro índico*. Austin, TX: Benson Latin American Collection, University of Texas.

Holland, William. 1961. "Relaciones entre la Religión Tzotzil Contemporánea y la Maya Antigua." *Anales del Instituto Nacional de Antropología e Historia* 13: 113–131.

Holland, William. 1963. *Medicina Maya en los altos de Chiapas*. México: Instituto Nacional Indigenista.

Jansen, Maarten. 1982. *Huisi Tacu. Estudio interpretativo de un libro mixteco antiguo: Codex Vindobonensis Mexicanus I*, 319–325. Amsterdam: Centro de Estudios y Documentación Latinoamericana.

Jonghe, Édouard de. 1906. "Thevet, mexicaniste." *Internationaler Amerikanisten Kongress*, Vierzehnte Tagung [1904], vol. 1, 223–240. Stuttgart: Kohlhammer.

Klein, Cecelia. 1982. "Woven Heaven, Tangled Earth: A Weaver's Paradigm of the Mesoamerican Cosmos." In *Ethnoastronomy and Archaeoastronomy in the American Tropics*, ed. Anthony F. Aveni and Gary Urton, 1–35. Annals of the New York Academy of Sciences 385. New York: New York Academy of Sciences.

Klein, Cecelia. 2000. "The Devil and the Skirt: An Iconographic Inquiry into the Prehispanic Nature of the *Tzitzimime*." *Ancient Mesoamerica* 11 (1): 1–26.

Klein, Cecelia. 2008. "A New Interpretation of the Aztec Statue Called Coatlicue, 'Snakes-Her-Skirt.'" *Ethnohistory* 55 (2): 229–250.

Köhler, Ulrich. 1995. *Chonbilal Ch'ulelal—Alma Vendida: Elementos fundamentales de la cosmología y religión mesoamericanas en una oración en maya-tzotzil*. México: Instituto de Investigaciones Históricas, UNAM.

Köhler, Ulrich. 2000. "Los llamados Señores de la Noche, según las fuentes originales." In *Códices y Documentos sobre México, 3° Simposio Internacional*, ed. Cosntanza Vega Sosa, 507–522. México: INAH.

Knab, Timothy. 1991. "Geografía del inframundo." *Estudios de Cultura Náhuatl* 21: 31–58.

Knab, Timothy. 2004. *The Dialog of Earth and Sky*. Tucson: University of Arizona Press.

Krickeberg, Walter. 1950. "Bauform und Weltbild im Alten Mexiko." *Paideuma* 4: 295–333.

Kubler, George, and Charles Gibson. 1951. *The Tovar Calendar*. Memoires of the Connecticut Academy of Arts & Sciences. Vol XI. New Haven, CT: Yale University Press.

León Portilla, Miguel. 1966 [1956]. *La filosofía náhuatl estudiada en sus fuentes*. México: UNAM.

León Portilla, Miguel. 1994 [1968]. *Tiempo y realidad en el pensamiento Maya*. México: UNAM.

López Austin, Alfredo. 1973. *Hombre dios: Religión y política en el mundo náhuatl*. México: UNAM.

López Austin, Alfredo. 1984 [1980]. *Cuerpo humano e ideología*. México: Instituto de Investigaciones Antropológicas, Universidad Nacional Autónoma de México.

Magaloni, Diana. 2011. "Painters of the New World: The Process of Making the *Florentine Codex*." In *Colors between Two Worlds: The Florentine Codex of Bernardino de Sahagún*, ed. Gerhard Wolf and Joseph Connors, 47–76. Florence:

Kunsthistorisches Institut in Florenz, Max-Planck Institut, Villa I Tatti, The Harvard University Center for Italian Studies.

Matos Moctezuma, Eduardo. 1987. "Symbolism of the Templo Mayor." In *The Aztec Templo Mayor*, ed. Elizabeth H. Boone. Washington, DC: Dumbarton Oaks.

Matos Moctezuma, Eduardo. 2008 [1975]. *Muerte a filo de obsidiana: Los nahuas frente a la muerte*. México: Fondo de Cultura Económica.

Mendieta, Gerónimo de. 2002. *Historia Eclesiástica indiana*. México: Consejo Nacional para la Cultura y las Artes, Cien de México.

Mikulska, Katarzyna. 2008. "El concepto de *ilhuicatl* en la cosmovisión nahua y sus representaciones gráficas en códices." *Revista Española de Antropología Americana* 38 (2): 151–171.

Mikulska, Katarzyna. 2010. "¿Cuchillos de sacrificio? El papel del contexto en la expresión pictórica mesoamericana." *Itinerarios* 12: 125–154.

Mikulska, Katarzyna. 2015. "Los cielos, los rumbos y los números: Aportes sobre la visión nahua del universo." In *Cielos e inframundos una revisión a la Cosmología Mesoamericana*, ed. Ana Díaz, 109–173. Mexico: Instituto de Investigaciones Históricas, Fideicomiso Felipe Teixidor y Montserrat Alfau de Teixidor.

Milbrath, Susan. 1989. "A Seasonal Calendar with Venus Periods in *Codex Borgia* 29–46." In *The Imagination of Matter: Religion and Ecology in Mesomaerican Traditions*, ed. Davíd Carrasco. International Series 515. Oxford: British Archaeological Reports.

Milbrath, Susan. 1999. *Star Gods of the Maya: Astronomy in Art, Folklore and Calendars*. Austin: University of Texas Press.

Milbrath, Susan. 2013. *Heaven and Earth in Ancient Mexico: Astronomy and Seasonal Cycles in the Codex Borgia*. Austin: University of Texas Press.

Nielsen, Jesper, and Toke Sellner Reunert. 2009. "Dante's Heritage: Questioning the Multi-Layered Model of the Mesoamerican Universe." *Antiquity* 83 (320): 399–413.

Nowotny, Karl A. 2005 [1961]. *Tlacuilolli*. Norman: University of Oklahoma Press.

Olivier, Guilhem. 2004. *Tezcatlipoca, burlas y metamorfosis de un dios azteca*. Mexico: Fondo de Cultura Económica.

Olivier, Guilhem. 2006. "The Sacred Bundles and the Coronation of the Aztec King in Mexico-Tenochtitlan." In *Sacred Bundles: Ritual Acts of Wrapping and Binding in Mesoamerica*, ed. Julia Guernsey and F. Kent Reilly III, 199–225. Barnardsville, CA: Boundary End Archaeology Research Center.

Prechtel Martin, and Robert S. Carlsen. 1988. "Weaving and Cosmos amongst the Tzutujil Maya." *Res: Anthropology and Aesthetics* 15: 122–32.

Quiñones Keber, Eloise. 2003. "Tonacatecuhtli and Tonacacihuatl: Azteca Deities of Trascenence and *Tonalli*." *Latin American Indian Literatures Journal* 9 (1): 172–186.

Ruiz de Alarcón, Hernando. 1984. *Treatise on the Heathen Superstitions That Today Live among the Indians Native to This New Spain, 1629.* Trans. J. Richard Andrews and Ross Hassig. Norman: University of Oklahoma Press.

Ruiz Medrano, Ethelia. 2011. *Mexico's Indigenous Communities: Their Lands and Histories, 1500 to 2010.* Boulder: University Press of Colorado.

Sahagún, Bernardino de. 1979. *Códice Florentino.* Edición facsimilar. México: Secretaría de Gobernación.

Sahagún, Bernardino de. 2002. *Historia General de las Cosas de Nueva España.* 3 vols. Study and paleography by Alfredo López Austin and Josefina García Quintana. México: Conaculta, Cien de México.

Santos-Granero, Fernando. 2009. "Introduction. Amerindian Constructional Views of the World." In *The Occult Life of Things*, ed. Fernando Santos-Granero, 1–29. Tucson: University of Arizona Press.

Schele, Linda, and Mary Ellen Miller. 1986. *The Blood of Kings: Dynasty and Ritual in Maya Art.* New York: Braziller; Fort Worth: Kimbell Art Museum.

Schwaller, John F. 2006. "The *Ilhuica* of the Nahua; Is Heaven Just a Place?" *The Americas* 62 (3): 391–412.

Seler, Eduard. 1961 [1920]. "Mythus und Religion del Alten Mexikaner." *Gesammelte Abhandlungen* IV: 1–167. Graz, Austria: Akademische Druck- und Verlagsanstalt.

Seler, Eduard. 1963 [1904]. "El viaje de Venus a través del infierno." In *Comentarios al Códice Borgia*, 2: 9–61. Mexico: Fondo de Cultura Económica.

Spitler, Susan. 2005. "Colonial Mexican Calendar Wheels: Cultural Translation and the Problem of 'Authenticity.'" In *Painted Books and Indigenous Knowledge in Mesoamerica*, ed. Elizabeth Hill Boone, 271–288. New Orleans, LA: Middle America Research Institute.

Stuart, David, and Stephen D. Houston. 1994. *Classic Maya Place Names.* Studies in Pre-Columbian Art and Archaeology, no. 33. Washington, DC: Dumbarton Oaks Research Library and Collection.

Taube, Karl. 2009. "The Womb of the World: The Cuauhxicalli and Other Offering Bowls of Ancient and Contemporary Mesoamerica." In *Maya Archaeology 1*, ed. Charles Golden, Stephen Houston, and Joel Skidmore, 86–106. San Francisco, CA: Precolumbia Mesoweb Press.

Tena, Rafael. 2002. "Introduction" to the *Histoire du Mechique*. In *Mitos e historias de los antiguos nahuas*, 115–122. México: Consejo Nacional para la Cultura y las Artes, Cien de México.

Thompson, J. Eric S. 1934. "Sky Bearers, Colors, and Directions in Maya and Mexican Religion." In *Contributions to American Archæology*, volume II (10): 209–242. Washington, DC: Carnegie Institution.

Thompson, J. Eric S. 1954. *The Rise and Fall of Maya Civilization*. Norman: University of Oklahoma Press.

Thompson, J. Eric S. 1970. *Maya History and Religion*. Norman: University of Oklahoma Press.

Tozzer, Alfred. 1907. *A Comparative Study of the Mayas and the Lacandones*. New York: Macmillan Company.

Vite, Alfonso. 2017. "De la Mesa a los Siete Cerros. Paisaje ritual entre los nahuas de Huautla, Hidalgo." Tesis de Maestría en Estudios Mesoamericanos. Universidad Nacional Autónoma de México.

Part II
Inventiveness

Reshaping Experience in Colonial Cosmologies

4

The Colonial Encounter

Transformations of Indigenous Yucatec Conceptions of K'uh

Gabrielle Vail

Research in the late nineteenth and early twentieth centuries first demonstrated structural parallels between descriptions of calendrical ceremonies recorded in the *Relación de las cosas de Yucatán* attributed to the sixteenth-century Franciscan friar Diego de Landa and almanacs depicting similar ceremonies in the Postclassic Maya codices. More recently, breakthroughs in the decipherment of the Maya hieroglyphic script have enriched our understandings of Maya conceptions of the supernatural agents that Landa encountered, whereas advances in ethnohistorical methodologies have improved our ability to reconstruct and understand transformations that occurred in indigenous cosmologies during the missionizing efforts of the sixteenth through eighteenth centuries.

This chapter examines the concepts of *ángel* "angel," *demonio* "demon," and *diablo* "devil" introduced in the *Relación* and included in later Yucatec-language texts such as the *Books of Chilam Balam*, which were written in the Latin alphabet from an array of source materials of both Maya and European origin. Conceptions of the underworld (Metnal or Xibalba) are also examined to draw comparisons with prehispanic sources.

Landa's descriptions of supernaturals are grouped under various rubrics, including the term *dios*, in addition to designations such as *imagen* "image," *ídolo* "idol," and *estatua* "statue." This chapter seeks to disambiguate these concepts and recontextualize them within current understandings of supernaturals based on

DOI: 10.5876/9781607329534.c004

hieroglyphic captions from Postclassic Maya codices. This, in turn, allows us to document how conceptions of supernaturals changed from the fourteenth through eighteenth centuries and influenced scholarly interpretations of prehispanic cosmology. What appears to be the earliest documented expression of this transformation involves Landa's attribution of the term *ángel* to two of the deities involved in rituals associated with the start of year on the day K'an in the Yucatec 260-day sacred calendar, or *tzolk'in*.

A comparative analysis of deity complexes from prehispanic codices and colonial manuscripts provides a unique window into some of the processes of culture change that occurred as a result of the engagement between Spanish ecclesiastics and educated members of the Maya nobility who acted as "cultural interpreters" for the Spanish missionaries. As scholars have recently demonstrated (e.g., Bricker and Miram 2002; Knowlton 2010), indigenous Maya were actively involved in the processes that led to these transformations, as the result of their attempts to make sense of the European worldview in light of their own understanding of the natural and supernatural worlds.

SOURCE MATERIALS

Textual sources from the Postclassic and colonial periods in Yucatan form the basis of this analysis, in combination with images from the codices that relate in a very explicit way to the hieroglyphic captions that accompany them. Three prehispanic manuscripts believed to be from the northern Maya lowlands on the basis of their style, the material culture depicted, types of calendars used, and later documentary evidence are today housed in European collections; the manuscripts are known by the name of the cities in which they are housed: the *Dresden*, *Madrid*, and *Paris* codices. They are each made from sheets of paper composed of the inner bark of the ficus tree that are joined together and then folded in a screenfold format and painted on both sides. They contain almanacs based on the 260-day *tzolk'in* calendar (in which 13 numbers are paired with 20 day names), tables of celestial observations and predictions, and ritual texts based on other calendrical cycles (including the 360-day *tun* and the approximately 20-year *k'atun*). Pictured and named in the hieroglyphic captions are a series of figures, many of whom can be identified explicitly as deities based on the epigraphic evidence.

Texts from the colonial period that are important to this analysis include the *Relación de las cosas de Yucatán* and a series of manuscripts known as the *Books of Chilam Balam* that include compilations of texts from both indigenous

Maya and European sources. The extant versions of these manuscripts date to the late eighteenth century, but there is clear evidence that the mythological portions of the manuscripts attributed to Chumayel, Tizimín, and Maní (the latter known as the *Códice Pérez*) can be dated much earlier, likely to the late seventeenth century (Knowlton 2010, 33–34). The *Books of Chilam Balam* were written in the Latin alphabet by indigenous Yucatec scribes (or redactors), as was another manuscript of interest to this study—the Morley Manuscript. Like the *Books of Chilam Balam*, the Morley Manuscript is written by hand in a leather-bound copybook dating to the eighteenth century. Evidence internal to the manuscript suggests that it derives from a late sixteenth-century original text (Knowlton 2008); Knowlton (2010, 48) describes it as a "missing link" between the *Books of Chilam Balam* and doctrinal works of Franciscan origin, based on the fact that it contains sermons composed in Yucatec that were later included in a publication dated to 1620, as well as a collection of European cosmogonies and cosmological texts that were translated into Yucatec and later copied into the *Books of Chilam Balam of Kaua* and *Chan Kan* (Knowlton 2008).

Landa's *Relación* must be regarded with special consideration, as the extant version is "an arbitrary collection by three or four compilers, probably made at various times but all after Landa's death, of excerpts of what may have been a large multi-volume work . . . or a collection of writings by Landa" (Restall and Chuchiak 2002, 663). Additionally, it is not clear how much of the text is Landa's, since it appears that the compilers drew from Landa's collection of papers, which included writings not only by him but also by his contemporaries and informants. One of the likely authors of certain sections of the manuscript is Antonio Gaspar Chi, the son of a Maya nobleman who was himself a priest-scribe. Chi was educated by the Franciscans and served as an interpreter; he was also extremely knowledgeable about Yucatec history and worked closely with Landa (Restall and Chuchiak 2002, 663). Restall (1998, 144–168) notes that there are some striking similarities between some of Chi's writing and sections of the *Relación*. Originally taken to be evidence of Chi's influence, it may instead be the case that the passages in question were copied from Chi's writings and not Landa's. There are, nevertheless, cultural biases present in the sections of the *Relación* that form the basis of the current study that suggest they stem from the pen of a European author or one raised in the Catholic tradition. Whether they are Landa's own observations or those of his contemporaries (or redactors) remains unknown.

THE CONCEPT OF *K'UH* IN THE MAYA CODICES

The term *k'uh [kuu]*, which is glossed as "dios" in colonial-period dictionaries (Arzápalo Marín 1995, I, 433), has a broad range of uses in the Maya codices (table 4.1, figure 4.1). It occurs

> as a title preceding the names of various deities;
> in place of the name glyph of the rain god Chaak, the creator Itzamna, the god of death (whose glyph may read Kimil "Death"), God C (the embodiment of *k'uh*), the jeweled flower god (God H, possible named Nik "Flower"), and the aged deity Pawahtun;
> conflated with the names of the maize, flower, and earth deities;
> and as an effigy (or what the Spanish termed "idol") representing both specific deities and "deity" more generally (with the specific referent remaining unspecified).

The concept of *k'uh* is personified in the codices not only by deities related to the natural world (i.e., pertaining to the domains of rain, agriculture, maize, creation, etc.), but also by an entity whose very appearance is a physical representation of the *k'uh* glyph in anthropomorphic form (see figure 4.1a). This figure, named God C by Paul Schellhas (1904), may be identified with a specific deity, based on which appellative glyph appears in the text caption (that of the death god appears in figure 4.1a, for example), or through the use of a particular epithet that can be linked to a specific deity. In a small number of cases, however, the referent appears to be K'uh (as epitomizing divinity) himself. An interesting example appears on page 18a of the *Madrid Codex*, where Itzamna and the death god may be seen emerging from the open mouth of K'uh's head. This forms the first of a series of three almanacs in which deities are otherwise shown emerging from the open jaws of serpents (generally conceived of as portals or passageways leading to or from the earth's interior). Blue-painted K'uh figures also serve as the protagonists of all but one frame of the almanac on *Madrid* 10c–11c (Chaak appears in the remaining frame). It is likely that K'uh was intended to represent Chaak in this almanac, as Chaak is the subject of the cognate almanac on pages 65b–69b of the *Dresden Codex*.

Based on this patterning, it appears that the figures pictured in the Maya codices can be identified emically as deities, although care must be taken when applying Western definitions (god or *dios*) to indigenous concepts.[1] The mapping of indigenous to European conceptions is not likely to involve a one-to-one correspondence, as may be seen for example by the encompassing of what Western cultures would interpret as effigies, rather than deities themselves, in the emic category. Moreover, where we might expect to see ritual specialists

TABLE 4.1. T14.1016c (*k'uh*) in the Maya codices

A. As a title associated with the following deity names.
God A: the death god
God A': Akan (a god of alcoholic beverages)
God B: the rain god Chaak
God CH: Yax Balam "First Jaguar"
God D: the creator Itzamna
God E: the maize god
God G: the sun god K'in Ahaw
God H: the god of wind and flowers
God K: the god of sustenance K'awil
God M: the merchant deity
Goddess O: the creator deity Chak Chel
God Q: the punisher god Kisin
God R: the earth god
Lahun Chan: 10 Sky
K'ak' Sip: Fire Peccary
Sak Nik: White Flower
Ix Kab Chel: Lady Earth Rainbow?
Kimil Chel: Lady Death Rainbow?
B. In combination with the names of the following deities.
God E (T24.1016c "maize god")
God H (T147.1016c "wind and flower god")
Goddess I (T58.1016c, "white god")
God M (T95.1016c, "black god")
God A (T1016c[738] "death god")

or human beings as the subject of the codical almanacs, instead the deities are themselves portrayed as the ritual actors and are depicted undertaking the acts that would normally be assumed by human agents (e.g., ritual specialists and those they are directing; see figure 4.2).

A particularly compelling example of the distinctions made by Western observers that play no role in indigenous Maya worldviews occurs on the *Dresden* yearbearer pages, which portray the rituals associated with the end of one year (a five-day liminal period) and those marking the inauguration of the next. These rituals are described in considerable detail in Landa's *Relación*,

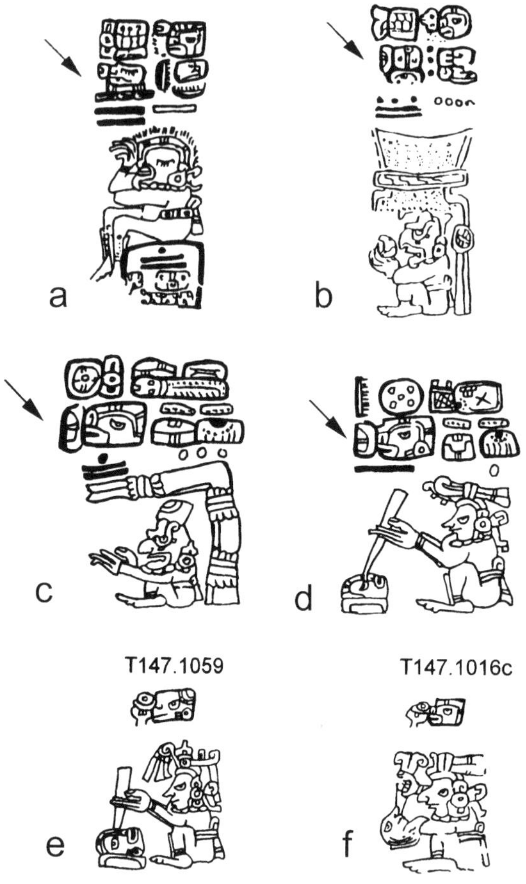

FIGURE 4.1. *Examples of k'uh (T1016c) in the Maya codices: (a) paired with the name of the death god indicated with the arrow; (b) paired with the title of Itzamna indicated with the arrow; (c, d) with the prefix T24 to name a complex of deities associated with the maize god; (e) naming God H, the god of wind and flowers, whereas in (f) T1016c substitutes for God H's portrait glyph. (Redrawn after Villacorta and Villacorta 1976.)*

where they are discussed in terms of processions of and offerings to the "statues" (*estatuas*) and "images" (*imágenes*) of particular deities (referred to most commonly in Landa's text as *demonios*), which were carried out by "the lords and the priest and the men of the town" (Tozzer 1941, 140).

On the yearbearer pages of the *Dresden Codex* (see figure 4.3), we can observe a conflation between the various participants mentioned in the *Relación*: the "priests" involved in performing the ceremonies, the deities themselves, and the effigies that formed an important component of the rituals. In the bottom register of each page, for instance, the rituals (including the scattering of incense and the sacrifice of a turkey) are being performed by the deities described as "statues" in Landa's text. In the top and middle registers, however,

146 GABRIELLE VAIL

FIGURE 4.2. *Ceremony involving sanctifying deity effigies. Itzamna and God H sprinkle sacred water on vessels containing* k'uh. *(*Madrid Codex *100d; after Brasseur de Bourbourg 1869–1870.)*

these same figures (or aspects of them) are portrayed being acted upon (carried in one instance and seated within a house or temple and receiving offerings in the other), suggesting that they most likely represent effigies of the deities mentioned in the *Relación*, rather than the deities themselves. This overlapping of roles and functions offers a fascinating glimpse into conceptions of "deity" among the Late Postclassic Maya of Yucatán. Clearly, it was a fluid category that included supernatural agents as actors, physical representations in various media (primarily wood or clay), and perhaps also human agents—actors who became, for a brief period (and under the appropriate circumstances), divine. With the advent of Christian missionizing efforts in which such notions would have been neither comprehended nor accepted, new ways of categorizing these different ritual actors came into being.

A series of almanacs that picture K'uh's glyph as an object being carved, sanctified, or ritually disposed of can be related to a description in the *Relación* referring to the making of wooden images ("idols") of the deities during the months of Mol or Ch'en, which corresponded to December and January in the mid-sixteenth century (Gates 1978, 76). A ceremony that may have involved animating these effigies (*k'uh*) is depicted on *Madrid Codex* 100d (figure 4.2); the two frames with pictures show the creator Itzamna and God H holding serpent scepters, which would have been used to sprinkle the vessels containing the K'uh essences with sacred water. The hieroglyphic

FIGURE 4.3. *Yearbearer ceremony. (*Dresden Codex *25; after Förstemann 1880.) The upper register shows an opossum performer carrying the deity of the outgoing year, named in the caption as K'awil. This scene takes place in the south. K'awil receives offerings in the middle register, before being "retired." In the bottom register, the sun god (on the right) performs a ritual involving scattering incense and sacrificing a turkey in front of an effigy of the god Chaak (left of scene). The ritual action has shifted from the south (in the previous registers) to the east in this one.*

captions remain opaque, so it is unclear how this particular ritual action is being described. The fact that the almanac begins on the day 4 Ahaw—a date with particular ritual significance in the *Madrid Codex*—is of interest,[2] as is the fact that the act of blessing the *k'uh* is being performed in each of the four world quadrants: by an unnamed deity in the east, Itzamna in the north, a death god in the west (this is mistakenly written in the text as *lak'in* "east," rather than *chik'in* "west"), and God H in the south. In this particular almanac, both Itzamna and God H are dressed as ritual specialists, or as what Landa would describe as "priests" (*sacerdotes*); they wear "miters" in addition to holding the serpent aspergillums.

Moving from prehispanic contexts into the colonial period, shifts in patterning and usage of the term *k'uh* can be clearly documented. Additionally, the frequency with which particular deities are mentioned and how they are designated also undergoes a significant transformation during the sixteenth to eighteenth centuries.

DIOS, DEMONIO, AND *ÁNGEL* IN COLONIAL CONTEXTS

In accounts of the yearbearer ceremonies from the mid-sixteenth-century *Relación de las cosas de Yucatán*, the term *dios* is reserved for the four named Bakabs (Bacabs; to be discussed below), Chaak (Chac), and an aspect of Itzamna called Yax Kok Ah Mut (Yaxcocahmut).[3] The statues and images of deities that formed part of the yearbearer ceremony, including both the color-directional Way (*uay*) and the year "patrons" named Bolon Tz'akab (Bolonzacab), K'in Ahaw (Kinichahau), Itzamna, and Wak Mitun Ahaw (Wac Mitun Ahau), were all referred to as *demonios*.[4] Moreover, a series of color-directional stones that received blood offerings, called the Ah Kaan Tun (*acantun*), were also categorized as demons.

The latter categorization is of particular interest in that it contrasts with the term *ángel*, which took on an important role in indigenous colonial-period manuscripts. A review of sixteenth-century texts suggests that the term *ángel* made its first appearance in the discussion of the yearbearer ceremonies in Landa's manuscript, where it is used in two different contexts (both relating to years beginning on the day K'an). The first occurs following a description of a series of rituals performed at the southern entrance to town, at which point the "image" of the Way (Kan-uvayeyab) was placed on a wooden standard in preparation for being carried to the house of the *principal*. Before the ritual procession began, however, the participants placed "on his shoulders an angel as a sign of water and of a good year, and these angels they painted so as to make them frightful in appearance" (Gates 1978, 62). A comparison with page 25 of the *Dresden Codex* (see figure 4.3), which depicts the ceremonies associated with southern years, suggests that the *ángel* may relate specifically to Bolon Tz'akab (Bolonzacab) in the *Relación* (Vail and Knowlton 2012).

The name (Ah) Bolon Tz'akab is translated as "something perpetual" in colonial-period dictionaries (Ciudad Real 1995, 92) or as "he of the nine generations" (Thompson 1970, 227); other possibilities that can be proposed are "nine changes [or successions]" (*bolon tz'ak*) or "nine changes [successions] of the earth" (*bolon tz'ak kab*). The deity pictured in the *Dresden Codex* that corresponds to the figure named as Bolon Tz'akab in the *Relación* is K'awil (Kauil), the god of sustenance and lightning. K'awil's name consists of *k'aa* and *wi'il* (Barrera Vásquez et al. 1980, 359, 922), together meaning "an abundance of food" (Thompson 1970, 289), which fits the description of the *ángel* mentioned in Landa's manuscript particularly closely.

Although the term *bolon tz'akab* is absent from the *Dresden Codex*, two possible glyphic references to this deity may be found in the *Madrid Codex*. The first, which reads *bolon tz'ak* "nine changes," appears in the caption to a frame picturing the creator deity Itzamna on page 19b. It is preceded by *u taan* "its face or presence," referring to the face or presence of Bolon Tz'ak(ab). The other example occurs in an almanac occupying the lower register of pages 95–96. Again, Itzamna is pictured, although in this instance he wears K'awil's nose above his own (figure 4.4). The appellative naming this deity reads *bolon yax* "nine green/blue."[5] In this regard, it is of interest that there is a reference to Yax Bolon Tz'akab (Dzacab) on page 43 of the *Chilam Balam of Chumayel* (Knowlton 2010, 61).

The term *ángel* also appears in the *Relación* in reference to the ceremonies performed in order to ensure that K'an years were favorable. To accomplish this, an "idol" called Itzamna K'awil (Itzamna-kauil) was made and placed in the temple.[6] A sacrifice was performed in the court of the temple that involved throwing a victim from a height onto stones piled below and then extracting his heart, which was "raised to the new idol, and offered between two plates . . . They say that an angel descended and received this sacrifice" (Gates 1978, 63). There is evidence, again from the *Chilam Balam of Chumayel*, to suggest that the *ángel* referenced is none other than Bolon Tz'akab. In the *Chumayel* text, Bolon Tz'akab (Dzacab) is described as descending from the sky for his rebirth ceremony, which takes place on the day K'an (Kan), when "his burden had been tied" (*Chumayel* 45; translated by Knowlton 2010, 76).[7] The fact that Landa's description and that in the *Chumayel* both refer to a ritual of descent taking place at the start of K'an years suggests a relationship between them, implying that the *ángel* and Bolon Tz'akab are one and the same. Evidence in support of this assertion may also be found on the page referring to K'an yearbearer ceremonies in the *Madrid Codex* (page 35), where a variant of the maize god with feathers on his arm is shown "diving" towards

FIGURE 4.4. *Itzamna wearing K'awil's nose in frame 3 and named as Bolon (Nine) Yax (at glyph E2). (*Madrid Codex *95d; after Brasseur de Bourbourg 1869–1870.)*

a sacrificial victim who has fallen head first onto some type of structure or altar (figure 4.5, upper right). There can be little doubt that the description in Landa's Relación correlates with this scene, suggesting that here the maize god is the physical embodiment of Landa's "angel."

In the Maya codices, there is a close association between the maize god and K'awil; both serve as embodiments of sustenance, and indeed, one of the meanings of *k'awil* is "Maya god of maize" (Barrera et al. 1980, 387). In this respect, it is of interest that the hieroglyphic caption to the K'an yearbearer page mentions the god K'awil, although whether this is in reference to the maize god remains unclear. As we have seen, the deity described as descending in the *Chumayel* manuscript on a day K'an is Bolon Tz'akab, suggesting a link between the codical maize god K'awil and the colonial-period deity (or deity complex) Bolon Tz'akab.[8]

To summarize, important changes can be documented when comparing the Postclassic Maya codices to colonial-period sources, revolving primarily around the prehispanic deity K'awil, who became known as Bolon Tz'akab in the colonial period, and also as Landa's *ángel*. Additionally, the prehispanic relationship between K'awil and the maize god leads to the merging of the

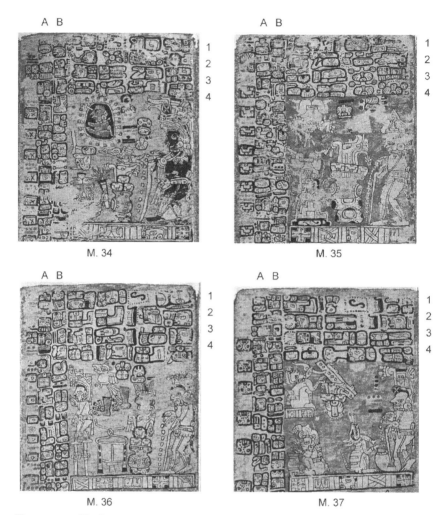

FIGURE 4.5. *Yearbearer ceremonies, corresponding to Kawak (M. 34), K'an (M. 35), Muluk (M. 36), and Ix (M. 37) years. References to four directional Chaaks appear in the captions (at A2–B2 on each of the pages), and Pawah deities are shown planting on the right side of each page. The "diving" figure pictured in the upper left of M. 35 has feathers on his arms, a characteristic that might inspire its interpretation as an "angel." (Madrid Codex, 34–37; after Anders 1967.)*

two, who both take on the persona of *ángel*/Bolon Tz'akab in colonial-period contexts. Another related term (*cangel*, or "archangel") was also adopted by indigenous Maya authors in the colonial period. It was used in relation to two other complexes of deities discussed by Landa—Chaak (Chac) and Pawahtun (Pauahtun)—who were in turn related to the deities known as the Bakabs (Bacabs), four "brothers" said to have been placed by God to hold up the heavens (Tozzer 1941, 135).[9]

THE *CANGEL* AND RAIN DEITY COMPLEXES

The Bakabs are not explicitly mentioned in the yearbearer almanacs but the quadripartite Chaaks and Pawahtuns are (although the latter are referred to as *Pawahk'in*, rather than *Pawahtun*).[10] The four color-directional Chaaks are named in the hieroglyphic captions to the *Madrid* yearbearer pages, beginning at the left with the black Chaak of the west (see figure 4.5). The Pawah deities are explicitly represented in the upper register on each page, where they are shown planting seeds (note that the first one is painted black, to match its association with Kawak/western years).[11]

A somewhat different role for the Pawah deities (specifically, the Pawahtuns) is evident in the Chilam Balam of Chumayel, where they are identified as the "angels" or "archangels" of the winds (Knowlton 2010, 92; Tozzer 1941, 137n638). The association of the Pawahtuns with the winds in the Chumayel text likely stems from the quadripartite nature of these deities in the prehispanic period, as well as their close relationship to the Chaaks. At the time of the conquest, the Europeans conceived of the winds as having directional associations, with a total of twelve winds related to both the cardinal and intercardinal points. Those stemming from the cardinal directions were highlighted and identified as "archangels" (Bricker and Miram 2002, 15). This same title is applied to Pawahtuns on page 51 of the *Chumayel* manuscript, where Red Pawahtun (Chac Pauahtun), White Pawahtun (Sac Pauahtun), Black Pawahtun (Ek Pauahtun), and Yellow Pawahtun (Kan Pauahtun) are equated with the *cangeles yk*, or "Wind Archangels."

In an agricultural ritual observed in 1813, the four Pawahtuns were identified with the gods of rain, fertility, and the four winds. The red Pawahtun was associated with the east and Santo Domingo, the white one with the north and Santo Gabriel, the black one with the west and Santo Diego, and the yellow one with the south and with both Santa Maria Magdalena and the Maya goddess X Kanleox (Baeza 1845 [1815], 170, as cited in Knowlton 2010, 90). In the contemporary Yucatec community of Yalcobá, the four Chaaks are identified with the four Babatun (likely derived from *pawahtun*) deities at the "four corners and limits of the earth" (Sosa 1989, 134), a concept that echoes Landa's assertion that the Bakabs were four brothers located at the four points of the world (Tozzer 1941, 135).[12] Like the prehispanic Chaaks, the Babatun deities are each assigned a color, which is said to correspond to the colors of the clouds associated with that particular direction. Although using a different set of color-directional associations, it is of interest that the *Madrid* yearbearer pages, which depict the four directional Pawah deities planting, also include references to color-directional clouds (Vail and Hernández 2013b, 356). From this, we can see the influence of both the indigenous prehispanic tradition, but also European influences such as those evident in Landa's manuscript.

The prehispanic associations of the Pawah deities are of particular importance in determining how later conceptions of this figure may have been altered by the mapping of prehispanic deities onto Catholic entities (archangels, saints, etc.).[13] In the Maya codices, the primary association of deities named with the Pawah title is with agriculture[14]—specifically, the task of planting (12 examples occur, seen on pages 26a–27a, 26b–27b, and 34a–37a of the *Madrid Codex*; see figures 4.5 and 4.6).[15] It is of interest that this figure (also called God P) is associated with the four world directions, whereas Chaak represents the center, or fifth direction, on *Madrid* 26a–27a and 26b–27b. This provides evidence in confirmation of the connection described in the *Relación* between the Pawahtuns and the Chaaks.

In light of colonial-period identifications, it is of considerable interest that none of the figures labeled with the Pawah title in the codices can be identified as a wind god, with the possible exception of the Pawahtun pictured in the upper register of page 48 of the *Dresden Codex*. Vail and Hernández (2013b, 288, 291) suggest the possibility that this deity takes on the aspect of a wind god (see their discussion for the reasons behind this identification). More generally, however, God H, who represents the god of the number three and of flowers, is thought to have an association with wind, and also with breath and life (Taube 1992, 63). As this figure does not have a quadripartite aspect, it was the Pawahtuns and Chaaks—based on their associations with rain, agriculture, and the four world quadrants—who took on the roles of the directionally based winds found in European sources at the time.

As part of this same process, the Chaaks also merged with the archangels. Page 43 of the *Chumayel* manuscript (see Text B in appendix 4.1) states: "Then his [Oxlahun Ti K'uh's] *cangel* was tugged and shall be removed" (Knowlton 2010, 57), or alternately "Then his [Oxlahun Ti K'uh's] *cangel* was tugged and his face was sooty" (Vail and Hernández 2013b, 85, 96n57). Knowlton identifies

FIGURE 4.6. *The four Pawahtuns (in frames 1–4) and Chaak (in frame 5) planting seeds. (Madrid Codex 26a–27a; after Brasseur de Bourbourg 1869–1870.)*

the *cangel* referenced in the text as the rain god aspect of Oxlahun Ti' K'uh (Oxlahun Ti Ku),[16] a plurality of deities (also sometimes represented in singular aspect) with celestial associations (see Vail and Hernández 2013b, 80–81, 295).

The deity known as Chauk or Anhel among the highland Maya of Chiapas may likewise be related to the prehispanic Chaaks (rain gods) but also to K'awil in his role as a god of lightning. In this regard, it is of interest that the Tzotzil identify this figure as "Younger Brother," and Balankil—a god associated with the healing power of tobacco—as "Elder Brother" (Groark 2010). John Carlson (2007, 11) has recently proposed that this relationship mirrors that between the two prehispanic deities known as God L and K'awil (see discussion below). Vail and Hernández (2013b, 74, 205) suggest that the colonial-period counterparts of these deities (or deity complexes) include Bolon Ti K'uh ("Nine as God," the Elder Brother) and Oxlahun Ti' K'uh ("Thirteen as God," the Younger Brother). Throughout the Maya region, stories of brothers (either two or three in number) abound. In the majority of these, the elder of the two brothers has clear associations with Venus and the underworld realm, and the younger with the sun and the celestial sphere (Carlson 2007; Milbrath 1999; Thompson 1930, 1970).

The association of archangels (*cangels*) with the rain god has continued to the present day in Yucatan as well, as may be seen in the writings of Redfield and Villa Rojas (1934, 116) in reference to their fieldwork in the Yucatec community of Chan Kom. They note that the eastern (or senior) Chaak is called Saint Michael Archangel and that the Chaaks are today "visualized as old men who ride on horses which are seen as clouds."[17] The horses, which are of different colors, are known as *santo cangel*. Bricker and Miram (2002, 78), following Donald Thompson (1954), suggest that this syncretism stems from the congruence of the prehispanic system of color-directional Chaaks and references in the Book of the Revelation to the four riders of the Apocalypse, whose horses are respectively red, white, black, and pale. The close association between these colors and those associated with the prehispanic Chaaks (red, white, black, and yellow) was clearly a contributing factor in "the final coalescing of Chaaks with archangels" (Thompson 1954, 13).

Like the process involved in assigning the term *ángel* to two related but distinct prehispanic Maya deities that resulted in the merging of their individual characteristics within a single framework, a similar process occurred as a result of linking the Chaaks and Pawahtuns. Although the two deities (or deity complexes) occasionally take on each other's names and attributes in the Maya codices, they remain distinct entities with their own particular functions, characteristics, and appellatives during the Postclassic period. In the *Relación*, however, they are described as largely interchangeable (both are subsumed under the category Bakab), a trend that continues with the mapping of both the Chaaks and Pawahtuns onto the Christian archangels, or *cangels*, seen in the Books of Chilam Balam. Indeed, by the nineteenth century, the Pawahtuns and Chaaks appear to have been thought of as merely two different names for the same complex of beings—rain and agricultural deities that had been at one time separate, albeit related, deities.

Bakabs in the Maya Codices and Landa's *Relación*

Of the deities discussed in the *Relación*, perhaps the most emphasis is given to the Bakàbs (Bacabs). They are described as follows:

> Among the multitudes of gods which this nation worshipped they worshipped four, each of them called Bacab. They said that they were four brothers whom God placed, when he created the world, at the four points of it, holding up the sky so that it should not fall. They also said of these Bacabs that they escaped when the world was destroyed by the deluge. They gave other names to each one of them and designated by them the part of the world where God had placed him bearing up the heavens . . . And they distinguished the calamities and fortunate events which they said must happen during the year of each one of them. (Tozzer 1941, 135–136)

The importance of the Bakabs in Landa's text appears at odds with their relative unimportance in the Maya codices, despite the fact that other aspects of these deities—namely, the Chaaks and Pawahtuns (see above)—were prominent. In the 210 pages that compose the extant Maya codices, there is only a single reference to Bakab, which occurs on page 74 of the *Dresden Codex*

FIGURE 4.7. *God L, possibly named as Bakab, in a scene associated with the ascent of the earth crocodilian to the sky. This creature belches forth a torrent of water at the top of the page. Its body is in the form of a "skyband," with glyphs for star or Venus, sky, darkened sun, and night or darkness. At the center of the scene is the goddess Chak Chel above the black God L with his own headdress, spear, and darts.* (Dresden Codex, 74; after Förstemann 1880.)

(figure 4.7), in a scene that some scholars have related to a mythological event involving the ascent of a crocodilian being to the sky in order to bring forth a flood to destroy a race of wooden beings (Knowlton 2010; Taube 1988, 143; 1993; Vail and Hernández 2013b, 155; Velásquez García 2006).[18] Two deity figures are pictured in the accompanying scene: the creator goddess Chak Chel, and the black deity God L (or a merging of this deity with Chaak). The hieroglyphic caption includes references to Bakab, Chaak, and Chak Chel (the second, third, and fourth glyphs in the bottom row of the text). The former two are related to "black sky" and "black earth," whereas the text preceding Chak Chel's glyph is eroded. It appears likely that the term *bakab* in the caption refers to God L, suggesting that he may represent one of the color-directional Bakabs referenced in the *Relación*, most likely Hosan Ek' (Hozanek), who is associated with black (*ek'*) years and the west.[19] It is also possible that God L and the black version of Chaak may be different aspects of the same underlying deity, as previous studies have suggested (Vail and Hernandez 2013b, 460n23). Nevertheless, the role of God L in the Maya codices and in earlier Classic-period contexts deviates significantly from the description of the Bakabs in Landa's account (i.e., as four brothers chosen by God to hold up the sky). In prehispanic

THE COLONIAL ENCOUNTER   157

contexts, God L is depicted as one of five manifestations of Venus, and he is associated very specifically with the underworld and its riches (Taube 1992, 79, 81; Vail and Hernández 2013b, 74, 284).

In the *Book of Chilam Balam of Chumayel*, the Bakabs play an important role as a plurality of deities (represented in their quadripartite aspect). On page 43 of the manuscript we read (Knowlton 2010, 65):

> Four stand as gods
> Four as Bacabs
> They caused their [the core-less people's] destruction
> And when the destruction of the world was finished
> They settled this [land].

This is followed by a description of the setting up of the world directional trees, one associated with each of the four quarters, and a fifth with the center (see text D in appendix 4.1).

The important role played by the Bakabs, and the fact that they are only one of a handful of deities labeled as gods in colonial-period texts, may stem from the positive qualities attributed to them by Spanish ecclesiastics, as evident from the account in the *Relación*. Moreover, refers to a flood (conceptualized as wiping out a former "evil") in both Maya and Christian traditions, was something highlighted in later (colonial period) writings (see text C in appendix 4.1).

It is very likely, however, that the Yucatec flood myth had prehispanic antecedents, as suggested not only from an examination of pages 74 of the *Dresden Codex* (and the cognate almanacs on pages 32 of the *Madrid Codex* and 21 of the *Paris Codex*; see Taube 1988, 150–152), but also from a Classic-period text from Palenque and Postclassic-period mural art from the sites of Mayapán and Tancah (Velásquez 2006). The Palenque text, like that in the *Chumayel*, refers to the decapitation of a crocodilian, followed by what has been interpreted as an event of world creation (Stuart 2005, 68).

However, important differences exist between the Postclassic-period sources and those related in the *Relación* and the *Books of Chilam Balam*—the most important being that none of the colonial-period texts that references these events discusses Chak Chel and her role.[20] There are likely several reasons for this. In the case of Landa's manuscript, the role of female deities was substantially deemphasized. Even more important, however, is the connection that is made between the four Bakabs (who merge with the Chaaks) and the four horsemen of the Apocalypse in later colonial-period manuscripts (see discussion in Bricker and Miram 2002, 78). If Landa (or the indigenous informants he was working with) similarly made this connection, it appears that the prehispanic episode involving the Bakab God L and the creator goddess Chak Chel was reframed in terms of the Christian text referring to the coming Apocalypse, as described in the Book of the Revelation. Instead of the four horsemen of the Apocalypse being responsible for this destruction, however, the author substituted the four Bakabs, thereby tying the Maya flood episode into both the prehispanic tradition (with the emphasis on the Bakabs), but also to Christian beliefs (with the emphasis on four "world destroyers").

Each of the four Bakabs discussed in the *Relación* is named and has a particular color-directional association. The southern Bakab (Hobnil) is further linked to the gods K'an Pawahtun (Kan Pauah Tun; *k'an* meaning "yellow," the color associated with the south) and to K'an Xib Chaak (Kan Xib Chac, or "yellow man Chaak"). The other three follow a similar pattern, with east being associated with Chak Pawahtun (Chac Pauah Tun) and Chak Xib Chaak (Chac Xib Chac; *chak* meaning "red"), north with Sak Pawahtun (Sac Pauah Tun) and Sak Xib Chaak (Sac Xib Chac; *sak* meaning "white"), and west with Ek' Pawahtun (Ek Pauah Tun) and Ek' Xib Chaak (Ek Xib Chac; *ek'* meaning "black").

The black Bakab (if that is indeed one of God L's manifestations) also appears as one of five heliacal rise aspects of the planet Venus on *Dresden* 46, where his speared victim is K'awil (figure 4.8). Vail and Hernández (2013b, 284) have recently suggested that this depiction can be related to a series of battles referenced in the *Books of Chilam Balam* involving Bolon Ti' K'uh (Elder Brother) and the Oxlahun Ti' K'uh (Younger Brother). According to this analysis, both K'awil and the sun god can be identified with the Oxlahun Ti' K'uh complex of deities (see Vail and Hernández 2013b, 447).

BOLON TI K'UH, OXLAHUN TI K'UH, AND CONCEPTIONS OF THE CELESTIAL AND UNDERWORLD REALMS

Although the deity complexes represented by the Bolon Ti K'uh ("Nine as God") and the Oxlahun Ti K'uh ("Thirteen as God") in the *Books of Chilam Balam* have clear prehispanic antecedents, there is only one explicit, documented reference (to the Oxlahun K'uh) in Yucatecan sources prior to the Spanish conquest (Knowlton 2010; Vail and Hernández 2013a). This occurs on page 101c of the *Madrid Codex*, in reference to a picture of K'uh seated in front of a bound stone stela or altar (Knowlton 2010, fig. 4.2). References to Bolon (Ti') K'uh all stem from Classic-period contexts, where this name appears as the first in a series of nine deities labeled the "Nine Lords of the Night" by Thompson (1929). Because of the context in which it appears (the lunar series), it is not associated with any explicit iconographic representations. In a recent study, Martha Macri (2005) has suggested that this series of nine deities refers to the nine "moonless" evenings (when the moon rises after midnight during its waning phase), whereas the series of 13 refers to the moon in its 13-day waxing phase.[21]

By the colonial period, however, the Bolon (Ti') K'uh were no longer linked to lunar cycles, but rather to agents of the underworld (in particular, Venus deities) and those of the celestial realm (including the sun god and deities associated with other celestial phenomena, such as lightning). These associations are made clear by reference to a text from page 14 verso of the *Chilam Balam of Tizimín*, which states that "then the dawn of Oxlahun Ku came because of Bolon Ti Ku / when he was born, engendered" (translation by Knowlton 2010, 73). The "dawn" of Oxlahun Ti' K'uh clearly refers to the first sunrise, which is attributed to the heliacal rise of Venus immediately prior to this in various indigenous texts, thereby suggesting a connection between Venus and the Bolon Ti' K'uh.

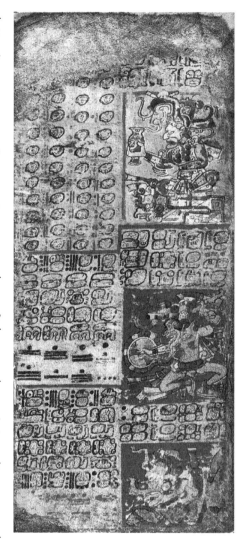

FIGURE 4.8. *God L as the Morning Star aspect of Venus (middle register) and his victim K'awil, who has been speared (lower register). The figure in the upper register is an anthropomorphized version of the earth crocodilian seated in the sky (*Dresden Codex, 46, Venus table; after Förstemann 1880.*)*

160 GABRIELLE VAIL

This is followed by a series of episodes in which first one and then the other of the two deities/deity complexes causes the downfall (through the blindfolding or blinding) of the other (for detailed discussions of this narrative, see text B in appendix 4.1, and Vail and Hernández 2013b, 448–451).

A comparison of colonial-period texts referring to these two deities/deity complexes with almanacs from the Maya codices allows the possibility of mapping the prehispanic deities who became known as Bolon Ti' K'uh and Oxlahun Ti' K'uh during the colonial period. God L is one of the former, whereas K'awil and Hun Ahaw (in his aspect as the sun) are two of the latter (Vail and Hernández 2013b, 447). Despite the link of the former set of deities (the "nine") to the underworld and the latter (the "13") to the celestial realm, there is no evidence from prehispanic contexts to equate the 13 with 13 layers of the upperworld and the nine with nine layers of the underworld (see chapter 1, this volume, by Nielsen and Sellner Reunert), as researchers have previously supposed (see, e.g., Thompson 1970, 280). Indeed, almanacs that reference 13 specific locations, such as the lower seasonal table in the *Dresden Codex* (65b–69b), move between the different realms. By way of example, Chaak is described in this table as being in the earth, in the foam (sea), on the road, in the sky, on a mountain, in the underworld (represented by the crossed-bones motif on which he sits), in a ceiba, in the white earth, in the water, in an unidentified location, with a wife, in the ?? (*kok?*, "turtle"?), and in the valley or grasslands.

A cognate to this almanac that appears in the *Madrid Codex* has nine frames instead of 13, thereby making reference to nine specific locations. The first five frames diverge from those in the *Dresden* almanac, whereas the last four are similar, including a reference to a watery place, a mountain, the sky (coupled with a road), and a valley or grasslands. This almanac also deviates from its counterpart in the *Dresden Codex* in picturing K'uh in place of Chaak in eight of its nine frames (with Chaak appearing in the fifth frame, seated within a house). The shift between 13 and nine frames is of considerable interest in light of previous interpretations of the layered upper and lower worlds. Nevertheless, while they highlight the importance of nine and 13 specific locations, they provide no evidence for linking them with the underworld and celestial realms. Instead, the codical data suggest that, for the prehispanic Maya, the chief distinctions were among the four world directions and a fifth, generally associated with the center, which was named or pictured as referring to the sky, the earth, a cave (or specifically the *kab'-ch'een*, or earth-cave, a location associated with the creation and emergence of humans), a cenote, a location within the earth, or a ceiba tree.

In prehispanic Maya contexts, many of the same deities are represented as being at times associated with the sky and at others within the earth. Chaak is most commonly associated with both of these regions; moreover, he is also portrayed as emerging from the latter, either through the open mouth of a serpent or directly from the water. In these scenes, he plays the role of a rainbringer or the planet Venus undergoing its heliacal rise (Vail and Hernández 2013b, 301–302). This evidence of fluidity suggests the importance of movement among these different regions, not only by Chaak but by a host of other deities. In certain cases, this relates to the appearance and disappearance of celestial objects, which are believed to be within the earth when they are not visible in the sky.

Deities that are pictured or described in the codices as being in the sky include not only Chaak but also the four directional Pawahtuns, Pawah Ayin (the crocodilian aspect of the Pawah deities), the moon goddess, Hun Ahaw, a hummingbird (the sun?), K'uh (likely representing Chaak in this context), the death god, the god of the number three, who is also associated with wind and flowers,[22] and the maize god. Similarly, deities described as being within the earth or pictured in a cave or cenote include Chaak, Chak Chel, the maize god, the death god, the god of punishment Kisin, Itzamna (and the black aspect of this deity), and the merchant deity God M. The fact that the same deity may have associations with the sky, the earth, or the region within the earth provides a compelling argument against making specific associations based on the region the deity inhabits in a particular example.

A mural from Structure 16 at the site of Tulum (figure 4.9), located on the Caribbean coast of present-day Quintana Roo, provides perhaps the most detailed depiction of Postclassic Yucatec conceptions of the realm within the earth. In this representation, the two lower registers with pictures are set off by their inclusion within the mouth of a toothed creature meant to symbolize the earth. Immediately below this are images of various water creatures (a fish and a stingray), likely intended to specify the watery nature of this realm. The two registers with pictures depicted as enclosed within the earth each show Chaak and Chak Chel, whose bodies sprout foliage, amidst entwined strands of vegetation or perhaps serpents. This imagery conveys the sense of this region (which can be identified with the *kab-ch'een*, or earth cave) as the birthplace of vegetation and likely also of the rains, given the presence of two deities who are rainbringers in the Maya codices (Vail and Hernández 2012). The upper register features the birth of the maize upon the earth's surface (where it is shown emerging from open-mouthed serpents), as well as a scene of Chak Chel grinding maize in association with Chaak. This has been

FIGURE 4.9. *Mural from Tulum Structure 16 showing Chaak and Chak Chel. (After Miller 1982, pl. 37.)*

previously interpreted as portraying the act that led to the creation of humans (Vail and Hernández 2012).

Chak Chel also appears in another important creation scene—that occurring in the central panel of the almanac on pages 75–76 in the *Madrid Codex* (figure 4.10). Here, she is paired with the male creator deity Itzamna, the two seated beneath a stylized version of a ceiba tree. Three *ik'* glyphs appear in front of Itzamna's outstretched hand, and another occurs as the middle glyph of three in front of Chak Chel; these may represent the maize seeds used to create human life (one of the associations of the *ik'* glyph is "breath" or "life").

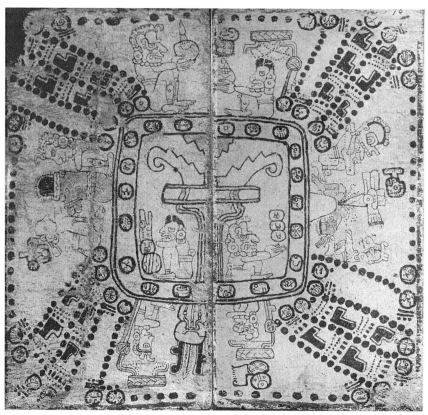

FIGURE 4.10. *The creator deities, Chak Chel and Itzamna, seated beneath a stylized world tree at the center of the scene. (*Madrid Codex, 75–76; after Gates 1933.)

It was perhaps based on this scene that the description in the *Relación* of a paradise for dead souls was drawn (see discussion below).

Landa's *Relación* makes a clear distinction between good and evil (omens, years, deities, etc.). As previously noted, one manifestation of this involves the use of the term *ángel* to describe a being symbolizing water and a good year that occurred as part of the yearbearer ceremony. This contrasts with another series of objects that were placed on the shoulders of the image of the black Way (Ek-uvayeyab) associated with western years. Before being carried from the western entrance of the community to the house of the *principal*, a skull, a corpse, and a carnivorous bird called *k'uch* ("vulture") were placed on the Way image "as a sign of great mortality, since they regarded this as a very evil year"

(Gates 1978, 67). As the image was being transported, various dances were performed, including one called *xibalba-okot*, which Landa translates as "the dance of the devil" (Gates 1978, 67). More appropriately, Xibalba refers to the underworld, or place of fright.[23] The "statue" (or patron) of Kawak (Cauac) years was moreover a deity named Wak Mitun Ahaw (Vacmitun-ahau), who can be associated with Schellhas's God A', a death or underworld deity whose name glyph may be read as *Akan* (the lord of drink or drunkenness; Grube 2004) or *Ah Kam* ("the dead one"; Vail 1996, 2000). Various translations have been offered for *Wak Mitun Ahaw*, which refers to Six "Mitun" Lord. *Mitun* may be a variant of *Mitnal* (also written as *Metnal*), one of the names of the underworld documented in colonial Yucatec sources (Thompson 1970, 302).[24] The prefix "six" is paired with the death god's glyph on several occasions in the Maya codices, and the deity of the number six is represented as a god of sacrifice (he has an axe in his eye, which has the value *ch'ak* "to cut" or "decapitate"). It may therefore be the case that the patron deity of Kawak years was an underworld lord of sacrifice.

A series of almanacs on pages 84–88 of the *Madrid Codex* offers an interesting parallel to the yearbearer ceremonies portrayed on *Dresden* 28 in relation to Kawak years. Each of the five almanacs pictures a deity representing a conflation of the death god and Kisin, the god of punishment. This figure is portrayed carrying a spear and torches, perhaps an indication of his role as a deity of castigation and punishment. In the final almanac in the series, the deity is shown seated within a thatched structure, holding an unidentified black-painted object (figure 4.11). The almanac's four previous frames include pictures and captions suggesting that they may be referring to yearbearer ceremonies. The last of these shows a bird perched on a skull, which recalls the description in the *Relación* of the objects placed on the shoulders of the *Way* during rituals associated with Kawak years (Bill, Hernández, and Bricker 2000). In light of this interpretation, it is possible to suggest that the conflation of the two underworld gods depicted here represents Wak Mitun Ahaw since *wak* may refer to sacrifice, which would fit well, given Kisin's roles as a god of punishment or the sacrificial god of the underworld.[25]

By the choice of vocabulary used in reference to the K'an and Kawak yearbearer ceremonies, the authors of the *Relación* are able to make allusions to heaven (associated with angels) and hell (associated with the dance of the "devil"), concepts that lacked meaning in the indigenous Yucatecan worldview. The text defines the paradise achieved by those whose conduct had been good as "a place where nothing would give pain, where there would be abundance of food and delicious drinks, and a refreshing and shady tree they called Yaxché,

FIGURE 4.11. *Possible depiction of Kawak yearbearer ceremonies, including a decapitated figure in frame 1, a vulture perched on a skull in frame 4, and an underworld deity, possibly the god of sacrifice, seated in a thatched structure in frame 5.* (Madrid Codex, 88c; after Brasseur de Bourbourg 1869–1870.)

the Ceiba tree, beneath whose branches and shade they might rest and be in peace forever" (Gates 1978, 57). There is good reason to suppose that this conceptualization derived from a comparison to images like that on *Madrid* 75–76, where the creator couple is seated beneath the ceiba tree. This is then contrasted with the torments suffered by the wicked, who were consigned to "an evil place below the other, and which they called Mitnal, meaning hell, where they were tormented by demons, by great pains of cold and hunger and weariness and sadness" (Gates 1978, 58). A figure called Hun Ahaw (Hunhau), who is identified as the "chief demon" in Landa's text, was said to preside over this region.

In the Maya codices, Hun Ahaw is linked explicitly to the sky (figure 4.12) and takes on the persona of the sun of the current era (Vail and Hernández 2013b, 279). Colonial-period texts—specifically, the mid-sixteenth-century K'iche' manuscript the *Popol Vuh* (Christenson 2007)—relate that this occurred after an arduous journey through Xibalba with his twin brother Xbalanque (known as Yax Balam in Yucatec).

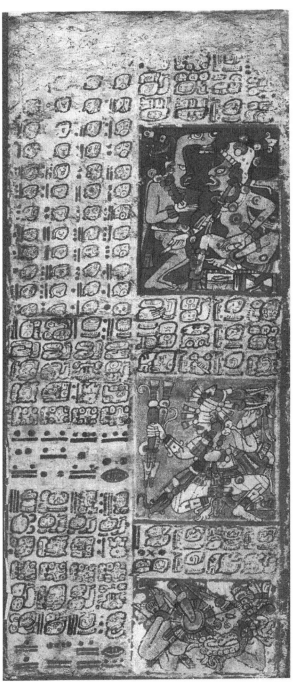

FIGURE 4.12. *Hun Ahaw seated on a skyband in the upper register, where he may serve as the sun of the new era. He is attended by the maize god. The warrior aspect of Kakatunal appears in the middle register as Venus as Morning Star, and his victim—a speared lord or stranger from the west—in the bottom register.* (Dresden Codex, 50, Venus table; after Förstemann 1880.)

The transformation of Hun Ahaw from a supernatural who vanquished the previous false sun and became the sun of the current era to that of the "chief demon" of the underworld, analogous with Lucifer, is one that remains difficult to interpret. In the Morley Manuscript, Hun Ahaw takes on the persona of the Christian Lucifer but also that of Seven Macaw, the false sun of the previous era from the *Popol Vuh* (Knowlton 2010, 114–117). It is of interest that this role was assumed by Hun Ahaw rather than one of the deities more commonly associated with the underworld by the prehispanic Maya, such as the lords of death (Kimil?) or castigation (Kisin). There is good evidence to suggest that the figure identified epigraphically as Hun Ahaw in the Maya codices is not the same Hun Ahaw as that described in Landa's manuscript or colonial-period indigenous texts. Rather, it seems more likely that this designation applies to the heliacal-rise aspect of Venus pictured in the middle register of *Dresden* 50 (see figure 4.12), who is said to emerge from the underworld into the eastern sky on the *tzolk'in* date 1 (Hun) Ahaw.[26] This figure, named Kakatunal (or Ka Acatl Tonal, "Two Reed Spirit") in the text, is a Mexican deity associated with punishment and castigation, who plays a role similar to that of the Maya god Kisin, a figure having clear associations with the underworld region.

Among the contemporary Lacandón Maya, Kisin is the adversary of the creator deity Hachakyum. In one particular episode, Kisin is said to have killed Hachakyum and buried him; Hachakyum, however, returned to life and, with the help of several other deities, created the underworld. He then caused the ground beneath Kisin to cave in, which sent him falling into the underworld, where he is relegated to this day. Hachakyum thereafter ascended to the sky in order to create the heavens (Thompson 1970, 344).

Based on this account, it seems likely that the "Hunhau" named in the *Relación* as the "chief demon of the underworld" is analogous to Kisin, as is also suggested by the *Dresden* Venus table, or to the black God L, another manifestation of Venus in its heliacal-rise aspect. Evidence in support of this latter identification stems from Classic-period depictions of God L, who, in the time before the present sun was created, presided over the underworld and its bountiful riches (maize, cacao, tobacco, etc.) (Martin 2006). In this respect, he has parallels to Seven Macaw in the *Popol Vuh*, who declared himself to be the sun and the moon for the wooden people of the previous era (Christenson 2007, 91–92). God L's arrogance, ascendancy, and later downfall likewise have parallels to the story of Lucifer in the Bible, suggesting a plausible connection to the colonial-period Hun Ahaw.

A merging of Lucifer, as the lord of the underworld, with Venus may be seen in various colonial-period sources (Thompson 1970, 303). This offers further evidence of the dialogic processes at work in the colonial period that are discussed in detail by Timothy Knowlton (2010)—in this case, a merging of indigenous concepts of the underworld and its presiding deity (Venus) with those of the Christian friars. Meanwhile, the figure named epigraphically as Hun Ahaw appears seated on a skyband in the upper register of *Dresden* 50 (see figure 4.12). There, he wears a skeletal costume, suggesting his emergence from the underworld region to the sky, where he becomes K'in Ahaw, or Lord Sun.

CONCLUSIONS

As a result of the encounter involving Franciscan missionaries and a literate group of Maya scholars engaged in the process of understanding the European worldview within the context of their indigenous beliefs (and vice versa), a number of transformations occurred with regard to how the class of beings identified prehispanically as *k'uh* was conceptualized. The adoption of European terminology—including the terms *ángel* and *cangel*—resulted in the merging of beings that had previously been understood as discrete deities: of K'awil and the maize god as *ángel* (omens of water and of a good year) and later as Bolon Tz'akab, and of the quadripartite Chaaks and Pawahtuns as "archangels" whose distinctive roles merged into a single set of deities having associations with both rain and agriculture during the colonial period.

Additionally, the practice of the Franciscan friars to characterize Maya *k'uh* as *demonios* or *dioses* led to the development of a dichotomy between the celestial and underworld regions and to the assigning of value judgments that are not in evidence in Postclassic Yucatec sources. Rather, the prehispanic codices conceptualize the natural world in terms of five regions: the four world quadrants and a fifth space in the center that may relate to the earth, the area within it, or the area above it. Deities are seen to have existed in multiple realms. To make this view more compatible with a Christian worldview, it became imperative to separate the three realms and assign specific supernaturals to each. It was through a process such as this that Hun Ahaw became identified with Lucifer—although, as we have seen, in this context Hun Ahaw likely referred very specifically to Venus deities undergoing heliacal rise, and not to the Hero Twin who is noted for vanquishing the Lords of the Underworld.

The transformations that can be seen in understandings of particular deities and complexes as a result of both intentional and unintentional processes led to a scholarly conception of prehispanic Maya cosmology that has often failed to differentiate between preconquest indigenous beliefs and those that became a part of the worldview of indigenous people as a result of the encounter between

European and Maya cultures that began with the conquest. This essay represents a small step towards disambiguating the two through a comparative analysis of conceptions of the supernatural in prehispanic and colonial-period texts.

Appendix 4.1

Selections from the cosmogonic narrative related in the *Books of Chilam Balam*: text A, text B, text C, and text D.

TEXT A

Ca tali uy ahal cab ti oxlahun ku tumen bolon ti ku
Then the dawn of Oxlahun Ku came because of Bolon Ti Ku

Ti ca sihi ch'abi
When he was born, engendered.

Ca sihi Ytzam Cab Ain
Then Itzam Cab Ain was born

Xoteb u kin balcah
That he may signal the day for the whole world.

Ca haulahi caan
Then the sky was turned face up

Ca nocpahi peten
Then the land was turned face down

Ca ix hop'i u hum oxlahun ti ku
And then Oxlahun Ti Ku's din began.

Ca uchi noh haicabil
Then the great destruction of the world arrived.

Ca liki noh Ytzam Cab Ain
Then great Ytzam Cab Ain ascended

Dzocebal u than u uudz katun lai hun ye ciil
That this deluge may complete the word of the katun series,

Bin dzocecebal u than katun
That the word of the katun might be complete.

Ma ix y oltah bolon ti ku i
But Bolon Ti Ku did not desire it

Ca ix xoti u cal Ytzam Cab Ain
And then Ytzam Cab Ain's throat was cut.

Ca u ch'aah u petenil u pach
So he sprinkled the island, its back

Lai ah uoh puc u kabae
This is its name: Calligrapher Hill.

Ma ix u toh pultah u kaba tiob
Neither did he really confess to them its name.

Ti kaxan tun u uich ualac y ahaulil lae
He had bound the eyes then of this current reign.

(Tizimín manuscript 14v.16–25; translated by Knowlton 2010, 73)

TEXT B

Tuchi yx ca dzoci vy ahalcabe
So when it finished dawning

Ma yx y oheltahob binil vlebal
They knew not that it would come to pass,

Ca ix chuci oxlahun ti ku tumenel bolon ti Ku
That Oxlahun Ti Ku was caught by Bolon Ti Ku.

Ca emi kak
When fire descended

Ca emi tab
When tumplines descended

Ca emi tunich y che
When stone and wood descended

Ca tali v baxal che y tunich
When his stick and stone came

Ca ix chuci oxlahun ti ku
Then Oxlahun Ti Ku was caught

y ca ix paxi u pol
And then his head was wounded.

Ca ix lahiv v uich
Then they put out his eyes.

Ca ix tubabi
Then he was spat upon.

Ca ix cuchpachhi xan
Then he was knocked down flat, too

Ca ix colabi v cangel y v holsabac
Then his archangel was tugged and shall be removed.

(CHUMAYEL MANUSCRIPT 42.26–43.3; TRANSLATION BY KNOWLTON 2010, 55–57)

TEXT C

Ca ix hu[t]lahi
And then it collapsed.

Ixma yumob y ah numyaob
Without the lords or the lowly

Ixma ychamob
Without spouses

Cuxanob ix ti minan u pucsikalob
That's how they lived without their hearts.

Ca yx mucchahij
And so were submerged

T u men v yam sus
By waves of sand

T u yam kaknab
And waves of sea.

Hun vadz hail
One fetching of rain

Hu[n] lom haail
One lancing of rain

Tij ca uchi col cangelili
Back then when only the Milpa Archangel arrived

Ti homocnac canal
It was tempestuous above

Homocnac ix ti cab
And tempestuous below [on earth].
(CHUMAYEL MANUSCRIPT 43.10–16; TRANSLATION
BY KNOWLTON 2010, 61–62)

TEXT D

Valic can tul ti ku
Four stand as gods

Can tul ti bacab
Four as Bacabs.

Lay hayesob
They caused their [the coreless people's] destruction.

Tuchij tun ca dzoci hay cabil
And then when the destruction of the world was finished

Lay cahcunah uchebal ca tzolic kan xib yui
They settled this [land] so that Kan Xib Yui puts it in order.

Ca ualhi sac imix che ti xaman
Then the White Imix Tree stands in the north

Ca ix ualhi y ocmal caan
and stood as the pillar of the sky,

V chicul hay cabal
The sign of the destruction of the world.

Lay sac imix che valic cuchic
This White Imix Tree stands there supporting it.

Ca yx ualhi ek ymix che
Then the Black Imix Tree stood

Cu [lic] ek tan pidzoy
[where] the Black-Bellied Pidzoy resides.

Ca yx ualhij kan ymix che
Then stood the Yellow Imix Tree

V chicul hay cabal
The sign of the destruction of the world.

Culic kan tan pidzoy
The Yellow-Bellied Pidzoy resides

Cumlic ix kan xib yui
And Kan Xib Yui sits.

Yx kan oyal mut
The Yellow Caller Bird.

Ca ix ualhij yax imix che t u chumuc
Then the Blue/Green Imix Tree stood in the center

U kahlay hay cabal
The history of the destruction of the world.

Culic uatal
It is erected.

(CHUMAYEL MANUSCRIPT 43.16–27; TRANSLATION BY KNOWLTON 2010, 65)

NOTES

1. It is also important to keep in mind that, in a large percentage of almanacs (80% or more), the term *k'uh* does not appear at all. It is presently unclear what this distinction means (i.e., why it was used in certain instances and not in others). Analyzing the contexts in which it occurs has not allowed this question to be resolved.

2. 4 Ahaw is the day in the 260-day calendar that corresponds to the start of the previous *bak'tun* cycle on 4 Ahaw 8 Kumk'u in 3114 BCE. 4 Ahaw is also the starting date of 28 almanacs in the *Madrid Codex*, most of which concern events associated with renovation and world renewal.

3. Colonial spellings appear in parentheses. Deity names remain in the orthography of the original source in all quoted material.

4. The term *way* refers to a spirit, or a nocturnal aspect of the soul, which may have sinister associations (Houston and Inomata 2009, 138). The discussion of these various categories in colonial contexts was originally outlined in Vail and Knowlton (2012).

5. This could also be a reference to the calendar date 9 Yax.

6. Note the relationship to the deity discussed previously (from *Madrid* 95d–96d), who represents Itzamna with some of K'awil's attributes.

7. The reference to his burden being tied on a day K'an suggests that this refers to a yearbearer date, when the beginning of the year (the month Pop) coincided with a day named K'an.

8. My study of the Maya codices indicates that deities were members of complexes that shared a particular name or title (Vail 1996, 2000). The most common of these

in the *Madrid Codex* is T24.1016c, which serves both to name the maize god (as a substitute for his more common appellative T24.1006b) and to designate a complex of deities with associations of abundance and fertility (see figure 4.1c–d).

9. The Maya deity Bakab and the central Mexican god Tezcatlipoca are both described in Spanish colonial sources in terms of four brothers. This was likely a means of simplifying their role as quadripartite deities with different aspects for a European audience.

10. The word *tun* means "stone" and is used in reference to the 360-day year, whereas *k'in* is the word for "sun," "day," and "festival." I interpret the suffix as *pawah* (meaning unknown), as originally suggested by Taube (1992). More recent interpretations as *itzam* (Stuart 2007) have good support in the Classic period, but evidence from the *Madrid Codex* (see Vail 2012) suggests that there was a shift in meaning to *pawah* sometime during the Postclassic period.

11. For an identification of these figures as Pawahs, see Vail (2012).

12. Note that *Pawahtun* is identified as another name for Bakab in Landa's manuscript (see discussion below).

13. Bricker and Miram (2002, 79) note that "the Maya paid little attention to the theological distinctions among God, the saints, and the angels, all of whom were regarded as deities and therefore conceptually equivalent to the gods in the Maya pantheon whose attributes they shared."

14. The Pawah title occurs with the following glyphs: *tun* ("year of 360 days" or "stone"), *haab* ("year of 365 days"), *k'in* ("sun" or "day"), *ooch* ("opossum"), *mak* ("turtle"), and *ayin* ("crocodile").

15. An agricultural association is also evident in the form that the Pawah prefix takes, since its central element likely refers to maize or to a tamale made of this substance (Taube 1992).

16. This identification is likewise supported by Vail's reading of *u holsabac* as "his face was sooty," a reference to the soot used by the rain deities to form the rain clouds (Vail and Hernández 2013b, 96n57). The term *sabak* also occurs as part of the name of the Lacandon rain deity Mensäbäk, or "Maker of Powder" (McGee 1990, 68), and refers to a black powder that Mensäbäk gives to his assistants to sprinkle through the sky to produce rain-laden clouds.

17. Note the close association of the Chaaks with clouds, much as was the case of the Babahtun deities in the previous discussion.

18. As several scholars have pointed out (see, e.g., Knowlton 2010, 62, 64; Velásquez 2006), the *Dresden* scene is closely paralleled by a text from the *Chilam Balam of Chumayel* (see texts A and C in appendix 4.1).

19. This attribution also fits well with Karl Taube's (1992, 84–85) proposal that God L may be associated with the western Maya area. He interprets the figure on *Dresden* 74 as a black version of Chaak, rather than God L, however (Taube 1988, 144).

20. A text occurring later in the *Chumayel* manuscript does, however, link the creator genetrix First Mother, described as "the mother of everything in the world," to creation by means of water or liquids (Knowlton 2010, 126–128).

21. This idea of the moon's visibility and invisibility may relate to later conceptions of these deity complexes, as seen in the *Books of Chilam Balam*, in that the former suggests a connection to the celestial sphere and the latter to the underworld region.

22. While this deity is named in the text, the figure pictured on the skyband is a deer in this instance.

23. Only one possible reference to Xibalba has been identified in the codices; this occurs in the almanac on *Madrid* 89a–90a (http://mayacodices.org/frameDetail.asp?almNum=179&frameNum=1). Each of its five captions refers to one of the world directions (south, east, north, west, and *kab'-ch'een*, or earth-cave) and to *xib beh* "road of fright." Although this term may refer to the journey that had to be made to reach the underworld, it may also be a reference to the Milky Way. Among the K'iche' Maya of highland Guatemala, the side of the Milky Way with the rift is called "ice road" or "underworld road" (*xibalba be*) (Tedlock 1992, 181), suggesting that the phrase in the *Madrid* almanac may be an abbreviation for this.

24. *Mitnal* (or *Metnal*) appears to be the Yucatec variant of the Nahuatl term *Mictlan* (Tozzer 1941, 132n617).

25. There is a reference to *wak lak'in wak taan kimil* "6 east, 6 is the face of death" in the caption to page 86c, one of the almanacs that pictures Kisin as a conflation of Gods A (Kimil?) and Q (Kisin).

26. It was as a result of Venus' sacrifice and heliacal rise that the present sun could be born (Vail and Hernández 2013b, 449).

REFERENCES

Anders, Ferdinand, coord. 1967. *Codex Tro-Cortesianus* (*Codex Madrid*). Graz: Akademische Druck- und Verlagsanstalt.

Arzápalo Marín, Ramón. 1995. *Calepino de motul: Diccionario maya-español*. 3 vols. Mexico: Universidad Nacional Autónoma de Yucatán.

Baeza, D. Bartolomé del Granado. 1845 [1815]. "Los Indios de Yucatan." In *Registro yucateco*, 168–178. Mérida: Impreso de Castilla y Cia.

Barrera Vásquez, Alfredo, Juan Ramón Bastarrachea Manzano, William Brito Sansores, Refugio Vermont Salas, David Dzul Góngora, and Domingo Dzul

Poot, eds. 1980. *Diccionario maya Cordemex: Maya–español, español–maya*. Mérida: Ediciones Cordemex.

Bill, Cassandra R., Christine L. Hernández, and Victoria R. Bricker. 2000. "The Relationship between Early Colonial Maya New Year's Ceremonies and Some Almanacs in the *Madrid Codex*." *Ancient Mesoamerica* 11 (1): 149–168.

Brasseur de Bourbourg, Charles E. 1869–1870. *Manuscrit troano: Études sur le système graphique et la langue des Mayas*. Paris: Imprimerie Impériale.

Bricker, Victoria R., and Helga-Maria Miram, trans. and notes. 2002. *An Encounter of Two Worlds: The Book of Chilam Balam of Kaua*. Middle American Research Institute, 68. New Orleans, LA: Tulane University.

Carlson, John. 2007. "From the Olmec to Columbus: The Maya." In *The Jay I. Kislak Collection at the Library of Congress*, ed. Arthur Dunkelman: 9–37. Washington, DC: Library of Congress.

Christenson, Allen J. 2007. *Popol Vuh: The Sacred Book of the Maya*. Norman: University of Oklahoma Press.

Ciudad Real, Antonio de. 1995. *Calepino de Motul: Diccionario maya–español*, ed. Ramón Arzápalo Marín. Mexico: Universidad Nacional Autónoma de México, Instituto de Investigaciones Antropológicas.

Förstemann, Ernst. 1880. *Die Maya Handschrift der Königlichen öffentlichen Bibliothek zu Dresden*. Leipzig: Verlag der A. Naumannschen Lichtdruckerei.

Gates, William. 1933. *The Madrid Maya Codex*. Publication 21. Baltimore, MD: Maya Society.

Gates, William, trans. with notes. 1978. *Yucatan before and after the Conquest*, by Diego de Landa. New York: Dover.

Groark, Kevin P. 2010. "The Angel in the Gourd: Ritual, Therapeutic, and Protective Uses of Tobacco (*Nicotiana Tabacum*) among the Tzeltal and Tzotzil Maya of Chiapas, Mexico." *Journal of Ethnobiology* 30 (1): 5–30.

Grube, Nikolai. 2004. Akan: The God of Drinking, Disease and Death. In *Continuity and Change: Maya Religious Practices in Temporal Perspective*, ed. Daniel Graña Behrens, Nikolai Grube, Christian M. Prager, Frauke Sachse, Stefanie Teufel, and Elisabeth Wagner, 59–76. Acta Mesoamericana, Vol. 14. Germany: Verlag Anton Saurwein, Markt Schwaben.

Houston, Stephen D., and Takeshi Inomata. 2009. *The Classic Maya*. New York: Cambridge University Press.

Knowlton, Timothy W. 2008. "Dynamics of Indigenous Language Ideologies in the Colonial Redaction of a Yucatec Maya Cosmological Text." *Anthropological Linguistics* 50 (1): 90–112.

Knowlton, Timothy W. 2010. *Maya Creation Myths: Words and Worlds of the Chilam Balam*. Boulder: University Press of Colorado.

Macri, Martha. 2005. "A Lunar Origin for the Mesoamerican Calendars of 20, 13, 9, and 7 Days." In *Current Studies in Archaeoastronomy: Conversations across Time and Space*, ed. John W. Fountain, and Rolf M. Sinclair, 275–287. Durham, NC: Carolina Academic Press.

Martin, Simon. 2006. "Cacao in Ancient Maya Religion: First Fruit from the Maize Tree and Other Tales from the Underworld." In *Chocolate in Mesoamerica: A Cultural History of Cacao*, ed. Cameron L. McNeil, 154–183. Gainesville: University Press of Florida.

McGee, R. Jon. 1990. *Life, Ritual, and Religion among the Lacandon Maya*. Belmont, CA: Wadsworth.

Milbrath, Susan. 1999. *Star Gods of the Maya: Astronomy in Art, Folklore, and Calendars*. Austin: University of Texas Press.

Miller, Arthur G. 1982. *On the Edge of the Sea: Mural Painting at Tancah-Tulum, Quintana Roo, Mexico*. Washington, DC: Dumbarton Oaks.

Redfield, Robert, and Alfonso Villa Rojas. 1934. *Chan Kom, a Maya Village*. No. 448. Washington, DC: The Carnegie Institution of Washington.

Restall, Matthew. 1998. *Maya Conquistador*. Boston, MA: Beacon Press.

Restall, Matthew, and John F. Chuchiak IV. 2002. "A Reevaluation of the Authenticity of Fray Diego de Landa's *Relación de las cosas de Yucatán*." *Ethnohistory* 49 (3): 651–669.

Schellhas, Paul. 1904. *Representations of Deities of the Maya Manuscripts*. Papers of the Peabody Museum of American Archaeology and Ethnology, vol. 4 (1). Cambridge: Peabody Museum of American Archaeology and Ethnology.

Sosa, John R. 1989. "Cosmological, Symbolic and Cultural Complexity among the Contemporary Maya of Yucatan." In *World Archaeoastronomy*, ed. Anthony F. Aveni, 130–142. New York: Cambridge University Press.

Stuart, David. 2005. *The Inscriptions from Temple XIX at Palenque*. San Francisco, CA: Pre-Columbian Art Research Institute.

Stuart, David. 2007. "Old Notes on the Possible ITZAM Sign." Posted on *Maya Decipherment: A Weblog on the Ancient Maya Script*. http://decipherment.wordpress.com/2007/09/29/old-notes-on-the-possible-itzam-sign/.

Taube, Karl A. 1988. "The Ancient Yucatec New Year Festival: The Liminal Period in Maya Ritual and Cosmology." PhD diss., Yale University, Department of Anthropology.

Taube, Karl A. 1992. *The Major Gods of Ancient Yucatan*. Studies in Pre-Columbian Art and Archaeology 32. Washington, DC: Dumbarton Oaks.

Taube, Karl A. 1993. *Aztec and Maya Myths*. Austin: British Museum Press and University of Texas Press.

Tedlock, Barbara. 1992. *Time and the Highland Maya*. Rev ed. Albuquerque: University of New Mexico Press.

Thompson, Donald E. 1954. "Maya Paganism and Christianity." In *Nativism and Syncretism*, ed. Margaret A. L. Harrison, and Robert Wauchope, 1–35. New Orleans, LA: Tulane University, Middle American Research Institute.

Thompson, J. Eric S. 1929. "Maya Chronology: Glyph G of the Lunar Series." *American Anthropologist* 31 (2): 223–231.

Thompson, J. Eric S. 1930. *Ethnology of the Maya of Southern and Central British Honduras*. Chicago, IL: Field Museum of Natural History.

Thompson, J. Eric S. 1970. *Maya History and Religion*. Norman: University of Oklahoma Press.

Tozzer, Alfred. 1941. *Landa's Relación de las Cosas de Yucatán: A Translation*. Papers of the Peabody Museum of American Archaeology and Ethnology. Cambridge, MA: Harvard University.

Vail, Gabrielle. 1996. "The Gods in the Madrid Codex: An Iconographic and Glyphic Analysis." PhD diss., Tulane University, Department of Anthropology.

Vail, Gabrielle. 2000. "Pre-Hispanic Maya Religion: Conceptions of Divinity in the Postclassic Maya Codices." *Ancient Mesoamerica* 11 (1): 123–147.

Vail, Gabrielle. 2012. "Embodying the Time of Creation: Yearbearer Rituals in the Dresden and Madrid Codices." Paper presented at the symposium New Perspectives on the Dresden Codex, organized by Nikolai Grube.

Vail, Gabrielle, and Christine Hernández. 2012. "Rain and Fertility Rituals in Postclassic Yucatan Featuring Chaak and Chak Chel." In *The Ancient Maya of Mexico: Reinterpreting the Past of the Northern Maya Lowlands*, ed. Geoffrey Braswell, 285–305. London: Equinox Publishing.

Vail, Gabrielle, and Christine Hernández. 2013a. *The Maya Hieroglyphic Codices Database, Version 5.0*. http://www.mayacodices.org/.

Vail, Gabrielle, and Christine Hernández. 2013b. *Re-Creating Primordial Time: Foundation Rituals and Mythology in the Postclassic Maya Codices*. Boulder: University Press of Colorado.

Vail, Gabrielle, and Timothy Knowlton. 2012. "From *Ku* to *Ángel*: Changing Maya Conceptions of Supernatural Agents before and after Landa's *Relación*." Paper presented at the session "The Friar and the Maya" at the International Congress of Americanists, Vienna, July 15–20.

Velásquez García, Erik. 2006. "The Maya Flood Myth and the Decapitation of the Cosmic Caiman." *Precolumbian Art Research Institute Journal* 7 (1): 1–10. http://www.mesoweb.com/pari/publications/journal/701/flood_e.pdf.

Villacorta C., J. Antonio, and Carlos A. Villacorta. 1976 [1930]. *Códices mayas*. Guatemala: Tipografía Nacional.

5

Zapotec Travels in Time and Space

The Correlation between the 260-Day Cycle and a Multilevel Cosmological Model

David Tavárez

Sometime in late 1704 or early 1705, a ritual specialist surrendered an alphabetic manuscript in Northern Zapotec, its folios sewn together as a European-style fascicle or booklet. On its last pages, besides other diagrams, this document contained a most curious design: an arrangement of three sets of five circles drawn as quincunxes—four directional points, one center. The graphic details were simple, the labels terse: each of the quincunxes was labeled, from top to bottom, House of the Sky, House of Earth, and House of the Underworld (see figure 5.3, discussed in detail below). This drawing stands as an elegant portal not only into the Zapotec cosmos it depicted, but also into one of the most spectacular and least-studied colonial Mesoamerican corpora: a group of 102 manuals and four collections of ritual songs preserved at the Archivo General de Indias in Spain.[1] These texts, written in Northern Zapotec variants, document beliefs of precolumbian origin regarding time, the cosmos, auguries, and proper offerings. The corpus provides us with the best documentation available for a Mesoamerican society regarding the management of the 260-day divinatory cycle and its social impact in colonial times.

This essay focuses on two important features of this corpus. The first one is the careful documentation of cosmological beliefs regarding what we may call a spatiotemporal correlation: that is, a belief in the rotation of each of the 260 *lani*, or "feasts," in the Zapotec divinatory count through a three-tiered cosmos with nine levels above and nine below earth. The second one is a clear narrative about how these feasts were associated with a particular tier in the cosmos (earth, sky, or underworld) or a particular intervening level in between these three tiers. The definition of this movement also splits the *trecena* into two parts, as is done also in the Borgia Group. In spite of the importance of the Northern Zapotec corpus for a full understanding of Mesoamerican calendrical systems, this essay is the first attempt to analyze the relationship between the 260-day cycle and cosmological realms that was recorded by colonial Zapotec daykeepers. It also emphasizes how ritual specialists developed what amounts to a set of instructions for other users that deployed a dynamic relationship between alphabetic and pictographic texts, and which relied on the presentation of calendrical information as a list. These shifts in the presentation of cosmological data were probably informed by a shift from formats employed in precolumbian divination texts to one that combined the transcription of oral performances with the literacy skills of Zapotec scribes.

A UNIQUE COLONIAL CORPUS: THE ZAPOTEC ANCESTRAL COUNT MANUALS

The aforementioned texts, authored by northern Zapotec ritual specialists sometime in the second half of the seventeenth century and circulated among specialists and believers, provides us with an exceptional opportunity to study Mesoamerican beliefs about time and space, as understood by practicing colonial specialists. The nature of the corpus places multiple challenges to its analysis: as an indigenous text meant only for other indigenous readers, an interpretation of its alphabetic texts requires considerable philological and linguistic knowledge about written colonial Northern Zapotec texts, and its multiple diagrams make basic assumptions about shared knowledge about the cosmos. While a first monograph on the corpus helped publicize its importance and contents, this work focused on decoding the names of the 260 feasts, and on a descriptive analysis of Zapotec collective ceremonies, but only from the standpoint of Spanish-language documents, as the author did not read Zapotec (Alcina Franch 1993; see also Alcina Franch 1998). Subsequently, various aspects of this complex corpus have been examined in detail in several articles and book chapters (Miller 1991, 1998; Justeson and Tavárez 2007; Tavárez and Justeson 2008; Tavárez 2009) and in two chapters of a monograph (Tavárez 2011). Based on the aforementioned works, Lind (2015) has also addressed this corpus.

This corpus was produced in the *alcaldía mayor* (province) of Villa Alta, a jurisdiction to the northeast of Oaxaca City in New Spain inhabited by speakers of Chinantec, Mixe, and residents of three distinct sociocultural regions who spoke Northern Zapotec variants: Cajonos on the south, Nexitzo on the north and west, and Bijanos on the east (Burgoa 1989a; Chance 1989). Between September 1704 and January 1705, the elected authorities of 104 *pueblos de indios* (indigenous communities) journeyed to the jurisdictional seat of San Ildefonso to register confessions in exchange for general immunity for engaging in "idolatry"—a terse term that, in this case, referred to ceremonies both private (*de particulares*) and collective (*del común*) that honored Zapotec deities and ancestors. This offer, made by the bishop of Oaxaca, Fray Ángel Maldonado, came only two years after the momentous execution of 15 Zapotec rebels from in the aftermath of a legendary riot that took place in the town of San Francisco Cajonos. By early 1705, the officials of 15 Bijanos Zapotec, 27 Cajonos Zapotec, 26 Nexitzo Zapotec, 29 Mixe, and seven Chinantec communities had deposited brief written confessions, denounced their ritual specialists, and turned in copies of their ritual texts (Archivo Histórico Judicial de Oaxaca 2004; Gillow 1978; Tavárez 2011).

This bureaucratized effort, which directly or indirectly involved the 60,000 indigenous inhabitants of Villa Alta, was the most ambitious idolatry extirpation attempt carried out in New Spain. This campaign must be set aside from other similar eradications in colonial Spanish America due to the amount of documentation that came into the open at that time: 107 separate textual units by officials from 40 Bijanos, Cajonos, and Nexitzo Zapotec communities. Among them, one finds four collections of transcribed ritual songs, two of which are devoted to Christian entities, while the remaining two celebrated Zapotec deities and founding ancestors from the standpoint of local cosmological theories (Tavárez 2009, 2011). The remaining 102 units contain partial or full lists of the 260-day Zapotec ritual calendar, bound into 99 *cuadernos* (notebooks).[2] Thus, the corpus contains 92 complete lists of the 260-feast cycle, seven calendars with at least 75 percent of the 260 feast names, and two calendar fragments, as well as two calendars with incorrect day orders. These documents, along with the Maya books of Chilam Balam (see, e.g., Bricker and Miram 2002), are the two largest extant corpuses of clandestine ritual texts authored by native specialists in colonial Spanish America.

While most ecclesiastical judges would burn effigies and ritual implements in a public disciplinary act, these divinatory texts were spared that fate, as Maldonado employed them as evidence of the uneven results obtained by members of the Dominican order after almost 170 years of residence in Villa

Alta. These texts demonstrate that, as late as the end of the seventeenth century, two separate time counts of precolumbian origin were in constant use in at least 40 Northern Zapotec communities. The foundational cycle was a 260-feast-day divinatory count called *biyee*, or "time period." The 260 time periods in the *biyee* were designated as *lani* "feast," in terms of devotional practice; from an astronomical perspective, each of them was also one solar day. Córdova glosses *piyè* as "time, or interval," and in the Villa Alta corpus *biyee* refers, in general terms, to time periods with divinatory or ritual significance, including the 260-day cycle but comprising other counts as well.

The second precolumbian time count depicted was a vague solar-year count of 365 days, called *yza*. A recent interpretation of the evidence in the Villa Alta time corpus shows that both the 260-day and the 365-day counts had a fixed correlation with the Gregorian year, and that colonial Zapotec specialists did not employ a correction to keep the yza current with the true solar year of 365.24 days. Moreover, the *yza*, like its Nahua counterpart the *xihuitl*, a vague solar year, had 18 periods, plus a final period designated with an ominous label, *ritola reyeni* "it is incapacitated, it is angry," or *quicholla queainij*, days that "will be disconcerted, will be angry." This period corresponds to the Nahua *nemontemi*, the five days that "are full in vain." The structure, correlation statements, and some period names of the *yza* in Book 85-1 are discussed in Justeson and Tavárez (2007). An important departure from the *xihuitl*, noted in the only complete surviving description of the *yza* in Book 85-1, is that the *cocii* (period) called Zachi (Fat) had 19 days instead of 20, and that the period at the end of the year had six days, rather than the five in the Nahua system. The names and order of the 18 *yza* periods and the final six-day period are listed below (AGI-Mex 882, 1405r, Book 85-1):

1. *Toohuà* (Entrance, or Maguey)
2. *Hui ttao* (Great Humidity, or Great Illness)
3. *Tzegag* (It Sprouts)
4. *Lohuee* (Parrot Feathers)
5. *Yagqueo* (1-Soaproot)
6. *Gabenàa* (They Will Keep a Vigil)
7. *Golagoo* (Nurturer)
8. *Cheag* (It Is Cold, or It Goes)
9. *Gogaa* (Nine)
10. *Gonaa* (Offering)
11. *Gaha* (Fruit)

12. *Tina* (It Will Be Cleaned of Sooth, or It Will Wrinkle)
13. *Zaha* (Beans)
14. *Zachi* (Fat, Grease)
15. *Zohuao* (It Can Eat, or It Can Buy)
16. *Yetilla* (It Will Fight Again)
17. *Yecho* (Blister)
18. *Gohui* (Buyer, or Reverse Side)
    Final period: *Quicholla quieainij* (They Will Be Disconcerted, They Will Be Angry)[3]

The divinatory count of 260 *lani*, or feasts, integrated two primary cycles: a count from one to 13, and a cycle of 20 words that referred to plants, animals, or forces of nature. This count also had four *cocijo*, or major subdivisions of 65 days, each of which subdivided into five *cocii*, or *trecenas*, 13-day periods (Córdova 1578b, 117r-122r), for a total of 20 trecenas. Here, I use "feast" to translate the colonial Zapotec term *lani*, also spelled *lanij*. *Lanij* is defined in Córdova (1578a, 384r) as "solemn occasion, feast." *Lanij* also appears as the main term in compounds glossed as "festivity" or "day of obligation" (Córdova 1578a, 196r), and also in connection with Spanish terms for major Christian observances. Hence, the Zapotec calendars organized festivities or observances correlated with a period of time measured by the passage of one solar day. However, two clarifications are in order. First, the label for these observations, *lani* or *lanij*, was independent from the notion of "day," which was glossed as *chij*, *chèe*, and *copijcha* (Córdova 1578a, 138v). More important, Córdova notes that the time period that corresponded to one *lani* was not measured by Zapotecs from midnight to midnight, or from dawn to dawn. In his *Arte*, after listing each of the 260 *lani*, Córdova (1578b, 122v) states: "Those who were born were called according to the aforementioned names, each with the name of the day. And the day was counted from midday until another midday." While the Dominican lexicographer subsumes here the notion of *lani* under the Spanish term *día*, "day," it is clear that he is referring not to a solar day, but to the passage of each of the 260 *lani* in the Zapotec sacred count.

According to Córdova, a Valley Zapotec ritual specialist who focused on counting and interpreting the 260-day count was designated as *peni colanij*, or *agorero* "augury maker" (Córdova 1578a, 13v), or merely as *colanij*, glossed both as *Echador de las fiestas o docto en ellas* "one who casts the feasts, or who is knowledgeable about them" (Córdova 1578b, 116v), and less neutrally as *sortilegos o hechizeros* "conjurers or sorcerers" (Córdova 1578b, 117r). *Huechijlla* or *huechilla* was often translated as *adiuino* or *sortílego* "diviner, conjurer" (Córdova 1578a, 10r, 13v, 325r, 387r–v), while *pijzi* was an "augury," often a bad one (Córdova 1578a, 13v, 376v, 329v), although it also appeared in other terms related to disgrace or injury.[4] In Northern Zapotec, a term written as *(bene) guechea* was employed to refer to ritual specialists in colonial times.[5] In the present, Northern Zapotec communities commonly employ a variant of the same term, now *bene washa*, a label that people employ for powerful ritual specialists who may also engage in malevolent practices (Ricardo Ambrosio, personal communication, 2008). While a comparison between Zapotec counts is outside the scope of this essay, the structure of the 260-day count in colonial Northern Zapotec communities is different from that of a day count that is still in use in some Southern Zapotec communities (Emiliano Cruz, personal communication, 2008; Pergentino Cruz, personal communication, 2010, 2017; for published work, see Meer 2000; Weitlaner et al. 1958; Weitlaner and De Cicco 1961).

What, exactly, is the label that best fits these texts? The specialists who prepared them often began the list of the 260 days with a sentence that presented them as *biyee xoci xotao reho* "the periods/time count of the fathers and ancestors of us all" (Tavárez 2011). Although not all documents employ this phrasing, this is certainly a designation that seems closely tied to how Zapotec specialists conceptualized these documents—although some are more creative, such as *libro quichi tia queani xotao* (1368r), "book, paper of the lineages, of the ancestors." As noted above, the word *biyee* was employed to refer to various counts, and thus is not exclusive to the 260-day cycle, strictly speaking. Specialists used these texts as highly portable, handwritten manuals that could be corrected, amended, or expanded following pragmatic criteria. Seizing on this fact, idolatry eradicators identified these works as *cuadernos* "notebooks," a word that captures how these texts were generated through multiple annotations but that minimizes their importance as sacred texts. Moreover, these documents contain not only the 260 feasts, but also a multiplicity of other cycles, auguries, cosmological diagrams, drawings that explore correlations between time and space, lists of deities worshipped at the beginning and end of the divinatory cycle, instructions for ritual practices segregated by gender and age, eclipse annotations, correlations with the Gregorian calendar, well as the occasional cosmological narrative. Given the heterogeneous and highly pragmatic nature of such contents, and the specialists' preference, as noted above, for stressing the ancestral nature of these cycles, I refer to these texts as "ancestral time counts," while the textual units themselves are "books" or "manuals."

THE ORGANIZATION OF CALENDRICAL AND COSMOLOGICAL INFORMATION IN THE ZAPOTEC MANUALS

For Postclassic Mesoamerica, one of the most succinct depictions of the interdigitation of space and time appears in the first page of the *Codex Fejérváry-Mayer*. This illustration manages to incorporate a representation of cosmological space, a set of nine deities usually designated as the Nine Lords of the Night, and the 260-day cycle. At the center of a quincunx, Xiuhteuctli[6] holds a spear thrower and arrows, and he is surrounded by the four cardinal points, each of which is associated with a pair of deities, a sacred tree with a bird, and a color. The boundaries defining the cardinal points seem to be delineated by the blood that flows from a dismembered Tezcatlipoca. Two separate iterations of the 260-day count, both proceeding in a counterclockwise direction, are fused with the cardinal points, each trecena represented by its first day's sign. The first iteration traces the edges of the cardinal points, so that each cardinal point is associated with four trecenas in sequence; every fifth trecena, whose first day corresponds with one of the four central Mexican yearbearers, is depicted as a sign on the back of a descending bird. The second iteration is represented by day signs placed between a cardinal point and a year-bearer trecena, with a sequence that describes a centrifugal spiral around the center of the diagram. Indeed, this is a "complex and masterful presentation" of a spatiotemporal diagram, as noted by Elizabeth Boone (2007, 114–116).

A comparable, but much simpler diagram from Book 23 of Villa Alta (figure 5.1), from the town of Yatzona, sketches out a cosmological space comparable to the one depicted in the *Fejérváry-Mayer*.[7] In the Zapotec diagram, which is also a quincunx, the four cardinal directions bear five lines each, except for the direction on top, which has six, unexpectedly, perhaps due to a scribal error. If each of these lines represent one trecena, then we would have a distribution of trecenas across cosmological space in a manner that echoes the much more detailed scheme in *Fejérváry-Mayer* 1. However, there is a striking difference as well: Book 23 depicts a division of cosmological space into eight units, formed by dividing the main rectangle into which the quincunx is inscribed into eight even and equally sized rectangular sections. These eight units are further subdivided by the diagonal lines that mark intercardinal directions, and by other lines that depart from each of the four corners of the main rectangle. While the *Fejérváry-Mayer* does have four cardinal and four intercardinal directions, it lacks the division of cosmological space into the eight equally sized units in Book 23.

However, the organization of cosmological information in the Northern Zapotec corpus did not follow the pictographic conventions of central Mexican codices. Instead, ritual specialists followed what I term a "recitation

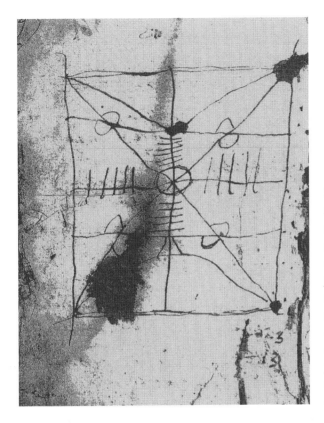

FIGURE 5.1. *Zapotec diagram of cosmological space, with four cardinal and four intercardinal directions indicated, plus a center and eight rectangular units. (AGI México 882, 527v, Book 23. Ministerio de Educación, Cultura, y Deporte, Archivo General de Indias, Spain.)*

format": that is, in every single manual, they wrote down the sequence of the 260 feasts, ordered as a descending list. As in other Mesoamerican divinatory counts, the list contained a cycle of 20 names (table 5.1).

In the Zapotec case, these 20 words were compounded with other lexical items that, while no longer semantically identifiable as numbers, stood for a cycle of 13 positions. In these manuals, the list of 260 feasts, which served as the central axis in terms of content, was sometimes preceded or followed by other lists, usually one containing the names of the 52 Zapotec years. In many notebooks, each of the feast names was followed by notes referring to deities, cycles within the 260-day count, observances, and auguries.

Given this strong tendency towards presenting information as a list, which may have reflected in part a recitation by ritual specialists, these time count manuals could drift towards what was, essentially, a paragraph format, which was also employed by Zapotec writers in other textual genres they practiced—such as petitions, letters, or formal statements. This orientation

TABLE 5.1. Colonial Northern Zapotec day names, extracted and interpreted by Terrence Kaufman, and adapted with some modifications from Justeson and Tavárez 2007[a]

	Córdova, Arte	Colonial Northern Zapotec	Meaning in colonial Zapotec	Other meanings for day names in Mesoamerica
1	+chiilla	+chila	cayman	cayman
2	+ii ~ laa	+ee ~ +laa	wind (+ee); lightning (+laa)	wind
3	+EEla	+Ela / ala	night	night, house
4	+Echi	+Echi	big lizard	lizard
5	+zii	+çee	omen (from snake or bird)[b]	snake
6	+laana	+lana	(to smell like) flesh or a carcass[c]	death
7	+china	+china	deer	deer
8	+laba	+laba	harvest?[d]	rabbit
9	+niça	+niza	water	water
10	+tella	+tela ~ +dela	to be tied in a knot	dog
11	+loo	+lao	monkey	monkey
12	+piia	+biaa	soaproot	tooth or twist
13	+ii ~ laa	+ee ~ +laa	reed	reed
14	+Eche	+Echi	jaguar	jaguar
15	+nnaa	+ina	sown field	eagle
16	+loo	+lao	crow or eye[e]	sun or buzzard
17	+xoo	+xoo	earthquake	earthquake
18	+opa	+opa	cold or dew	flint
19	+aappe	+Epag	drop[f]	storm
20	+lao	+lao	face	macaw, flower

[a] E stands for a sound transcribed either as *e* or *i*; EE is for *ee* or *ii*; + marks the morphemic boundary at which these words acquire prefixes with a numerical function.

[b] Córdova (1578a): *Agorar tomar aguero. tizàca pijcia . . . tàca pijzia; Agorar lo que hace el aue, que canta o culebra. toni pijcia, toçàca pijzia; Agoero.l. aguero. pijzi, pijze, peezi; 141r: Dios de los agueros. pitòo pijzi.* This interpretation is based on the work of Javier Urcid, whom I thank for discussing it with me (see Urcid 2005, figure 1.20).

[c] Córdova (1578a, 73v): *Carne cosa que hiede a carne o carnaza. hualàna, nalàna; carne heder algo a carnaza. tillàa nalana.*

[d] Butler (2000, 382): *cosecha gwlap*; Long and Cruz (2000, 338): *cosechar, chelap* (*mazorcas*); Córdova (1578a): *cocij collàpa* (95v), *cosecha de pan el tiempo* (95v), *frutos el tiempo de ellos* (201v), *otoño tiempo del año* (296r), *tiempo de miesses frutas o desiega o de algo* (401r).

[e] See Urcid (2005, figure 1.20).

[f] Ibid.

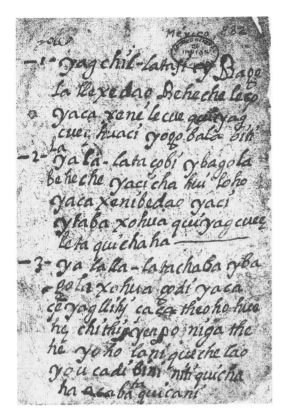

FIGURE 5.2. *Notes on the first three days in a divinatory count. (AGI México 882, 160r, Book 1; Ministerio de Educación, Cultura, y Deporte, Archivo General de Indias, Spain.)*

is showcased in Book 1, whose first folio is taken up completely by a series of notes that refer to each of the first three days in the divinatory count: 1-Caiman, *yagchil[a]*; 2-Wind, *yala*; and 3-Night, *yalalla*, as seen on figure 5.2.

SPACE AND TIME TOGETHER: THE RELATIONSHIP BETWEEN THE *BIYEE* AND THE LEVELS OF THE COSMOS

Like the *Fejérváry-Mayer* and other central Mexican codices, the Villa Alta corpus contains many references to a complex continuum in which temporal cycles were closely linked to cosmological spaces. However, rather than depicting this continuum in a single diagram, as was often done in precolumbian texts, these manuals often divide information into the depiction of cosmological realms, and annotations closely linked with the lists of 260 feasts. Calendars state that each of the 20 trecenas in the *biyee* originated in each of the three main cosmological levels.

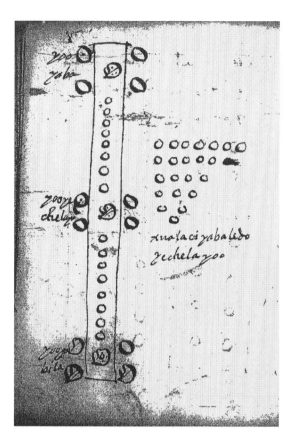

FIGURE 5.3. *Zapotec cosmological diagram with nine levels above and nine below earth. (AGI México 882, 384v, Book 11; Ministerio de Educación, Cultura, y Deporte, Archivo General de Indias, Spain.)*

On the other hand, diagrams such as the one in Book 23 often summarize an important amount of cosmological details. Hence, one of the best graphic representation of the Zapotec cosmos, as understood by colonial specialists, appears as a simplified sketch drawn by the anonymous author(s) of Book 11 of Villa Alta (see figure 5.3).

To elaborate on the brief description given previously, this drawing depicts the universe as a structure with three main tiers (sky, earth, and underworld), and with nine levels above earth and nine levels below. Eight levels, represented by circles, existed between the House of the Underworld (*yoo gabila*) and the House of Earth (*yoo yeche layo*; literally, "House of the Town on Earth"), and there were eight more levels between the House of Earth and the House of the Sky (*yoo yaba*). Four circles surround each of the three houses; as discussed below, they turn out to be essential components of the Zapotec cosmos. These four circles surround a fifth circle at the center; as shown below, they are cosmological locations where feasts were said to "be sitting" or to "fall into." Similar depictions of a three-tiered cosmos with nine levels above and nine below earth, and houses depicted as quincunxes, also appear in seven other manuals: 5, 6, 36, 62, 66-1, 94, and 97.[8]

Remarkably, the Book 11 artist inserts some unusual pictographic details in the depiction of these three central circles: lines, dots, and what may be a quick sketch of an animal or human head for the underworld circle. The exact model for these pictographic representations remains unknown. On the right side, flanking the space between earth and sky, an arrangement of 23 circles grouped into six levels, with one circle crossed out, bore the following label:

xua	laci	yaba	le	do	yeche	layoo
lord	small section/allotment	sky	next	to	town	earth

This phrase may be glossed as "the lords of a small section of sky next to earth," or "the lords of a small section of sky [that is] on the whole earth." It refers to the 22 circles drawn above, which do line up with the House of Earth and the first four levels between earth and sky.[9]

One of the most remarkable manuals in the corpus for its simple and spare presentation of calendrical data is Book 85-1 (figurre 5.4), which was written by one of three ritual specialists from the Nexitzo Zapotec town of San Miguel Tiltepec: Miguel Hernándes, Juan de Luna, and Juan Velasco. Tiltepec was one of the communities with the largest concentration of Zapotec ritual specialists: although inhabited by only 736 families, it had more than 10 specialists, which placed it on the same level with some communities that were twice as populous, as was the case for Yalálag (Tavárez 2011, 220). A measure of the calendrical knowledge contained in 85-1 is that it contains the only extant list of the periods in the *yza*, the 365-day Zapotec year (see Justeson and Tavárez 2007).

A set of annotations in Book 85-1 follow a pattern that can be termed "trecena initial," as they refer to one of the three cosmological houses at the start of the trecena. On the folio depicting the first trecena, the gloss on either side of the house image drawn atop the list of the first thirteen days in the count (*yagchilla*, 1-Cayman, to *queçee*, 13-Reed) states that the *trecena*, called *cocii* in Valley Zapotec and *lani* or *llanij* in Villa Alta, is associated with the House of Earth (*yoho lleo; lleo*, and *leo* are variants of *yeche layo*), and that the first of four 65-day divisions (*goçio i*) in the calendar "takes the turn" (*ricij laza*) here.

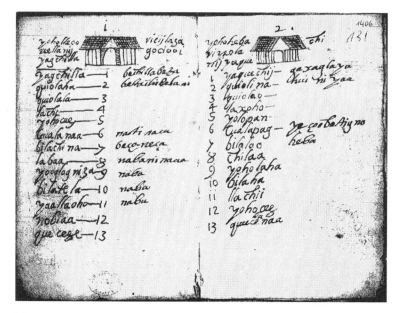

FIGURE 5.4. *The first two trecenas in Book 85-1. (AGI México 882, 1415v–1416r [pagination corrected]; Ministerio de Educación, Cultura, y Deporte, Archivo General de Indias, Spain.)*

Trecena 1 (1415v)

1. | yoho lleoo | ree | llanij | yagchilla | ricij | laza | gocioo i |
   |---|---|---|---|---|---|---|
   | yoho lleoo | ree | llanij | yag+chilla | ri+cij | laza | gocioo i |
   | house [Earth] | STA.be[10] | feast | 1+Cayman | HAB+take | turn | [65-day period] 1 |

   "The feast of 1-Cayman is [resides] in the House of Earth; Gocio One takes a turn"

Trecena 2 (1416r)

2. | yoho yeba | riyeo | lanij | yaquechij |
   |---|---|---|---|
   | yoho yeba | ri+yeo | lanij | ya+quechij |
   | house Sky | HAB+give[11] | feast | 1+Jaguar |

   "House of Sky gives away the feast 1-Jaguar"

Trecena 3 (1416v)

3. | yoho leo | reeyexog | | lanij | yagchina |
   |---|---|---|---|---|
   | yoho leo | re+ey+exog | | lanij | yag+china |
   | house [Earth] | HAB+RST+fall.into/come.out[12] | | feast | 1+Deer |

   "Into House of Earth falls the feast of 1-Deer"

Trecena 4 (1417r)

4.	yoho gabila	rijyexog	lanii	yaglao
	yoho gabila	rijy+exog	lanii	yag+lao
	house Underworld	HAB+fall	feast	1+Face

"Into House of Underworld falls the feast of 1-Face"

Trecena 5 (1417v)

5.	yoho leo	riyyeo	lanij	yagquee
	yoho leo	ri+yyeo	lanij	yag+quee
	house [Earth]	HAB+give	feast	1+Reed

"The House of Earth gives away the feast 1-Reed"

The design of the pages in Book 85-1 is spare and systematic: each folio contains one trecena, a realistic sketch of a Zapotec house with a straw roof appears on the top, flanked by references to the transit of the first date in each trecena across the houses of the cosmos. The verbs used for the feasts seem to refer to living beings, as 1-Cayman is said to "reside" (*ree*) in the House of Earth, while a movement down the cosmological levels is stressed when the ritual specialist states that 1-Deer "falls into" (*reeyexog*) the House of Earth, and then 1-Face "falls into" (*rijyexog*) the House of the Underworld. In the end, in Book 85-1 the first date of Trecena 1 is associated with the House of Earth, that of Trecena 2 with the House of the Sky, that of Trecena 3 with the House of Earth, that of Trecena 4 with the House of the Underworld, that of Trecena 5 with the House of Earth, and so on in a circuit that goes up to Sky, returns to Earth, descends into Underworld, and comes up again, until Trecena 20. This arrangement yields ten trecenas associated with earth, and two groups of five trecenas each linked to either sky or underworld.

Moreover, each of the 13 days in a trecena was also associated with a particular level or location in the cosmos. Going back to the diagram in Book 11, each of the three houses has five places that correspond to one level (four supports, plus the house itself, which together also constitute a quincunx). There are also eight separate levels between the House of the Sky and the House of Earth and eight more levels between the House of Earth and the House of the Underworld. Since various calendars report that the trecenas "leave from" and "arrive or fall" into the various houses, these annotations imply that the first five feasts in the trecena make the rounds around the house in question (Earth, Sky, or Underworld) and its four supports (five locations = five feasts), and that the *trecena*'s last eight days ascend or descend, one by one, the eight

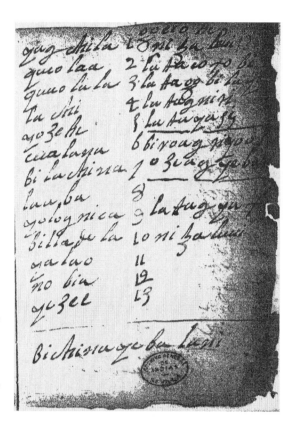

FIGURE 5.5. *Annotation concerning the sixth feast in Trecena 1. (AGI México 882, 282r: Book 6; Ministerio de Educación, Cultura, y Deporte, Archivo General de Indias, Spain.)*

intervening levels between houses (eight levels = eight feasts). This would mean that each of the three cosmological levels is three quincunxes: that is, four different directions surrounding a central location.

This interpretation is supported by some strategically placed annotations in Calendars 5 and 6. These calendars are copies of each other, and come from the Nexitzo Zapotec town of San Juan Yatzona. Book 6 was signed by local specialist Don Juan de Bargas, and Calendars 5 and 7 may have been owned, respectively, by the specialists Don Gaspar de Bargas and Juan Ximénez.[13] Both Calendars 5 and 6 divide the trecena into a five-day circuit through a quincunx and an eight-day cycle up or down eight cosmological levels.

Indeed, several annotations focus on the transit of the sixth feast in each trecena from one cosmological house into another. In Book 6 (282r), the following annotation is aligned with Day 6 of Trecena 1 (see also figure 5.5):

Day 6 of Trecena 1

6. | biroag ti | yoo | lleo | zeag | yebaa |
   |---|---|---|---|---|
   | bi+roag=ti | yoo | lleo | z+eag | yebaa |
   | CMP+be.taken/picked[14]=EMC[15] | House | [Earth] | PGM+go[16] | Sky |

"It [Day 6] has been taken from the House of Earth, it is going to the Sky"

On a line below the last day of the first trecena, we read, *bichina quieba lani* "the feast arrived in the sky." Then, near Day 6 of Trecena 2, we find the following phrase:

Day 6 of Trecena 2

7. | beroag ti | yoo | yeba | zeta | leo |
   |---|---|---|---|---|
   | be+roag=ti | yoo | yeba | z+eta | leo |
   | CMP+be.taken=EMC | Hous | Sky | PGM+come[17] | Earth |

"It [Day 6] has been taken from House of the Sky, it is coming to Earth"

A note placed at the end of Trecena 2 reads *bechina lao yoo niga* "[it] arrived here on Earth/the ground." Then, right next to Day 6 of Trecena 3, we read:

Day 6 of Trecena 3

8. | biroag ti | yoo | leo | ceag | cabila |
   |---|---|---|---|---|
   | bi+roag=ti | yoo | leo | c+eag | cabila |
   | CMP+ be.taken=EMC | House | [Earth] | PGM+go | Underworld |

"It [Day 6] has been taken from the House of Earth, it is going to the Underworld"

Another line at the end of this trecena states, *bexog cabila*, "[it] fell into Underworld." Finally, near Day 6 of Trecena 4, we read:

Day 6 of Trecena 4

9. | biroag ti | yoo | cabila | ceheta | leo |
   |---|---|---|---|---|
   | bi+roag=ti | yoo | cabila | c+eh+eta | leo |
   | CMP+be.taken=EMC | House | Underworld | PGM+RST+come | Earth |

"It [Day 6] has been taken from the House of the Underworld, it is coming to Earth"

We know that the days eventually come back to the House of Earth because, on 278v, Book 5 states that, by the end of Trecena 20, *bechina lao yoo lani* "the feast arrived on Earth."

While some Zapotec specialists simply placed a sentence near Day 6 indicating the movement of the feasts, some documents specify that the movement in question takes place on the sixth day of the trecena, as occurs with Book 70. A note in the middle of its second trecena reads:

AGI México 882, 1270v

10.	niga	lani	xopa	birog	yoho	yeba
	niga	lani	xopa	bi+rog	yoho	yeba
	here	feast	six	CMP+be.taken	house	Sky

"Here, Feast Six was taken from the House of the Sky"

Hence, the annotations discussed above suggest that, after the circuit of five places—each cosmological house, plus the four places represented by circles surrounding it, as depicted in Book 11—the last eight days of the trecena ascend or descend towards another cosmological house, represented in Book 11 as the center of a quincunx.

Zapotec specialist also linked their temporal cycles with European time counts. The specialist who composed Book 51 had a particular interest in recording the European days of the week, as reflected by his notations regarding the first and last day in each trecena:

AGI México 882, 1070r

Trecena 1, Day 6:				Trecena 1, Day 13:			
11. piroa	ioho	çe	epa	penita	iho	ieba	mierecole
pi+roa	ioho	çe	epa	pe+nita	i[o]ho	ieba	mierecole
CMP+be.taken	house	POT.be.aloft	Sky	CMP+be.placed[18]	house	Sky	Wednesday

"It was taken from the house, it will be in Sky"   "It was placed in the House of the Sky on Wednesday"

AGI México 882, 1070v

Trecena 2, Day 6:				Trecena 2, Day 13:			
12. peroa	ieba	marti		penita		ioho	marti
pe+roa	ieba	marti		pe+nita		ioho	marti
CMP+be.taken	Sky	Tuesday		CMP+be.placed		house	Tuesday

"It was taken from the Sky on Tuesday"   "It was placed in the house on Tuesday"

AGI México 882, 1071r

Trecena 3, Day 6:					Trecena 3, Day 13:			
13. peroa	ioho	çe	cabila	loni	penita	ioho	cabila	loni
pe+roa	ioho	çe	cabila	loni	pe+nita	ioho	cabila	loni
CMP+be.taken	house	POT.be.aloft	Underworld	Monday	CMP+be.placed	house	Underworld	Monday

"It was taken from the house, it will be in the Underworld on Monday"   "It was placed in the House of the Underworld on Monday"

By using references to the days of the week, the author of Book 51 was not alone, since correlation statements involving weekdays and the use of Dominical letters as a point of reference appear in other Zapotec calendrical manuals. What seems important here, however, is that this ritual specialist recognized that a subdivision of the trecena, the seven-feast period between Days 6 and 13, could be correlated to a fixed interval in the European week: Tuesday to Tuesday, Monday to Monday, and so on.

So far, the evidence discussed here concerns the cosmological diagram in Book 11, and various annotations from five manuals: 5, 6, 51, 70, 85-1. In fact, the movement of feasts up and down the Zapotec cosmos was carefully followed and commented on through annotations in the majority of the notebooks in this corpus, as shown in table 5.2.

Table 5.2 renders visible two important features of the spatiotemporal correlation in this corpus of 102 calendar manuals. First, the idea of a spatiotemporal correlation was essential to an understanding of the 260-day count, as it appears in 89 manuals, or 87.25 percent of the corpus. Even notebooks that contain little beyond a list of the day names (such as Book 70) nonetheless refer to this correlation. Additionally, the table shows a range of theories about the details of the correlation. To portray a division of the trecena into two parts, an annotation could be placed on either Day 5 or Day 6 to signal the beginning of the movement of the feast across cosmological space. The authors of 55 of the notebooks, or more than half of the corpus, agreed that the significant day was Day 6, as on that day the movement up or down the eight levels (as depicted in Book 11) began. The most popular solution, which appears in 44 manuals, was to indicate the start of a movement on Day 6, and to indicate the arrival of Day 13 before one of the cosmological houses. While most calendrical texts that indicate the movement of days described it to annotations, a minimalist solution was to draw a line separating the trecenas into two parts, as was done in seven manuals.

On the other hand, the authors of 27 manuals, or more than one-quarter of the corpus, preferred to indicate an association between one day in the trecena and a house in the cosmos. Twenty-four manuals follow a trecena-initial pattern: as was done in many central Mexican codices, the trecena is represented by its first day. A more unusual solution, employed in only three cases, is associating a cosmological house with the last feast in the trecena.

It should not be a surprise that diagrams in eight Zapotec manuals (Books 5, 6, 11, 36, 62, 66-1, 94, and 97) and annotations made on 87.25 percent of this corpus depict a cosmos divided into sky, earth, and underworld, which was firmly intertwined with cosmological theories of precolumbian origin. Such theories

TABLE 5.2. Annotation patterns for the movements of the 260 feasts across cosmological space

Trecena division	Days on which annotation is placed	Manuals that use this system	Manuals by AGI numeration
Days 1–6, 7–13	6 and 13	44	5, 6, 8, 12, 13, 14, 15, 17, 18, 19, 20, 21, 22, 23, 24, 25, 26, 30, 31, 32, 34, 36, 37, 42, 47-2, 51, 53, 57, 60, 61, 62, 63, 65, 66-1, 66-2, 67, 68, 70, 72, 76, 84, 90, 94, 99
	1 and 6	5	29, 49, 54, 73, 89
	1, 6, and 13	1	4
	Only on 6	2	7, 62
	Only on 1, line between 6 and 7	3	71, 77, 91
Days 1–5, 6–13	Only on 5	2	78, 95
	Only on 1, line between 5 and 6	3	58, 59, 98
	1 and 13, line between 5 and 6	1	74
	5 and 13	1	9
Trecena-initial; division not indicated	Only on 1	24	1, 10, 11, 28, 33, 35, 38, 43, 44, 45, 46, 48, 50, 52, 64, 69, 75, 81, 85-1, 85-2, 86, 88, 92, 93
Trecena-final; division not indicated	Only on 13	3	16, 27, 41
No reference to houses or feast movement	N/A	13	2, 3, 39, 40, 47-1, 55, 56, 79 and 80 (aberrant), 82, 83, 96, 97

Source: AGI-Mex 882.

included the 260-feast cycle, the *trecenas* and half-*trecenas*, and quincunxes. After all, extensive archaeological and iconographic evidence shows that the sky and the underworld were crucial elements in the cosmology of ancient Oaxacan societies. Monte Albán, the most important Classic-period site in the Valley of Oaxaca, was built as a model of the cosmos with a northern platform associated with the sky and ruling lineages, and a southern platform linked to underworld motifs (Joyce 2004). Furthermore, iconography related to the sky and the sun appears in genealogical registers carved in several Classic-period slabs (Urcid 2005, 120), and a nine-step staircase in Classic-period Zapotec tombs has been interpreted as a direct reference to the nine levels between the earth and the underworld (Urcid 2005, 94).

CONCLUSIONS

This essay provides the first thorough analysis of a correlation between the feasts in the 260-day count and nine levels above and nine below earth, as articulated in the largest extant corpus of colonial Mesoamerican calendars. In summary, the Northern Zapotec calendars established a spatiotemporal continuum that links time in the 260-day count with space in a three-tiered cosmos. Each of these tiers is a quincunx—four supports surrounding a central "house" of earth, heaven, or underworld—and there were eight levels in between the underworld and earth, and another eight between earth and the sky. Hence, counting the underworld and the sky as the first and last levels, there are nine levels below and nine above the place where humans live, the House of Earth. Such a depiction appears explicitly as a diagram in eight manuals. Indeed, this striking cosmological theory provides further evidence about the multiplicity of precolumbian calendrical systems and cosmological theories in Mesoamerica, which cannot be subsumed under a single, unified calendrical scheme (Prem 2008, Tena 1987).

In order to understand this spatiotemporal cycle, the cosmological model described above must be combined with the annotations that describe the movement of feasts across the cosmos. Such a circuit is described explicitly through annotations in half of the corpus (53 manuals), which alert about the displacement of Day 5 or 6 in each trecena. According to this description, the first day of each trecena leaves from one of the three cosmological houses; Days 2–5 travel through each house's four cosmological supports; Days 6–13 go up or down the eight levels between the house of departure and the house of arrival; and the first day of the next trecena leaves again from the house of arrival. In the end, ten trecenas (1, 3, 5, 7, 9, 11, 13, 15, 17, and 19) are associated with the House of Earth, as the first feast of each of them departs from earth and initiates an upward or downward movement across the cosmos. Five trecenas (2, 6, 10, 14, and 18) are associated with the House of the Sky, as the first day of each of them is "taken from" or "falls from" the sky. Five trecenas (4, 8, 12, 16, and 20) are associated with the House of the Underworld, as the first day of each of them "comes down" to the underworld.

It should be stressed that ritual specialists discuss the movement of each of the 260 feasts across cosmological levels almost as if they were living entities: they were "sheltered" in *gonita*, or even "fall into" *riyexog* ("cosmological houses"). Indeed, the Zapotec cosmos that is described in the Villa Alta corpus comes across as a vibrant spatiotemporal continuum that is in motion itself, and that admits the motion of various entities, as two cycles of ritual songs (AGI-Mex 882, Books 100 and 101) contain acts of propitiation for the return of foundational ancestors to the Northern Zapotec community of Lachirioag, who come down from various locations in the cosmos.

While a discussion of the origin of the Zapotec spatiotemporal correlation goes well beyond the scope of this essay, there is little evidence to suggest that this Zapotec cosmological model is based on a non-Mesoamerican cosmogony, as it has strong parallels with other Mesoamerican vertical models of the cosmos with nine levels (see, e.g., López Austin 2016). A recent but already influential interpretation has posited that the depiction of a layered cosmos in the *Vaticanus A* and in a few other central Mexican pictographic sources may reflect the influence of European cosmological models (Díaz 2009; chapter 3, this volume; Nielsen and Sellner Reunert 2009; chapter 1, this volume). Certainly, none of these works should have been expected to address the dense layers of evidence inscribed in the Northern Zapotec corpus, which still remains poorly known. A more plausible scenario holds that Mesoamerican cosmological theories were pliable and contingent on historical and political changes. Thus, according to Ross Hassig (2001), the transition in Mexica cosmology from a nine-layer to a 13-layer upper realm, and a change in the date for the celebration of the New Fire marking the beginning of a period of 52 vague solar years, may have been introduced early in the reign of Moteuczoma Xocoyotzin.

During the seventeenth century, Northern Zapotec ritual specialists lived beyond the reach of educational projects that thrived among metropolitan indigenous elites. As the Dominican chronicler Francisco de Burgoa (1989b, 96) acknowledged, in Villa Alta, "the errors were harsher, and barbarity was less civilized." Indeed, Dominican deficits in terms of basic catechetical instruction in the region—which did not even consider the possibility of an education that comprised European cosmological and astrological teachings—were at the root of Bishop Maldonado's attempts to confiscate the Zapotec divinatory manuals and establish new parishes managed by secular clergy.

A concise *reductio ad absurdum* is in order here. If the Northern Zapotec cosmos were based on a European model of a three-tiered, multilayered cosmos, then it follows that virtually everything that colonial Zapotec specialists recorded, as discussed in this essay, was the result of a radical act of reverse engineering. In order to fit a putative European innovation, Zapotec specialists would have had to labor, rather swiftly and over a period barely exceeding a century, to retrofit into Dante's cosmos preexisting Mesoamerican theories about the spatiotemporal order, the use of a thirteen-day period as a fundamental unit in the divinatory calendar, the division of these units into half trecenas, and the significance of quincunxes in the cosmos—without even considering cardinal points, local and pan-Zapotec deities, and ancestors, which were also closely associated with the Zapotec spatiotemporal continuum. Such cosmogonical labors would have had to take place in order to achieve the perfect fit that a thirteen-day period has with a theory about three cosmological houses, their structure as quincunxes, and a series of eight cosmological levels above and below Earth.

Hence, we must look elsewhere to find the cosmogonical principles painstakingly recorded in these Zapotec manuals. We know from the work of Eduard Seler (1983) that plates 75 and 76 of the *Codex Borgia* depict the various lords of the half trecenas, in which the trecenas are divided into groups of seven and six days. Elizabeth Boone (2007, 132) has stressed that these lords seem to emphasize cardinal directions. There is an obvious parallel between the half-trecenas in the *Borgia* and in the Northern Zapotec corpus. Rather than proposing an explicit continuity here, this essay focuses on providing a detailed account of the spatiotemporal cycle in the Zapotec manuals, while the continuities and differences between the Borgia Group and the colonial Zapotec *biyee* are discussed elsewhere (Tavárez n.d.). In the end, future work on this formidable corpus of Zapotec manuals that recorded the "time of the ancestors" will result in insights regarding the structure, mechanics, and meaning of this assemblage, which may in turn help elucidate other travels in time and space depicted in central Mexican codices.

NOTES

1. This essay is a revised version of the first analysis of the relation between the Zapotec 260-day count and cosmological layers, which appeared in Tavárez (2006, 428–29), Justeson and Tavárez (2007, 71), and Tavárez (2011, 196–198). Earlier versions of this analysis were presented at the 2007 and 2011 meetings of the American Society for Ethnohistory, and also at three presentations at the Universidad Nacional Autónoma de México, which were sponsored by the Instituto de Investigaciones Filológicas (2011) and by the Instituto de Investigaciones Históricas (2014 and 2015). I am grateful to those who organized and attended these events. The input and suggestions of Ricardo Ambrosio of Lachirioag have been extremely helpful over the years. I also thank Ana Díaz, John Justeson, Pergentino José Cruz, the late Emiliano Cruz Santiago, Javier Urcid, Brook Lillehaugen, Aaron Broadwell, Federico Navarrete, and Elizabeth Boone for providing valuable feedback. All errors are my own. Archive

abbreviations: AGI: Archivo General de Indias; -Mex: Audiencia de México; AHJO-VA: Archivo Histórico Judicial de Oaxaca, sección Villa Alta. My analytical translations from the Zapotec follow, for the most part, Smith-Stark (2008), and use "+" to mark proclitics, "—" morpheme boundaries, and "=" for enclitics. Abbreviations: 3: third person, CMP: completive; EMC: emphatic particle associated with the full completion of an action; HAB: habitual; POT: potential; PGM: progressive movement; RSL: resultative; STA: stative.

2. Alcina Franch (1993) numbered the Villa Alta calendars 1–99, and this numeration is used by the Archive of the Indies. Nevertheless, some of these *cuadernos* contain two different calendars, or split the same calendar into two cuadernos. There are, thus, 102 separate calendars bound into 99 manuals.

3. The interpretation of the names of the 18 *cocii* and their relationship with the Nahua *xihuitl* is discussed in greater detail in Tavárez (n.d.). Most of the glosses above are based in entries in Córdova (1578a): (1) 253r: Maguey. tòba; or the word for "mouth, edge"; (2) 415v: Vmidad de tierra ovmido para sembrar. nàhui; 165v: Enfermedad o dolencia. huij; (3) 342r: Rebentar la planta o semilla. zegàaya; 359r: Reuerdecer. zegàaya; 423v: Verdecerse. zegàaya; (4) 317r: Plumas las ordenes dellas que tienen los papagayos en si. lohuè; (6) 21r: Alerta estar yterum, tipeennáayà; 34r: Aprestado assi . . . tipeeñaa; 420v: Velar de noche esperando algo, tipèennàaya; (7) Analyzed as the agentive *go* plus *lago* (food, sustenance); 258r: Mantenimiento. làgo; 257v: Manjar generalmente todo lo que se come. làgo; (11) 201r: Fruto de arbol o otro qualquiera. xigàha; (12) 128v: Deshollinado ser . . . titiña; Deshollinado assi . . . pitiña; (13) 200v. Frisol legumbre. pizàa; (14) 206v: Gordura o grassa, zàchi; 209v: Grossura en cuerpo o lardo, zàchi; (Butler 2000, 345): zaš´ gordo (de carne); (15) analyzed as the possibilitative aspect prefix *zohu-* (Smith-Stark 2008, 409–10) and the verbal root *-ao*, "to eat" or "to buy"; (16) analyzed as the restorative *ye* (Smith-Stark 2008; see Foreman and Dooley 2015 for contemporary data) plus the verb *-tilla*; 361r: Ryxoso. natílla; (17) 61v: Buba o grano assi; (18) *còhui* is glossed as: comprador (83v), color de morado (80r), enues de algo (176v), reboluedor tal (343r). The verb *go+hui*, "he/she/they gave," appears in some Zapotec wills (see, for instance, AHJO-VA Civil 96 L7 E5, 6r).

4. *Colani* is composed by the agentive *co-* and the nominal root *lanij* "festivity," and it can be glossed as "maker of feasts." For a detailed discussion of data in Córdova regarding deities and ritual specializations, see Smith-Stark (1999).

5. The term *guechea* was used by a Spanish visitor to the town of San Miguel Tiltepec when referring to a number of local residents regarded as *brujos*, or malevolent specialists (AHJO-VA 2004, Criminal 129, 21v).

6. The name of this deity is also spelled *Xiuhtecuhtli* in Nahuatl sources. *Xiuhteuctli* is used here, as this spelling better reflects the labialized /k/ in the name, transcribed with a "c" in both spellings.

7. A more detailed comparison between the cosmological diagrams in several central Mexican codices and those in the Villa Alta corpus appears in Tavárez (n.d.).

8. Tavárez (n.d.) proposes a detailed analysis of the extraordinary parallels between the diagram in Book 11 (Figure 5.3) and two illustrations in Codex Vaticanus B whose cosmological significance was first addressed by Eduard Seler.

9. See Córdova (1578a), where *laaci* and *làaci* are glossed, respectively, as "Derecho o parte q[ue] me cabe como de ofrenda" (118r), and "Suerte por obra o destajo" (390v). On the other hand, *laci* may be "small part" ibid.; 307r: "Pedaçito . . . *làci*"; 317v: "Poco nombre, adiectivo . . . *làci*"; "Poco iterum . . . *lacì=ca, làci tào*." *Le do* may be a cognate of the Colonial Valley Zapotec locative phrase *le to bi*, which Córdova (1578a) glosses as "Todo cantidad continua" (404r) and "Toda vna cosa entera toda vna pared" (403v), while *le to bi yza* meant "ogaño por todo este año o casi todo este año" (288r). Butler (1980, 240, 243) reported that, in Yatzachi el Bajo, *do le'e* is used to indicate approximate location, "more or less in one place"; in this construction, Butler stated that *do* means "somewhere," while *le'e* is the body part locative "inside, in the middle of" (see Sonnenschein 2005). Ricardo Ambrosio of Lachirioag (personal communication, 2016) interpreted *le do* as "being next to" something. A first gloss of this annotation appeared in Tavárez (2011, 198).

10. Córdova (1578a) glosses *tèe=a* as "estar ser [es decir] sum est fuy" (188v); "sentado estar" (375v); "tener algo consigo o en su poder" (397r). See also Smith-Stark et al. (2008, 339). For a study of Colonial Valley Zapotec positional verbs, see Foreman and Lillehaugen (2017). In Northern Zapotec texts, the equivalent verb, *ree*, is often glossed not only as "to be" or "to be sitting," but also as "to reside." See AHJO-VA Civil 144, 3r: *tzela to lichiya ree sa[n] miguel beñe chi peso* (y por una casa mia en que vive Pedro of San Miguel dio dies pesos); AHJO-VA Civil 171, 4r: nigaa **ree** Juan ylesca (en donde vi[ve] Ju[an] Yllescas).

11. The verb *yeo* is translated as "to give" in some Northern Zapotec texts. See AHJO-VA Civil 866, 24v: **yeo** y bechia bisente nijaq[u]e yesia gatzag (le doy a mi hermano vicente para que yo coja la mita[d]); AHJO-VA Civil 29, 72v: cana tza catino **yeo=hui=no** xinia yoco quenaya (Y que quando se muera se lo a de dar todo a mis hijas todo lo que tubiere mio); see also AHJO-VA Civil 25, 1r: ximenes **yeo** na tzona p[es]os (ximenes a de dar tres pesos).

12. In Northern Zapotec texts, the verb *+exog* is often rendered by court translators as "to fall [into]." See AHJO-VA Civil 277, 10r: lani batti **b+exog=aa** ylaao guichi huee (ya e caedo en esta enfermedad); AHJO-VA Civil 29, 16v: **b+exog**=te quiqueag quiag belog (Y ba a caer a la caveza de la piedra y cueba); AHJO-VA Civil 304, 2r: beyetag lidia **b+ey+exog** yiquiag (y baja dr.o a caer por la cabesa). This meaning may be preserved in an apparent cognate from Yatzachi el Bajo, in Butler (2000, 109): *ch+exjw*, bajar, disminuir. At least in one instance, *be+xog* is glossed as "to come out": AHJO-VA

Civil 98, 1r: yalag b+exog lichi guia frasco (hasta que salio de la carcel el frasco). See also Butler (2000, 97): *ch+echoj*, salir (de cualquier lugar ajeno, pero no de la casa de la persona); Butler (2000, 84): *ch+choj*, salir.

13. AGI-Mex 882, 255r–256r.

14. Córdova 1578b, 78v, glosses *ti+roa*, with the completive *pi+*, as "cogido ser assi algo," while *ti+ròa=ya* is "coger flores o fruta de los arboles," or for fruit to be cut (94v). See also AHJO-VA Civil 196, 11v: **gui+roag** ttomi atta noseea quie xana alguasil mayor (para que salga el dinero que debo a mi amo el Alguasil m[ay]or).

15. The morpheme *ti* does not appear to be either the morpheme *=ti*, which is used widely in Colonial Valley Zapotec in negative constructions (Anderson and Lillehaugen 2016), or a demonstrative. Instead, I propose that *ti* is an emphatic morpheme that stresses completion (EMC), and that it is a variant of *=te*. This enclitic appears after several verbs in Colonial Northern Zapotec texts, and emphasizes that the movement or action was fully completed. An example appears in AHJO-VA Civil 361, 4r: *bi+china=te na+chaga fran.co her.s* (y llega **hasta** juntar con fransisco hernandes). Another one is in AHJO-VA Civil 451, 14r: *yaghe-zoa=te xana la-yoo* (y ba a parar **hasta** avajo de la tierra). In Lachirioag (Ricardo Ambrosio, personal communication, 2017), the enclitic *=te* indicates that the action encoded by a verb has been completed: for example, *w+dita* (played) and *w+dita=te* (finished playing). Butler (2000, 294) glosses *=te* as an emphatic, and also as "later."

16. Pickett (1976), continuing a discussion of Zapotec verbs of motion that began in Speck and Pickett (1976), noted the presence of an aspect marker *z+* and *zi+*, which was informally denominated as "on-the-wayative." Based on these works, Smith-Stark (2008, 408) identified the aspect marker *z+* and *c+* in Colonial Valley Zapotec as *progresivo de movimiento*, "progressive movement." An alternative account exists: Colonial Northern Zapotec forms such as *zeag* could be interpreted as verbs in the stative aspect, following the analysis of contemporary Zapotec variants spoken in Yatzachi el Bajo (Butler 1980), Yalálag (López and Newberg 2005), and Zoogocho (Sonnenschein 2005). These three grammatical descriptions identify only four verbal aspect markers (continuative, stative, completive, and potential), and analyze verbs of motion such as *zej* as statives, as it appears that there is no habitual versus progressive contrast in those variants (Broadwell 2015). In addition, there is little evidence in Colonial Northern Zapotec texts for the andative and venitive forms attested in those variants. However, I tentatively adopt Smith-Stark's analysis here, based on the following facts. First, some Colonial Northern Zapotec texts employ *na+yeag* as the stative form of "to go" and thus indicate a distinction in aspect marker between this form and *z+eag*. Moreover, Reyes (1891, 40), following Córdova (1578b, 66r), identified *zee+* as denoting "perseverance" (*perseveransia*). This *z+e* may be composed by the progressive marker *z+*, plus Smith-Stark's *+e* as resultative. Hence, the present chapter tentatively identifies

the aspect marker z+ in Colonial Northern Zapotec as PGM, "progressive movement," as it appears mostly in verbs of motion. Progressive movement z+ is quite different in shape and function from the progressive *ca-*, whose historical development is exhaustively discussed in Broadwell (2015).

17. AHJO-VA Civil 108, 5r: leni **s+eda** canij la dioha (y beniendo para aca); AHJO-VA Civil 227, 12r: **z+eeh+etag** lij vechina esquina (ba bajando hasta llegar a la esquina).

18. Butler (2000, 143): chnitə', colocar, poner, compl. bnitə'.

REFERENCES

Alcina Franch, José. 1993. *Calendario y religión entre los zapotecos*. Mexico City: UNAM.

Alcina Franch, José. 1998. "Mapas y calendarios zapotecos; siglos XVI y XVII." In *Historia of the Arte en Oaxaca*, vol. 173–191. Oaxaca: CONCA, Gobierno of Oaxaca.

Anderson, Caroline J., and Brook D. Lillehaugen. 2016. "Negation in Colonial Valley Zapotec." *Transactions of the Philological Society* 114 (3): 391–413.

Archivo Histórico Judicial de Oaxaca. 2004. *Los documentos of San Francisco Cajonos*, transcribed by Claudia Ballesteros. Oaxaca: Tribunal Superior de Justicia, IAGO, Proveedora Escolar.

Boone, Elizabeth. 2007. *Cycles of Time and Meaning in the Mexican Books of Fate*. Austin: University of Texas Press.

Bricker, Victoria R., and Helga-Maria Miram. 2002. *An Encounter of Two Worlds: The Book of Chilam Balam of Kaua*. Middle American Research Institute, Publication 68. New Orleans, LA: Tulane University.

Broadwell, George A. 2015. "The Historical Development of Progressive Aspect in Central Zapotec." *International Journal of American Linguistics* 81 (2): 151–85.

Burgoa, Fray Francisco de. 1989a [1674]. *Geográfica descripción*. 2 vols. Mexico City: Editorial Porrúa.

Burgoa, Fray Francisco de. 1989b [1670]. *Palestra Historial*. Mexico City: Editorial Porrúa.

Butler, Inez. 1980. *Gramática zapoteca: Zapoteco de Yatzachi el Bajo*. Mexico City: Instituto Lingüístico de Verano.

Butler, Inez. 2000. *Vocabulario zapoteco de Yatzachi el Bajo*, 2nd ed. Mexico City: Instituto Lingüístico de Verano.

Chance, John K. 1989. *The Conquest of the Sierra*. Norman: University of Oklahoma Press.

Córdova, Fray Juan de. 1578a. *Vocabulario en lengua Çapoteca*. Mexico City: Pedro of Ocharte y Antonio Ricardo.

Córdova, Fray Juan de. 1578b. *Arte en lengua zapoteca*. Mexico City: Pedro Balli.

Díaz, Ana. 2009. "La primera lámina del *Códice Vaticano A*: ¿Un modelo para justificar la topografía celestial de la antigüedad pagana indígena?" *Anales del Instituto de Investigaciones Estéticas* 31 (95): 5–44.

Foreman, John, and Sheila Dooley. 2015. "Causative Morphology in Macuiltianguis Zapotec." In *Valence Changes in Zapotec: Synchrony, Diachrony, Typology*, ed. Natalie Operstein and Aaron Sonnenschein, 237–280. Amsterdam: John Benjamins Publishing Company.

Foreman, John, and Brook D. Lillehaugen. 2017. "Positional Verbs in Colonial Valley Zapotec." *International Journal of American Linguistics* 82 (2): 263–305.

Gillow, Eulogio. 1978 [1889]. *Apuntes históricos sobre la idolatría e introducción del cristianismo en Oaxaca*. Graz: Akademische Druck-u Verlagsanstalt.

Hassig, Ross. 2001. *Time, History, and Belief in Aztec and Colonial Mexico*. Austin: University of Texas Press.

Joyce, Arthur. 2004. "Sacred Space and Social Relations in the Valley of Oaxaca." In *Mesoamerican Archaeology*, ed. Julia Hendon and Rosemary Joyce, 192–216. New York: Wiley-Blackwell.

Justeson, John, and David Tavárez. 2007. "The Correlation of the Colonial Northern Zapotec Calendar with European Chronology." In *Skywatching in the Ancient World: New Perspectives in Cultural Astronomy*, ed. Clive Ruggles and Gary Urton, 17–81. Niwot: University Press of Colorado.

Kubler, George, and Charles M. Gibson, ed. 1951 *The Tovar Calendar: An Illustrated Mexican Manuscript*. Connecticut Academy of Arts and Sciences, Memoir 11. New Haven: Connecticut Academy of Arts and Sciences.

Lind, Michael. 2015. *Ancient Zapotec Religion: An Ethnohistorical and Archaeological Perspective*. Boulder: University Press of Colorado.

Long, Rebecca, and Sofronio Cruz. 2000. *Diccionario zapoteco of San Bartolomé Zoogocho, Oaxaca*, 2nd ed. Mexico City: Instituto Lingüístico de Verano.

López Austin, Alfredo. 2016. "La verticalidad del cosmos." *Estudios de Cultura Náhuatl* 52: 119-150.

López, Filemón, and Ronaldo Newberg. 2005. *La conjugación del verbo zapoteco: Zapoteco de Yalálag*, 2nd ed. Mexico City: Instituto Lingüístico de Verano.

Meer, Ron van. 2000. "Análisis e interpretación de un libro calendárico zapoteco: El Manuscrito de San Antonio Huitepec." *Cuadernos del Sur* 15 (6): 37–74.

Miller, Arthur. 1991. "Transformations of Time and Space: Oaxaca, Mexico, circa 1500–1700." In *Images of Memory: On Remembering and Representation*, ed. Susanne Küchler and Walter Melion, 141–175. Washington, DC.: Smithsonian Institution Press.

Miller, Arthur. 1998. "Espacio, Tiempo y Poder Entre los Zapotecas de la Sierra." *Acervos* 10 (3): 17–20.

Nielsen, Jesper, and Toke Sellner Reunert. 2009. "Dante's Heritage: Questioning the Multi-Layered Model of the Mesoamerican Universe." *Antiquity* 83 (320): 399–413.

Pickett, Velma B. 1976. "Further Comments on Zapotec Motion Verbs." *International Journal of American Linguistics* 42 (2): 162–164.

Prem, Hanns J. 2008. *Manual de la antigua cronología mexicana*. Mexico City: Centro de Investigaciones y Estudios Superiores en Antropología Social, Porrúa.

Reyes, Fray Gaspar de los. [1704] 1891. *Gramática de las lenguas Zapoteca-serrana y Zapoteca del Valle*. Oaxaca: Imprenta del Estado.

Seler, Eduard. 1983 [1904]. *Comentarios al Códice Borgia*. 2 vols. Mexico: Fondo de Cultura Económica.

Smith-Stark, Thomas. 1999. "Dioses, sacerdotes, y sacrificio: Una mirada a la religion zapoteca a través del Vocabulario en lengua Çapoteca (1578) de Juan de Cordova." In *La religión de los Binnigula'sa'*, ed. Víctor de la Cruz and Marcus C. Winter, 89–195. Oaxaca: Instituto Estatal de Educación Pública de Oaxaca, Instituto Oaxaqueño de las Culturas.

Smith-Stark, Thomas. 2008. "La flexión de tiempo, aspecto y modo en el verbo del zapoteco colonial del valle de Oaxaca." In *Memorias del Coloquio Francisco Belmar*, ed. Áurea López Cruz and Michael Swanton, 377–419. Oaxaca: Biblioteca Francisco de Burgoa, CSEIIO, Fundación Harp Helú, INALI.

Smith-Stark, Thomas, Áurea López Cruz, Mercedes Montes de Oca Vega, Laura Rodríguez Cano, Adam Sellen, and Alfonso Torres Rodríguez. 2008. "Tres documentos zapotecos coloniales de San Antonino Ocotlán (con Vicente Marcial Cerqueda and Rolando Rosas Camacho)." In *Pictografía y Escritura Alfabética en Oaxaca*, ed. Sebastián van Doesburg, 287–350. Oaxaca: Instituto Estatal de Educación Pública.

Sonnenschein, Aaron. 2005. *A Descriptive Grammar of San Bartolomé Zoogocho Zapotec*. Munich: Lincom Europa.

Speck, Charles H., and Velma B. Pickett. 1976. "Some Properties of the Texmelucan Zapotec Verbs *Go, Come*, and *Arrive*." *International Journal of American Linguistics* 42 (1): 58–64.

Tavárez, David. 2006. "The Passion According to the Wooden Drum: The Christian Appropriation of a Zapotec Ritual Genre in New Spain." *The Americas* 62 (3): 413–444.

Tavárez, David. 2009. "Los cantos zapotecos de Villa Alta: Dos géneros rituales indígenas y sus correspondencias con los Cantares Mexicanos." *Estudios de Cultura Náhuatl* 39: 87–126.

Tavárez, David. 2011. *The Invisible War: Indigenous Devotions, Discipline, and Dissent in Colonial Mexico*. Stanford, CA: Stanford University Press.

Tavárez, David. n.d. "Zapotec Cosmology" (unpublished manuscript).

Tavárez, David, and John Justeson. 2008. "Eclipse Records in a Corpus of Colonial Zapotec 260-Day Calendars." *Ancient Mesoamerica* 19 (1): 67–81.

Tena, Rafael. 1987. *El calendario mexica y la cronografía*. Mexico: Instituto Nacional de Antropología e Historia.

Urcid, Javier. 2005. *Zapotec Writing: Knowledge, Power, and Memory in Ancient Oaxaca*. Internet publication, FAMSI, http://www.famsi.org/zapotecwriting. Accessed on October 28, 2019.

Weitlaner, Roberto, and Gabriel De Cicco. 1961. "La jerarquía de los dioses zapotecos del sur." *Proceedings of the Thirty-Fourth International Congress of Americanists*, 695–710. Vienna.

Weitlaner, Roberto, et al. 1958. "Calendario de los zapotecos del sur." *Proceedings of the Thirty-Second International Congress of Americanists*, 296–299. Munksgaard.

6

A Cosmology of Water

The Universe According to the Ch'orti' Maya

Kerry Hull

The Ch'orti' Maya reside in southern Guatemala in the *departamento* of Chiquimula (figure 6.1). Despite commonly cited overestimates claiming 52,000 speakers (Cojti 1992), today there are closer to 12,000 fluent native speakers of the Ch'orti' language based on my fieldwork (cf. Metz 1998, 326). The Ch'orti' primarily live a subsistence, agricultural lifestyle, with the *milpa* (cornfield) being the heart of their existence (Dary, Elías, and Reyna 1998; López García and Metz 2002) (figure 6.2). The dominant religion remains Catholic, although manifesting itself at times as a hybridization with traditional Ch'orti' beliefs, symbols, and practices.

In the last one hundred years, a steep decline in native ceremonialism and traditional religious rites, including also clear ideas of native cosmology, is directly attributed to three primary factors: (1) missionary work by the Catholic church and by evangelical Christian groups, (2) fear that traditional practices might be labeled "witchcraft" (*brujería*), which often results in the murder of the practitioner, and (3) a "folklorization" of traditional Ch'orti' beliefs by the Guatemalan government (Hull 2003, 47–62; Metz 1998, 334; 2007). Indeed, the first two factors are in part related. Since the 1990s, however, there has been a reemergence of traditional ceremonialism to stem the tide of loss (Hull 2003). More to the point, these efforts toward language documentation and promulgation, traditional rituals, and a return to the "respectful" way of life of times past (Metz 2009, 171) are also seen as a means of redefining

DOI: 10.5876/9781607329534.c006

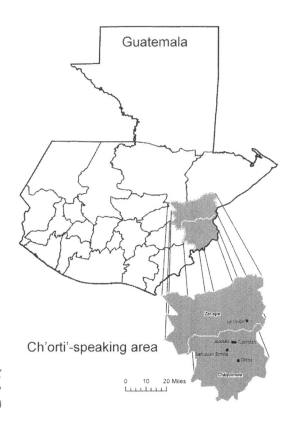

FIGURE 6.1. *Ch'orti'-speaking areas, Guatemala. (Map by Zachary Nelson.)*

Ch'orti' identity, securing land rights and human rights, improving education, and gaining political influence (Metz 2007). Some elderly rural Ch'orti', however, blame the famines and widespread poverty in the Ch'orti' area today on the loss of ritual practice, and specifically on not making "payments" (*tojma'r*) to the rain-making angels (Hull 2003, 55–56; Metz 2007, 334–335). A return to traditional forms of ceremonialism, in their view, would solve many of the problems facing the Ch'orti' today.

The third cause in the loss of traditional Ch'orti' cosmological understanding can be traced to what Metz describes as the "folklorization" of the Ch'orti', that is, reducing and redefining what it means to be "Ch'orti'" to antiquated festivals, dances, competitions, and so forth. As Metz has powerfully argued, "*Quizás no haya nada más efectivo en el ataque contra la cosmovisión tradicional ch'orti' como su folklorización por parte de los ladinos*" (Perhaps there is nothing more effective in attacking the traditional worldview of Ch'orti as its folklorization by Ladinos) (Metz 2007, 14, my trans.).

FIGURE 6.2. *Cornfields in the Ch'orti' area. (Photography by Kerry Hull.)*

Recent efforts to revitalize Ch'orti' language and culture have included cosmological instruction, although often in traditions from non-Ch'orti' regions of eastern Guatemala. For example, in 2002 a ceremony was performed in the hamlet of Suchiquer Arriba by a Mam-speaking priest who had been invited by a ritual revitalization group among the Ch'orti' (figure 6.3). The cosmological layout and symbolism of the ritual items (candles, cacao, corn, etc.) were largely foreign to Ch'orti' participants. The notion, however, of an emerging, externally defined and taught "cosmology" is problematic. In one of three seminars that were held in the Ch'orti' area (one in 1999 and two in 2004), Metz (2007, 11) noted the difficulty Ch'orti' field laborers had in articulating just what their "cosmology" was, especially in light of the complex definitions given by the visting leaders from other parts of Guatemala (Metz 2007, 10). Oddly, the few remaning Ch'orti' ritual priests today (known as *padrinos*) are not being included in efforts to "re-teach" Ch'orti' cosmology. Metz has notes that "the *padrinos*, with their strong localist vision, have not yet been made

FIGURE 6.3. *A ceremony performed in the Ch'orti' hamlet of Suchiquer Arriba, Guatemala, by a Mam-speaking priest. (Photography by Kerry Hull.)*

aware of the movement. Ironically, national leaders use Girard's books on *cofradias* [religious brotherhoods] to teach the Maya cosmovision to Ch'orti' peasants, while *padrinos* who continue to practice much of what Girard wrote are not invited."[1] Rafael Girard and Charles Wisdom certainly documented the purest form of traditional Ch'orti' ritual practice and cosmology available, yet some parts of that cosmology are just as foreign to modern Ch'orti' as are the eastern Maya cosmologies being taught in the Ch'orti' area today in seminars and ritual revitalization ceremonies.

Notions of heaven and hell, cause and effect, the nature of the gods, and others aspects of Ch'orti' cosmovision have been altered by 500 years of Christian influence in the area. However, the increased Christianization of the Ch'orti' people since the middle of the twentieth century has significantly contributed to the erasure of certain aspects of traditional worldview. Nevertheless, among many rural Ch'orti', as well as among the rapidly disappearing traditional healers and ritual specialists, a formidable retention of native understandings is still observable. The following description of some of the elements of Ch'orti' cosmovision is a collaborative-source effort, drawing upon the earliest extant Ch'orti' sources, together with my own fieldwork data, and that from other recent work of ethnographers, linguists, and anthropologists.

THE EARTH, THE HEAVENS, AND HELL

The earth in Ch'orti' is *e rum* or *u't e rum*, but ritual specialists usually refer to the earth as *Santa Magdelena/Magalena, Santa Madre* ("Holy Mother"), *Mundo Cazadora* ("Earth Hunter"), or *Katu' Mundo* ("Our Mother Earth").[2] Traditional Ch'orti' society views the earth as their "Mother," and they therefore treat it with great respect. For instance, before digging postholes for a new house, the owner prayers to *Katu' Mundo* and asks her permission because the earth will "feel" their shovels and "cry" from the pain (Girard 1949, 222). Also, when a traditional healer needs leaves of a plant for medicine, they first pray and ask if Mother Earth will allow them to take from her plants. In addtion, during the "*promesa*" field ceremony, where offerings are placed into four or five holes in the cornfield, "*Ya uputo'b' ujtz'ub' tya' ak'otoy ucho'b' entregar la limosna . . . I ucho'b' nombrar uk'ab'a' tama e Santa Magalena*" ("There they burn copal when they arrive to deposit the '*limosna*' offering . . . And they name the name of Santa Magalena"), asking for her to ensure a bountiful harvest. The earth is also thought to be the consort of *ijb'en (ikb'en)*, "the male spirit of the earth" (i.e., the god of agriculture, who is also *Katata'*, the sun god). The earth is also known as the "*Virgen doña María del Ombligo*" (the "Virgin Lady Mary of the Navel"), according to Girard (1995, 249), and the hole dug in the center of a cornfield (or other ritual spaces) is said to be her vagina as well as her umbilicus (Girard 1995, 185, 216; Hull 2009, 133). The earth also petitions the gods in heaven for the welfare of humans, whom the Ch'orti' say are "her children." In the informative words of one of Fought's consultants (Fought 1972, 516), "*Che'nob' e katu' mundo uyetojroner e diosob' tichan i ja'x ak'ajtmayan tamaron taka ajtichaner diosob'*" ("They say that the Our Mother Earth is the fellow speaker of the gods in heaven, and it is she who pleads on our behalf with those heavenly gods"; orthography altered, my trans.).

The earth was created before the sun, moon, and stars in Ch'orti' mythology. Water was involved in the creation at various stages and in curious ways. In a creation myth I recorded in 2002, an elderly Ch'orti' recounted that before creating the land, "*Padre Celestial 2000 año ch'a'r atz'i makwe' ja' che*" ("2000 years ago Heavenly Father was indeed lying within the water, they say"), an enigmatic reference, perhaps meaning God was in a watery environment as he created the land (Hull 2016, 12). Later in the narrative it reconfirms that Heavenly Father was "*maku' ja' ani ch'a'r ya*" ("lying inside water there"), either meaning God was creating from a watery place or was in the water himself on earth as he created dry land on top of the sea. God formed the land as "*kay xana uturb'a e rum najtnajt*" ("he went about placing dirt far and wide") (Hull 2016, 18). An account by a consultant of Fought provides more detail (1972, 378): "The Angel San Reimundo went, and took a little earth, and dropped it on the water. And they say that as the earth fell on the water, it swirled all over the water, and the water was covered over by it, and it immediately began to harden like foam on the water. They say that it went on floating on the water, until it thickened all the water of the sea."

From ancient times a core belief in Mesoamerica has been that the earth (thought to be either a turtle, a crocodile, or land on the back of one of them) floated on the sea (Houston et al. 2006, 95; Miller and Taube 1993, 83; Taube 1993, 67). The Ch'orti' likewise envision the earth as a land mass floating like "foam floats on the water when we are cooking" (see Fought 1972, 373, 375) on an enormous sea, or segmented, varicolored seas. The land is sometimes said to rest on the back of a large turtle or crocodile, two animals that are often interchangeable (together with the chijchan snake) in Ch'orti' mythology. The earth connects to the sky at an area of only water. The earth and sky are said to connect at one edge with tar in Ch'orti' thought, just as the wings of a wasp are attached at one point on its back. Fought's (1972) consultant stated, "because they say that where the edge of the world remains, which we live on, and at the edge, they say, there is just water—which is called—the sea of tar. Because they say that it is the sea of tar because they say that there—there they join, the edge of the world and the sky" (Fought 1972, 373). He continued, "because they say that there is a sealed heaven, which we call sky, and the earth. Because they say that the world and the sky are alike—like—the back of a wasp, coming together at a joint" (Fought 1972, 374).

A defining structural principal of Ch'orti' cosmology is the quincunx pattern of four corners in a square and the all-important center, which is considered *umujk e rum* ("the navel of the earth"). The world, indeed the universe as a whole, is said to be supported by four corner posts (*oyob'*), analogous to those found in common Ch'orti' houses. Earthquakes are the result of the earth or its supporting posts moving on the water. The corners of the world are the "resting places" or "seats" of four of the Older Brother Angels (*sakumb'irob' anxerob'*) after they finish bringing the rains for the *b'ajxan pa'k'ma'r* ("first planting season") (Hull 2000). A consultant of Fought (1972, 378) states that angels stand at the cornerposts "holding them" in supportive fashion. Certain high mountain peaks in the Ch'orti' area are also known individually as "the support of the angels" (Fought 1972, 414). These four mountain peaks are also said today to be resevoirs of water used by supernaturals in rain-making activities. Each corner (mountain) of the world is associated with a specific color, a guardian deity, and an enormous basin of water. Girard (1995, 159) notes that the four colors of the corners of the world are red, black, white, and yellow, and that the beasts[3] (called "dragons," "horses,"[4] or "snakes") on which rain-bringing angels ride share those corresponding colors.[5] Each of the four angels is assigned to one of the four corners of the world, although there is some disagreement today as to which angel is associated with which direction.

Home altars, standard features of traditional Ch'orti' houses, are also viewed as cosmograms (figure 6.4). Bowls or candles are commonly placed at the four corners of the altar along with items sacred to the family, such as saint figures, crosses, incense burners, important photos or artwork, and precious family relics (Wisdom 1940, 134). The altars are often covered in *conte* leaves (*Philodendron anisotomum*), a plant with considerable ritual significance among the Ch'orti'. The arch of Ch'orti' home altars is said to represent the sky, and the plain of the altar symbolizes the *milpa*, or cornfield.

A PARALLEL EXISTENCE

An elderly consultant of mine spoke of *upat e mar* ("the underside of the sea"), which designates the watery locus of myriads of evil spirits said to exist under the watery realm of the otherworld (Hull 2009, 189). Due to its watery environment, this zone is therefore a possible portal through which evil spirits can pass to enter this world to afflict humanity. Conversely, another location is said to exist at *upat e glorya* ("the underside of heaven") (Hull 2009, 189). These terms refer to a mirrored existence—worlds parallel to the one we live on. In the watery world below, an elderly consultant of mine stated: "*Kochwa' turu to'r e rum b'an ch'a'r ejmar malob'*" ("Just as they live on the earth also the evil spirits are below"). This "underside of heaven" (*upat e glorya*) that refers to an alternate world to our own is confirmed by data from Fought (1972, 362, 374) in the expression "*tu'pat e sielo tichan*" ("on the back side/underside of heaven above") (orthography altered; my trans.). In addition, the Ritual of the Bacabs contains a cognate concept and phraseology of *pach can* ("the back side of the sky") (Roys 1965, 64; original orthography), a name for one of the directional locations in the otherworld. Fought's Ch'orti' consultant describes *upat e sielo* ("the back side of the sky") as "a town like—just like we see here," except

FIGURE 6.4. *A Ch'orti' home altar in the hamlet of Amatillo above Jocotán, Guatemala. (Photography by Kerry Hull.)*

the inhabitants are all those "who die here" and so "go to live in that place" (Fought 1972, 362, 374). The lifestyle and occupations there are essentially the same as on this earth. Interestingly, above the mirrored, heavenly world above is yet another location where the gods are said to live, that is, "the sky which we see above is not the dwelling-place of God, but that the dwelling-place of God is farther up" (Fought 1972, 374).

Heaven in Ch'orti' is simply referred to as *tichan* (lit. "in the sky"), or by the Spanish term *gloria*. Christian notions of heaven have influenced the native Ch'orti' understanding of this realm, yet much remains of traditional conceptions. Heaven is much like life is here on earth. Antonio Hernández, a Ch'orti' consultant of Girard, stated that "whatever is in the heavens is also on the earth" (Girard 1995, 117). Daily activities continue unabated in the next world. According to a consultant of López García and Metz (2002, 234), "*Allí habían también en ese lugar albañiles y carpinteros, levantando casas, haciendo mesas, haciendo puertas y mujeres moliendo, mujeres tejiendo, mujeres allí haciento ollas*" ("There were also masons and carpenters there, building houses, making tables, making doors and women grinding, women weaving, women there making pots"; my trans.). One difference, however, is that heaven is said to be a gorgeous, aromatic place. "Heaven is described as a place of beautiful landscapes and gardens, with beautiful flowers and fragrant smells, or as a city in the hustle and bustle of constant construction to welcome new souls"[6] (López García and Metz 2002, 234). This indigenous, Ch'orti' notion of heaven as a fragrant, flowery place is well documented throughout Mesoamerica (Taube 2004).

The underworld (at times equatable with "hell") is not a singular location, rather it is conceived of a series of *estados*, or "levels." Evil spirits of differing types occupy these seemingly endless regions. Some of the most commonly mentioned levels in healing prayers are 5, 10, 11, 12, 13, 14, 15, 17, 18, 500, 1,500, 500,000, and 17,000,000. Level 1,500 of the underworld is said to be "*lejos, lejos*," "far, far away," and is where some of the worst, most dangerous ("*más terrible*") spirits live, such as the "bad eclipse spirits." It is, according to a consultant of mine, "the part of the world without end ['*sin fin*']. It is Hell, where San Manuel Rey Pilato is. This is Rey Koludo. They are in the sea there." Some of the more prominent names of evil spirits that dwell in the watery 1,500th level are Don Juan Fabriquero, San Pablo Fabriquero, and Don Pedro Puertero del Mundo, who are responsible for causing diarrhea, cramps, choking, fatigue, and so on by traveling to this world from the underworld by means of waterways.

WATERY WORLD

The Ch'orti' understanding of the universe envisions bodies of water under the earth, and above the earth in the heavens, especially at the four corners of the universe (figure 6.5). The notion of a quincuncial universe, with four corners and a central point, is pervasive in Ch'orti' mythology and is crucial for comprehending the worldview of the Ch'orti' as it relates to water.[7]

For the Ch'orti', the watery landscape of the otherworld is segmented into five enormous, colored lakes at the corners and center of the world, which are usually (but not always) black, white, red, blue-green, and brown (Girard 1995, 160). Two sets of lakes exist, one above the earth and one below it. These bodies of water that are located under the earth and are individually positioned at the four corners of the world, with another at the center (Fought 1972, 373). These colored lakes figure prominently into nearly all Ch'orti' curing ideology. *Curanderos*, or traditional healers, regularly make reference to these lakes in their

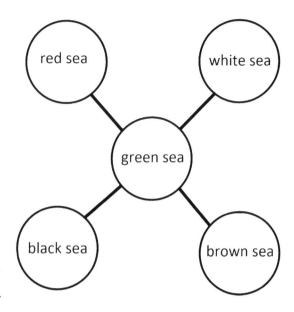

FIGURE 6.5. *The five colored lakes in Ch'orti' cosmology.*

curing rites as sources of disease because the Ch'orti' believe these lakes to be repositories of evil spirits who mill about, spreading illnesses to humans on earth.

The particular action attributed to these water-borne, menacing spirits that cause illnesses is always described as *a'syo'b'* ("they play"). "Play" in this context means "mischievously causing illness" or infecting (Hull 2000). In curing rites, Ch'orti' healers usually ascribe certain illnesses to the influence of spirits from these lakes. For example: "*a'si tamar enyax alaguna, ensak alaguna*" ("they come playing in the green lagoon, the white lagoon"). Another example of a reference to these colored lakes appears in a healing text recorded by Fought (1972, 118): "Perhaps this punishment judgment was dislodged there and came here, rising up in the blue lake, the white lake; in the blue water channel, the white water channel. It joined with the blue Lake Güija, the white Lake Güija." Similarly, "it came playing from the big lake underneath the earth, from the blue lake, the white lake, since all disease-causers play there" (Fought 1972, 169). Specific diseases are associated with different colored lakes. Dysentery (*k'uxnak'*) is said to originate in the *mar de sangre*, or "sea of blood." Fought's consultant describe it as a sea of "pure blood, perfectly red" (Fought 1972, 354). If one's feces are white in color, the evil spirit who caused it is thought to have originated in the *mar de leche* ("sea of milk"). Other forms of dysentery are said to come from the *mar negro*, the *mar de brella*, or the *mar de verde* (the "black sea," "sea of tar," and "green sea," respectively).

In addition to "bad seas," the Ch'orti' also envision a parallel set of seas in which "good" angels "play." The Ch'orti' label these with the triplet *mar divino, mar dichoso, mar sagrado* ("divine sea, lucky sea, holy sea"). In fact, the Ch'orti' believe that there are both "good" and "evil" parallel existences in the otherworld. Of this dual existence I have elsewhere written:

> For the Ch'orti', most Otherworld beings, including disease-spreading spirits, have both a male and a female counterpart. This duality can take the form of a good and evil equivalent for each heavenly being. An illustrative example of this dualism in Ch'orti' mythology is San Antonio. San Antonio, the God of Fire, while usually feared as one who causes fevers, burns down houses, and causes all sorts of other illness if one falls near a fire, has a good counterpart who is one of the four "princes of the angels of God" ("*principos[8] de ángeles de Dios*"). (Hull 2003, 96)

Examples of gender dyads appear everywhere in Ch'orti' ritual prayers. For example, "*A'si Don Leoniso, Doña Leonisa*" ("Don Leoniso and Doña Leonisa play"). This pair of beings resides in the ashes of fires and are said to be responsible for illnesses that afflict people who fall near a fire with ashes. Similar paired gender terms for otherworld being in Ch'orti' mythology illuminate our understanding of Ch'orti' gender conceptions and worldview (see Hull 2003, 145–148; 2004; 2009).

These malevolent pairs of disease-causing spirits are able to have an influence on humans in curious ways that relate directly to water. A consultant from Tunucó of Palma Ramos (2001, 132) warned: "That is why you have to be afraid of the water currents, because in a stream of water there are many murdered souls, drowned, killed, shot. And all that comes depending on water and the justice of the water. As it is raining now, when the angels of heaven come, there are thousands of plagues, so you have to protect yourself" (my trans.).

Evil spirits, in Ch'orti' thought, can manifest themselves (or their power) at numerous locations on the earth considered to be *vivo* ("alive") in a very special sense. This notion of *vivo* is of particular interest in these contexts. For the Ch'orti', certain items are imbued with a sacred, living force that make them dangerous to humans. Such objects are termed *vivo*, which refers to (1) their infusion with a harmful spiritual energy that emanates from the "playing" of malevolent spirits there, or (2) a spiritually active place where otherworld beings or their influence can exist. These areas are portals between realms through which these beings travel. In most cases, being *vivo* relates to the fact that one can get infected (*t'e'nsan*) by an evil spirit through contact with or close proximity to these objects or places. While in most instances the object is viewed as inherently *vivo*, it is also possible for a place to become infected with this type of spiritual energy (Hull 2003, 96–98, 164, note 6). Without mentioning the term *vivo*, Girard likewise enumerated various locations that are especially active with spirits, all of which are precisely places that my consultants today describe as *vivo*: "Legends that refer to incantations abound; the hills, water courses, pools, mountains that come together, legendary lagoons, enchanted caves, canyons, any irregular terrain is a mystical place populated with spirits for the native eyes" (Girard 1949, 333; my trans.).[9]

Houses and house compounds have several areas that are viewed as *vivo*. The eaves of a house are particularly dangerous since the evil spirits are said to "play" there. The eaves of a house become ever more dangerous when it begins to rain. Ch'orti' children are taught not to play under the eaves of a house during a rainstorm since they may get struck by lightning. Even on a sunny day, if a child falls while under the eaves of a house, the Ch'orti' say the child may fall ill.

Since in Ch'orti' mythology the otherworld is a watery place, all bodies of water on earth are thought to be potentially *vivo*. Lakes, rivers, water troughs, puddles of water, and even washing areas are highly charged and dangerous. While washbasins (*pilas*) and washing stones are items used daily by any Ch'orti' family, they are also respected and feared, since evil spirits are said to be constantly "playing" (*a'si*) in or around them. Specifically, *Doña María de Lavandera* (also called *Doña María de Estumeka Espumador*) is the evil spirit who resides in washbasins and is responsible for causing sicknesses related to them. *Niño Estumeka del Mundo* is a title for other menacing spirits who "play" in washbasins and other watery locations. If it begins to rain, a Ch'orti' woman will stop washing clothes, since there is a good chance she will be struck by lightning if she continues. Children who trip and fall near a washbasin are quickly taken to traditional healers since they will surely become sick within a few days. A washing board or washing stone is similarly dangerous. If a child falls near one, or if a woman becomes angry around one, a child could fall ill. The malevolent spirit that resides at the washing board carries the title *Espumador*, since this is where soap bubbles up, resembling a sickness called *sakb'ub'* or *sakb'uk*. *Sakb'ub'* is described as *un mal deseo, un calor de mal deseo* ("a bad desire, a heat of bad desire") and is a kind of lust that emanates from people who have been infected. The result of *sakb'ub'* is a sore throat that eventually causes a kind of frothing at the mouth in young children.

In addition, gullies and valleys are said to be *vivo* since they are conduits of water. Rivers, streams, and creeks are likewise *vivo* and therefore are prime locations for evil spirits at "play." The Ch'orti' believe that the spirits of those who have been murdered, called *mal aires matados*, and the spirits of those who drowned in water, called *tijtijutir ausente ahogado, espanto de ausente matado ahogado*, especially haunt areas of running water. According to the Ch'orti', if a person gets a *susto de agua* ("water fright") while near a river, it will cause a swelling of the stomach that could be fatal if it ruptures. In other cases, these spirits are said to grab and drown unsuspecting individuals who get too close to the deep sections of a river (Wisdom 1940, 425). The drainage ditches or gutters (*corrientes*) that are dug around Ch'orti' homes to redirect water away from the house during rains are also *vivo*, and one can get diarrhea from the evil spirits who play here. Even clouds can be dangerous, since the Ch'orti' say that malicious spirits commonly travel and "play" *dentro de las nubes verdes* ("within the green clouds").

GOD, THE SUN

The sun, *k'in* or *katata'* (lit. "our father"), is a major Ch'orti' deity, one to whom prayers are offered just as its rays begin to shine in the early part of the morning (Fought 1972, 527n34.5). A consultant of Metz (2006, 125) stated: "If we don't take care of him, he won't shine every day." The sun is said to be composed of "fire" (*k'ajk'*), and the only reason the earth is not consumed by its heat is that there is a "sacred wind" (*ik'ar sagrado*) in the sky, a dry wind distinct from the wind that brings rain, that forms "a great shelter" between the realms (Fought 1972, 376).

The sun is male, and its consort, the moon, is female in Ch'orti' thought. During the new moon each month the Ch'orti' understand that the sun and moon are "hidden" and are having sexual relations for several days (Hull 2000). A consultant of Fought (1972, 430–431) elaborated that the moon *unut'u ub'a taka e k'in* ("joins itself with the sun"; orthography altered). The sun is also an avatar of the fertility god. Therefore, it is no coincidence that the name of the sun, *Katata'*, according to López García and Metz (2002, 202; cf. Metz 2006, 125), also means *la bendición* (lit. "the blessing"), that is, the rains or clouds. In his agricultural aspect, the sun, as Girard's consultants explained, "impregnates" the earth when it reaches its zenith at the center of the sky, thus "making it fertile for cultivation" (Girard 1995, 58–59).

The sun god also bears the title "The Seven," a designation that Girard attributes to the fact that from "his celestial abode to the earth there are only seven steps" (Girard 1995, 59). The sun is commonly referred to by a myriad of appellations that include the number seven: for example, *El Señor de las Siete Pilas, de las Siete Palabras divinas, de los Siete imaginarios [sic], de los Siete Oidores del mundo, de las Siete disposiciones* ("The Lord of the Seven Fountains, of the Seven Divine Words, of the Seven Imaginaries [*sic*], of the Seven Listeners of the World, of the Seven Dispositions"; Girard 1962; 1995, 59). Today, traditional healers (*curanderos* in Spanish, *ajnirom* or *ajk'in* in Ch'orti'), generously employ these and other "seven" titles in healing prayers to the sun god—the patron god of healing (see Wisdom 1950, 399).

The sun has numerous other names in Ch'orti', many of which correspond to different stations or time segments relative to its "position," as one consultant of mine noted, in the sky. Girard notes that these "positions may be multiple" (Girard 1995, 59). I have documented names for the sun such as *San Gregorio, Niño San Gregorio, Niño San Gregorio de Cristo, Niño San Antonio Revistador Pastoreador, Niño Revistador del Mundo, Niño Jesús, Padre Jesús*, as well as a dozen others in healing prayers. Fought gives the clearest explanation of the meaning of these various epithets for the sun:

> They say that the Sun has not just one name. The one which he is best known by people continues to be Jesus Christ. They say that when it is just getting light its name is Child Redeemer of the World. One name is San Gregorio the Illuminator. One name is San Antonio of Judgment. One name is Child Guardian. One is Child Refuge. One is Child San Pascual. One is Child Succor. One is Child Creator. They say that at each hour, one of these is its name.
> (Fought 1972, 485)

Each "position" of the sun represents a different hour of the day with its individual name. Fought also describes a festival that took place on March 3 celebrating the "the birth of the sun" (*uk'uxna'r e k'in*), or the beginning of cycle of names (Fought 1972, 485). The first "position" is the "birth" of the sun, and by night the sun is said to "die" (Girard 1962, 266). The individual names of the sun given by Fought's consultant were "given in a fixed order by Damasio López," while Fought's consultant González "knows only that the Redemptor comes first, before the actual sunrise" (Fought 1972, 527n34.6–7). In my experience, however, the exact order of all the names associated with the individual stations has largely been forgotten by most Ch'orti' today.[10]

THE MOON IN CH'ORTI' CULTURE

While the sun and related solar deities are known to dominate the pantheons of multiple Maya groups, as well as with the ancient Maya, the moon also plays a pivotal role in many of these societies today. As noted above, the sun in Ch'orti' thought is male and is married to the female moon. The Ch'orti' overtly relate the strong rays of the sun (*uwarar e katata'*) to a stricter father-figure, and the soft beams of the moon (*uwarar e katu'*) to the more gentle personality of the female moon (Hull 2000). Indeed, the moon in Ch'orti' is known by several names, all referencing a female: *katu'* (lit. "Our Mother") and *kanana'* (lit. "Our Mother"). The sun, conversely, is referred to as *katata'* (lit. "Our Father"). It is the moon, however, that is closely tied to the production of rain.

The phases of the moon are carefully observed, as they have a direct bearing on agriculture and sexual reproduction. A new moon is referred to as either *satar e katu'* (lit. "the moon is lost") or *ch'ok e katu'* (lit. "the moon is young"). The Ch'orti' say that during a new moon, *la tuna tiene un repozo con el sol* ("the moon takes a rest with the sun"). A new moon also represents a dangerous time in Ch'orti' society. For instance, bathing can be harmful during this three-day period since bath water is said to become "contaminated" during a new moon. In addition, farmers know to not plant during a new moon, or, according to my consultant, "the roots of the plant will not be strong," which would result in poor plant growth. It is better to wait three to four days after a new moon during the waxing of the moon to ensure the plant's strength. Also, if one cuts plants during a new moon, the fruit of those plants will not grow well, and it can damage the plants in other ways. Waiting two days later is still dangerous, but three days is said to be probably safe.

There are differing opinions relating to having sex during a new moon. Some say that having sexual relations during a new moon is perfectly acceptable since the moon herself "is going to give birth at that time, too." In addition, they say that since the sun and moon are having sex during a new moon, humans can therefore also do it. Conversely, others say that having sex during a new moon will result in a child with birth defects because of this *descuido* ("lack of care"). López García and Metz (2002, 202–203) note that the moon typifies women in their reproductive cycles, and that the moon becomes *fría* ("cold") and without energy (that is, low on blood) while it is in the new moon phase, just as a woman is weak after blood loss after menstruation.

The waxing quarter moon is referred to variously as *t'erpi'x e katu'* (lit. "the moon is now titled"), *oraj*[11] *ch'ok e katu'* (lit. "the young/unripe moon is 'central'"), *oraj e katu'* (lit. "the moon is 'central'"), and *ch'okix e katu'* (lit. "the moon is now young/unripe"[12]). This is a time when planting is possible, and a respectable result can be expected. You can begin to cut palm plants anytime from the eighth day to the full moon, but closer to the full moon is better. Sexual relations, they say, are also acceptable during this time. Interestingly, at the time of the waxing quarter moon, castrating animals is prohibited. From the third to the seventh days after a new moon, castration is done; however, on the eighth day, the blood of the animals is said to be too "simple." Thus, castration would pose a particular danger to them on the eighth day due to possible uncontrollable blood loss.

There are three terms for "full moon" that I have found among the Ch'orti': *b'ut'ur e katu'* (lit. "the moon is filled"), *k'in katu'* (lit. "sun-moon"), and *ne'p katu'* (lit. "the moon is semi-ripe"). The rare verb *pechka* or *pechk'a*[13] is an intransitive verb meaning "be a full moon" (Hull 2016, 325). The full moon is a highly productive time for the Ch'orti', when planting of all kinds is at its best. Planting done under a full moon is said to guarantee the best possible growth of the plant. The moon, according to the Ch'orti', actually fortifies plants and helps them to grow stronger roots. Similarly, sexual relations during a full moon also ensure the strongest possible baby. However, any time from the tenth to the seventeenth day is good overall. The Ch'orti' say a child conceived during a full moon will grow to be strong and healthy (Palma Ramos 2001, 55).

THE MOON AND RAIN

The Ch'orti' rely on celestial phenomenon to understand and predict the cycles of rain. For example, the moon is one of the most reliable means for foretelling the coming rains. The moon is referred to as the Water Goddess (Girard 1979, 54) and as *katu'* ("Our Mother"). During the months of March and April the Ch'orti' look to see if the moon is level or tilted in the sky.[14] If the moon is tilted (*t'erer ch'u'r*), it is a signal that the moon will be "bringing" (*uk'eche*) the rain. If the moon does not appear in a tilted position, it is a bad sign that the rains will not come (Hull 2000). The moon is thought by the Ch'orti' to be like a pot of water that will spill rain onto the earth when tilted, a belief shared by other Maya groups (Milbrath 1999, 29). For example, among the Tz'utujil, when the moon tilts over Ch'umil volcano, it is said to be "bathing," and the drops from pouring water over herself fall to the earth as rain (Tarn and Prechtel 1986, 176).

ECLIPSES

Throughout Mesoamerica the event of an eclipse (*kilis*) is commonly interpreted as the sun or moon being "eaten" by a celestial beast. For example, a colonial Zapotec text from 1693 records an eclipse as *miercole tza niga bitago beoo bisabini 21 enero año de 1693* ("Wednesday. On this day, the moon got eaten [eclipsed]. It floated in the air. January 21, year of 1693"; Tavárez and Justeson 2008, 73). The Yukatek Maya likewise refer to a solar eclipse as *chi'bal k'in* ("eaten sun") (Barrera Vásquez et al. 1980, 93; see Redfield and Villa Rojas 1934, 206). The Ch'orti' say that during a solar eclipse *e katu' uk'uxi e k'in* ("the moon eats the sun"). Sometimes the eclipse is described as a dark cloud with *boca abierta* ("an open mouth"). More commonly, it is thought to be a beast, either a large snake or feline. A Ch'orti' consultant of mine stated that an eclipse is caused by *el diablo con cola* ("the devil with a tail"), and that this devil "eats" the moon during a lunar eclipse. During an eclipse, they say the moon *uchompati e pak'ab' e konoj twa' ma'chi ak'ujxob' umen e kilis* ("offers up humans in exchange so that it will not be eaten by the eclipse") (Fought 1972, 425; orthography altered, my trans.).

Wisdom records the name of eclipse as *ah kilis* (*ajkilis*) (lit. "He of the Eclipse"), a name still used today, which he defines as "a deity said to eat the sun and the moon (thus causing their eclipse)" (Wisdom 1940; 1950). Wisdom also notes that *ah kilis* eats the sun "when angry" (Wisdom 1940, 399). In my experience, a commonly heard name for the eclipse today among traditional healers is *Eclipse de Ángel*, meaning the "Eclipse Angel." A traditional healer and consultant of mine, Don Lauterio, named the eclipse evil spirit as *Ángel Kilisante de Mal Hombre*, or *Eclipse de Ángel*. This evil spirit associated with an eclipse (*ajkilis*) infects bodies of water during an eclipse so that one cannot bathe or drink water anywhere because it has been "contaminated." Other associated evil spirits that cause sickness during an eclipse are *Doña María de Eclipse*, *Don Juan de Eclipse de Ángel*, and *Don Pedro de Eclipse*.

A woman who is close to giving birth when there is an eclipse is said to be in considerable danger. Furthermore, if a child is born during or near the time of an eclipse, it risks being *kilisado* ("eclipsed"), which means it could be born without hands, with a deviated septum, or with malformed ears due to the evil spirits associated with eclipses.[15] Since women are closely associated with the moon in Ch'orti' society, lunar eclipses are particularly hazardous because the event creates a disruption in the "hot/cold" balance of the body just as women are said to be in their highest moment of *calor* ("heat") (López García and Metz 2002, 203). The moon, patron of women, is said to "lose partially her powers of fecundity" when eclipsed, which is why this is such perilous time for pregnant women (Wisdom 1940, 399). Therefore, miscarriages or birth defects for the child are said to be common results after a lunar eclipse.

An elderly consultant of mine from the hamlet of Amatillo related the following belief regarding the danger to a pregnant woman of an eclipse and what preventative measures are taken:

In older times the moon would be eclipsed. So what the ancestors would do is they would go outside with their candles. They prayed in order to not be harmed by the moon, they say, the moon, they say. Then whoever was pregnant would have a cord tied around her stomach so that they [the evil spirits] would not see it also, so that the child would not be born [with defects]. They would tie it so that they would not be harmed by the eclipse. So they did not see it [the baby]. "Go on, kids" they use to say if they looked at the sun when it was being eclipsed. They used to say it because it would eat them or the children would get an impediment. The now-deceased people, our poor grandparents used to say it. That's right. (my trans.)[16]

The *kordon* ("cord") mentioned above that is tied around the expectant mother's waist is always red (cf. Paul and Paul 1952, 177). The color red in Ch'orti' tradition has a mysterious protective power against the harmful influences of an eclipse, or that of a sorcerer's spell. My primary consultant, Hipólito Ohajaca Pérez, clarified the importance of the color red during an eclipse as well as the belief that the sun is "eaten" during the event (Hull n.d.b.; my trans.):

1 *K'ani ink'ajti inte' k'ub'seyaj twa' e pak'ab'ob' onya'n Ch'orti'.*
   I want to relate a belief of the old Ch'orti' people.
2 *E k'ajpesyaj ira watar konde e k'in iksijb'a o konde kawa're e k'in ak'ujxa umen e kilis.*
   This belief comes when the sun darkens, or when, as we say, the sun is eaten by an eclipse.
3 *Oni'x ani kanoya konde kuchur uyarob' konde uwiro'b' ani e k'in iksijb'a o e katu' iksijb'a ub'an umen e kilis ukachi unak'ob' wa'kchetaka tante' chakchak payuj twa' yer e chuchu' xe' kuchur umenerob' tu'nak'ob' ma'chi ak'ujxa uti' i ayo'pa inb'utzir uyerojir.*
   Long ago our grandmothers, when pregnant, when they would see the sun darken or the moon darken also because of an eclipse, they would quickly tie up their stomachs with a red cloth so that little baby that was carried by them in their stomachs would not be eaten, and that it would be born without facial defects.
4 *B'anto ani ucho'b' konde uwiro'b' ke' e katu' i e k'in war ak'ujxa umen e kilis.*
   Thus would they always say when they saw the moon and the sun were being eaten by an eclipse.
5 *Ja'x ukacho'b' wa'kchetaka e chakchak payuj tu'nak'ob'.*
   They would quickly tie a red cloth on their stomachs.
6 *B'anixto ani koche'ra uk'ub'seyajob' te' onya'n pak'ab'ob' tara.*
   Thus were the beliefs of the ancestors here.

This use of the color red to guard against the harmful effects of an eclipse is common throughout Ch'orti' communities today. A possible visual representation of both the mythic beast that causes the eclipse and the use of red cloth to protect against it may be found on a Late Classic polychrome Maya vessel (Kerr n.d., 5359).[17] The scene shows an open-mawed snake rising to the sky about to bite an "eclipse cartouche" enclosing the heads of the sun god and moon goddess. All four individuals in the scene hold their hands up towards the sky while wearing red loincloths, possibly in defense of the dangerous effects of the eclipse (Hull 2000).

In addition, as noted above, red is also effective in blocking acts of sorcery. The Ch'orti' say that the Eclipse Angel (*Eclipse de Ángel*, lit. "Angel Eclipse") is an evil being whom sorcerers (*brujos*) call upon in order to cast spells on others to kill them. However, the color red can protect them against such curses. For example, Wisdom states that the Ch'orti' place a red cloth over a peppermint plant in order to protect it from sorcery (Wisdom 1950, 151). Likewise, someone who has been cursed with an "evil eye" wears a collar of the red beans of the corral bean plant (*Erythrina rubrinervia*) for spiritual protection (Girard 1947, 353). A consultant of Palma Ramos (2001, 93) mentioned that when one has a type of bewitchment known as "evil eye" (*mal de ojo*), a traditional healer has them put *un trapo rojo en la cabeza* ("a red cloth on the head") for protection.

CLOUDS, THUNDER, AND LIGHTNING

Clouds

The production of rain is a multistepped process involving numerous supernatural beings in Ch'orti' thought. It begins with the Chijchan snakes, who are known as the *dueños de la lluvia* ("owners of the rain") (López García 2010, 89; see also Dary et al. 1998, 249–251). The *Noh Chih Chan*, or great Chijchan serpent that resides in the center of the earth, is the one who, according to Girard's consultant, "either releases or doesn't release the water" from the sacred spring (Girard 1995, 115–116). Once dispersed, the water goes up in the form of vapor to form clouds (*tokar*). Girard states that "Chortí gnostics" believe "that clouds are supplied with water from the cosmic 'basins'" (Girard 1995, 25), usually by the Chijchans. A cloud itself is also viewed as *la jarra de los Ángeles* ("the jar of the angels") (López García 2010, 113; see also Fought 1972, 413), which is why angels are continually petitioned to "pour" their cups of water onto the earth: "*Ángel San Gabriel, vierta, vierta el agua*" ("Angel San Gabriel, pour, pour out the water") (López García 2010, 113).

Thunder

Thunder is said to be the beating of the drums[18] of the rain-bringing angels (either one of the Elder Brother Angels or the Kumix Anxer). Girard adds that these celestial drums are "of gold and silver" (Girard 1995, 154). Wisdom likewise states that "the Working Men," the rain-making angels, "beat the clouds into rain and who produce thunder and lightning" (1940, 439). Thunder can also be understood as the shouts of the Chijchan snakes in different parts of the sky (Wisdom 1940, 394).

Lightning

On lightning among the Ch'orti' I have elsewhere written:

> Lightning is much feared and numerous beliefs are associated with it. The Ch'orti' see lightning as the "machete of god" (*umachit e katata'*) that is wielded by various angels (*anxerob' e katata'*) working under the auspices of God. Lightning is generally known as *jijb'ya'r*. However, special lightning bolts contain small, sharp stones on their tips so that when the angels of God throw them to the earth the Ch'orti' believe you can go to the strike spot and find the small flint point (also called "la hacha de dios"). These dangerous bolts are called *senteyo*, a term specifically used for the powerful, deadly lightning bolts that are accompanied by thunder (which, they say, is caused by the angels playing their drums or by angels chopping at the clouds with their machetes). (Hull 2000)

Girard records that thunderbolts are known as a "dagger of gold, dagger of silver," and Girard's consultant Hernández states the rain gods "temper the clouds and strike them so they will release the rains" (Girard 1995, 155). Furthermore, thunderbolts, according to Girard's consultants, "cut the layer of clouds" (Girard 1995, 162)—a direct reference to the slashing movement of the rain angels' swords. Lucío García Onofre from Tesoro Abajo, Jocotán, recounted that "the angels, mounted on the dark horses, formed dense, gray clouds; the men, mounted on the dappled gray horses, formed whitish clouds. Each time they handled their swords, they formed lightning and thunder" (Dary et al. 1998, 251, my trans.).[19]

Lightning is especially feared beginning on September 1 because the Younger Brother Angel, Kumix Angel (*Kumix Anxer*), begins his work, and so there is a much better chance one will get struck by a lightning bolt at that time. Lightning can also be used to target certain creatures on earth, such as snakes and scorpions (Hull 2000), as well as Chijchan snakes (Wisdom 1940, 396). Fought also notes that rain-bringing angels fire thunderbolts at evil spirits (Fought 1972, 436n28, 518–519).

More than simply the fear of being struck by lightning, the Ch'orti' also ascribe spiritual dangers to lightning. It is believed that one can get possessed by an evil spirit if lighting strikes nearby because it will "weaken your spirit" due to fear, making you susceptible to harmful spirits. In such cases a *curandero*, or traditional healer, must be called to rid one of this *espanto* (*b'ajk'ut*) or "fright" (see Wisdom 1940, 319; for Mesoamerica more generally, see Wisdom 1952, 130).

Lightning can also be a sign regarding rain. Excessive summer lightning flashes in the sky can signal that months of dry weather are at hand. However, if there is lightning in the north and south simultaneously, it means that two angels are fighting with their swords, the one in the south to bring the rain and the one in the north to stop the rain. Fortunately, they say, the southern angel will always win in this case, resulting in abundant rains.

RAINBOWS AND WATERY REALMS

According to Wisdom, a rainbow is the body of a Chijchan, a mythical horned serpent of Ch'orti' lore, "stretched across the sky" (Wisdom 1940, 394).[20] For the Ch'orti', rainbows (*makchan*, lit. "blocking-snake")[21] are also watery loci teeming with evil spirits[22] "playing" (*a'si*). A traditional healer I recorded once stated that *a'si tamar enyax arcoíris, ensak arcoíris* ("they play in the green rainbow, the white rainbow"). Rainbows were so feared in times past that elderly Ch'orti' today say their ancestors used to chop at a rainbow with a machete in an effort to kill the *mal brujo* ("evil sorcerer"). Also, persons finding themselves hexed through sorcery must *pagar un tributo* ("pay a tribute") to the rainbow, but all the while being careful never to point at it or *se le pudriría la mano* ("their hand would rot off") (Girard 1949, 323). A widespread belief among the Ch'orti' is that one should not point directly at a rainbow lest one's finger rot away. Rainbows are said to have such deforming powers because they are repositories of harmful spirits. For example, a consultant of mine from Amatillo related that if he ever dared point (*tuch'i*) at a rainbow, *o'k'oy o'r nik'ab'* ("my finger would rot off").

The tradition stating the danger of pointing at a rainbow has deep roots among various Maya groups and throughout Central America. For example, the Kaqchikel view the rainbow as a sinister snake (as do the Ch'orti'), and so, according to Thompson, mothers in the past "warned their children that if they pointed at it, they would have distorted fingers" (Thompson 2001, 138). The Tz'utujil similarly believe that if one points directly at a rainbow *se le tuerce el dedo con que se le señala o la mano completa* ("the finger with which one pointed or the whole hand will get twisted out of shape"; Tz'utujiil Tinaamitaal 1998, 68, my trans.). Some among the Tz'otzil Maya of Chiapas say one's bellybutton will rot if one points at a rainbow (Past 2014, 44), while others in Zinacantán say it is the tip of the finger that rots away (de León Pasquel 2005, 151n14). Similar traditions exist in Guerrero and Oaxaca, Mexico (Neff 1994; 1997, 37–38; Katz 1997, 117), and even as far as south as the Peruvian Amazon (Murúa 1986, 438–439).

Rainbows are sometimes a sign that the rain will stop soon (Fought 1972, 431), or are viewed as inhibitors of the rain: *e makchan ma'chi uyakta ak'axi e jaja'r* ("the rainbow doesn't let the rain fall") (Fought 1972, 383, 388; orthography altered). Another type of "rainbow" is said by the Ch'orti' to indicate the coming rains. Known in Spanish as the *cabañuelas*, this practice, ostensibly imported from Spain, predicts the weather for the coming months of the year based on counting[23] and comparing them to the first 8, 10, or 12 days of January. For the Ch'orti', as Girard observed, the *cabañuelas* are *los pronósticos anunciadores del invierno . . . durante las lloviznas de enero* ("the forecasters of the rainy season . . . during the drizzle of January") (Girard 1949, 335; my trans.). The Ch'orti' also relate the *cabañuelas* to a specific meteorological phenomenon during that time period—the formation of long, snake-like clouds. They say that when two *arcos* ("arches")—that is, long, white, bending clouds—go east onto the mountaintops, it means rain is coming. However, if the clouds veer more toward the middle of the sky or onto the west mountains, it means that the rains will not arrive. The Ch'orti' compare these arched clouds to a *makchan* or rainbow. One of my Ch'orti' consultants related more about this tradition:

> Understanding about the rainbow clouds that lie in the sky for a sign of the rain. I also want to tell about the belief about whether we know if it is going to rain soon, quickly, or if it is going to last a long time. And in the sky here we, where we live, we see when there are not a lot of clouds in the sky, we see a rainbow lying there that is circular. We see it double in its appearance lying in the sky, this rainbow lying. It brings white clouds. If we see that it is really white, or there are lots of clouds lying in the sky, we say that it is going to rain, that this rainbow brings the rain. Now the rain won't be far off. And it is going to rain a lot; it's just that the day will not arrive, we say. Then we are seeing that this rainbow lets us see that the rain is going to fall. And when, or when we see that it doesn't bring a lot of clouds, this rainbow is very simple. It lies very simple in its appearance in the sky. We say it is not going to rain quickly. It [the dry season] is going to last. It's going to be very dry. Now there is still no rain because we see that the rainbow doesn't bring the rain. Therefore, in this way was our belief, and we also see if it is lying on the mountain horizontally, we say that the rain is now coming. (Hull n.d.c.; my trans.)[24]

A third type of "rainbow" among the Ch'orti' is sun dogs (parhelion) or moon dogs (paraselene), the colored rings around the sun and moon. In Ch'orti' thought, they are watery zones that occasionally appear days before a rainstorm. In addition, however, Ch'orti' healing texts make constant reference to the watery rings as places where malevolent spirits are thought to "play," often before coming to this world (Hull 2000; 2003, 516–520). The red-colored part of this *pila* ("trough"), according to a consultant of Fought (n.d., field notes #300–354), is what causes bloody noses and dysentery. Note these examples from Ch'orti' curing prayers I recorded in Oquen, Guatemala:

Example 1:
A'syo'b' wato'b' uruedir te' Katata',
   tamar uruedir te' Reina.
They [evil spirits] come playing (in) the wheel of the sun,
   in the wheel of moon.

Example 2:
A'si takar e Niño San Gregorio de Cristo,
   takar e Niño San Gregorio de Cristo.
A'si tamar e mediante cielo,
   tamar e mediante de la gloria.
They play with the Child Saint Gregorio of Christ,
   with the Child Saint Gregorio of Christ.
They play in the middle of the sky,
   in the middle of the heavens.

In Example 1, the term *Katata'* ("God") is an appellative regularly applied to the sun. The mention of *Reina* refers specifically to the moon in her role as "Queen" of the sky. In addition, in Example 2, *San Gregorio* or *Niño San Gregorio* are both versions of the name most commonly attributed to the sun by traditional healers. In all of these examples, the evil spirits are said to be playing in the watery rings (variously referred to as lagoons, rainbows, or wheels) that at times surround the sun and the moon.

During a discussion with a Ch'orti' healer in 2000 about sun dogs he clarified, "God is lying down in a ring. There God is indeed being bathed. It is going to rain." Several other healers have made similar observations to me using this *exact* terminology referring to the sun: "God is being bathed, it is going to rain." The rings around the sun represent water in which God (the sun) is said to be "bathing." The same is true of the moon, according to a consultant of mine: *La luna cuando tiene la rueda alrededor, trae un señal que va llover fuerte, muy furioso* ("When the moon has a ring around it, it brings a sign that it is going to rain heavily, very furiously"). Our scientific understanding of these rings decrees they are the result of light being refracted on water or light ice crystals floating in the upper atmosphere in cirrostratus clouds that usually precede rainstorms. The Ch'orti' interpret these rings as "troughs" (*pilas*) whose presence clearly signals the coming rains (see also Fought 1972, 267).

The understanding of sun dogs and moon dogs as "troughs" of water may provide us with crucial details in understanding similar Classic-period Maya conceptions. In 1999 Søren Wichmann and I independently discovered that the ancient Maya considered sun and moon dogs the "bathing" of those astral bodies (Hull 2000; Wichmann 2004).[25] I refer specifically to the images on Jimbal Stela 1 (figure 6.6) and Ixlu Stela 3 (figure 6.7a). Both stelae show images of the Classic Maya deities known as the "Paddler Gods." I have argued that the Paddler Gods can represent an aspect of the sun and the moon (Hull 2000). Milbrath has also similarly concluded that the Paddler Gods "may represent a conceptual pairing of the sun and the moon" (Milbrath 1999, 130). This is particularly likely in my mind, since their name glyphs are commonly substituted for the signs *k'in* ("sun, day") and *ak'ab'* ("night, darkness"). In both scenes on Ixlu Stela 3 and Jimbal Stela 1, the Paddler Gods are shown surrounded by *muyal* ("cloud") markers in the form of a dotted-scroll motif. According to Stuart, Houston, and Robertson (1999, 169–170), the scenes on these two stelae may depict the Paddler Gods involved in a "bathing" (i.e., rain-making) ceremony or possibly a purification ritual. The glyph that often appears with the Paddler Gods in this context reads **ya-AT-i**, *y-ati* ("they bathed them")[26] (David Stuart, personal communication 2000; see also figure 6.7b).

Another possibility is that the compound is translated nominally as "it is the bathing of [the gods, the Paddler Gods]." Stuart et al.'s reading of the verb *at-* in these inscriptions as "bathe" is significant, and allows us to deduce from these texts and images that the Classic-period Maya saw the rings around the sun and moon as watery places in which deities were "bathed." In the words of a consultant of mine, a Ch'orti' traditional healer, who stated, *e Katata' war a'tesna. K'ani ak'axi e jaja'r* ("God is being bathed. It is going to rain"), we see that the same verb *ati* "bathe" is used to describe the watery rings ("basins" or "rainbows") around the sun and moon—which I would argue are the watery clouds that surround the actors on Jimbal Stela 1 and Ixlu Stela 3.[27] On Jimbal Stela 1, both Paddler Gods carry a caption, the leftmost of which reads *ubaah naah-ho'-chan chahk* ("it is the image of the Jaguar Paddler in Naah Ho' Chan, Chahk"). It is

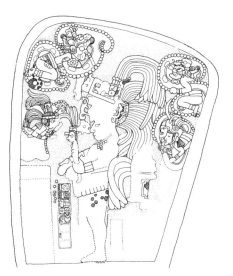

FIGURE 6.6. *The Paddler Gods and other deities being "bathed" in watery clouds. (Stela Jimbal 1 SD-2029; drawing by Linda Schele © David Schele.)*

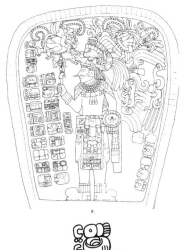

FIGURE 6.7. *(a) Two Paddler Gods in the act of "bathing" in watery clouds. (Ixlu Stela 3, SD-2054; drawing by Linda Schele © David Schele.) (b) The glyphic compound reading* yati, *related to "bathing." (Altar to Stela 1, Copan, CPN 39, Block L; drawing by Asa Hull, after drawing by Linda Schele.)*

of considerable interest that the Maya rain god, Chahk, is mentioned in both captions, confirming the close association of these watery rings with rain.

CONSTELLATIONS

Several constellations are recognized and have some cultural significance. The Pleiades are known as *Mormorak'* or *Mormorek'*, the latter meaning "group of stars" (*mormor* "group" and *ek'* "star"). Girard also notes the name *Cabrillos* ("Little Goats") for the Pleiades. Based on its position in the sky, this constellation figures into the calculations of when the first rainy period will begin (Girard 1995, 183; for a related belief among other Maya groups see Milbrath 1999, 258). Known and recognized among the Ch'orti', the Southern Cross serves a similar function (Girard 1995, 100; 1962, 78), although there are few

A COSMOLOGY OF WATER 233

associations with it today. Also known as "La Cruz de Mayo," it is visible in the month of May (*anumuy tama uyuxinar e jab'* "it passes in the middle of the year").

While little is known of the Milky Way today, Metz (2006, 130) does note that the Ch'orti' refer to it as *chan* ("snake") and believe it is an "unpropitious sign of dryness." Girard similarly affirms that the Milky Way is conceived of as a white snake (Girard 1979, 170). Girard states that Ch'orti' priests call the Milky Way *el camino de Santiago* ("Santiago's road") and that *Santiago gobierna tanto la Vía Láctea como el arcoíris, comparados ambos a gigantescas serpientes que se mueven en el firmamento* ("Santiago governs both the Milky Way as well as rainbows, both compared to giant serpents that move in the firmament") (Girard 1966, 205, my trans.).

STARS

In Ch'orti' creation mythology, the earth and sky were created before the sun, moon, and stars, leaving the first humans to live in complete darkness (Hull 2016, 19). It was during a later creation event that stars were set into the sky. Stars (*ek'ob'*) in Ch'orti' thought are viewed in two distinct ways. First, these astral bodies are said to be "fellow runners with the sun" that serve as "watchers over mankind."[28] Their permanent presence in the night sky is said to be a gift "left by God Our Father" (i.e., *Katata'*) so that "not even for a moment would man remain abandoned on the earth" (Fought 1972, 427).

The second and arguably the dominant notion today is that stars are physical manifestations of evil spirits. Not unlike early thirteenth-century Italian conceptions of the "influence" (*influenza*) stars could have on human illness (Fleming 2002, 124), the Ch'orti' believe stars are responsible at times for various sicknesses. Certain astral bodies, including some planets, which are also called *ek'*, can be employed in sorcery rites to curse individuals, often in highly injurious ways. Sorcerers (*ajb'a'x*) call upon various stars or planets to assist in infecting a person with a disease. Therefore, traditional healers divine to learn which star or planet is responsible for a particular illness in order to determine the proper cure (Hull 2000).

COMETS

For the Ch'orti', comets and falling stars[29] are inauspicious signs of a coming negative event. Falling stars are said to destroy crops and can also foretell death (Metz 2006, 130). López de Rosa and Chactún (2004, 48) describes a tradition wherein someone who sees a falling star has to immediately ask to put the palm of a child into his or her own palm *para que la estrella caída no se lleve la suerte* ("so that the fallen star won't take away their luck/destiny").

The only name I have found for comets is the Spanish *bolas de fuego* ("balls of fire"). Wisdom noted comets were called *k'ahk' tuut e k'in* (lit. 'fire in the sky'), "comet, falling star" or as *"yuch' e k'in"* (lit. "spark of the sun"), with either term meaning either "comet" or "falling star" (Wisdom 1950). An elderly Ch'orti' consultant once told that it was a comet (or falling star) hitting the earth that brought the first fire to humans. However, another consultant heard that it was a flying chunk of lava expelled from a volcano that introduced humans to fire.

THE PLANET VENUS

The Ch'orti' still retain names of some of the planets, but many are only known by their Spanish forms. Ch'orti' Mayan today does not have a word that strictly corresponds to "planet," rather they refer to them simply as *ek'ob'* ("stars"). Wisdom, however, recorded the term *noh yuch'* for "planet, bright star," where *noh* [*noj*] means "big" and *yuch'* is "point of light, star" (Wisdom 1950).

Venus is perhaps the most noted planet in contemporary Ch'orti' society. It is called by various names such as *Sakojb'ix Tob'ix Ek'* (lit. "it-just-dawned, it-just-jumped star") when it appears as the Morning Star. One elderly *curandera* (traditional healer) I recorded referred to Venus as *e luser xe' k'ob'ir* ("the Big Star"), *e nukir lusera—Prinsipo de Lusero* ("the Big or Leader Star"), or simply *e noj ek'* ("the Big Star"), which corresponds well with other Mayan languages that refer to it in similar terms, such as *noh ek'* in Yukatek, meaning *"lucero de la mañana"* ("Morning Star") (Barrera Vásquez et al. 1980, 573). However, it is most commonly referred to by a number of Spanish names, such as *San Pascual* (Fought 1972, 527n34), *San Pascual del Verónico, Lucero Pascual, Uviada, Niño San Pascual,* and *Don Pascual de Verónico*. A female counterpart to *Don Pascual* is *Doña Pascual*, an unidentified planet that is said to reside in the *mediante del cielo, mediante de la gloria* ("middle of the sky, middle of heaven"). Fought recorded the name of Venus as Evening Star as *San Ramón* (Fought 1972, 437n29.7). There does, however, seem to be some overlap in the Spanish names noted above for Venus's appearance as both Morning and Evening Star.

Venus may also have some relation to grinding corn, as it is called *la estrella moledor* ("Grinder Star"), according to a text from the ALMG (2005, 49), and *estrella chilatera o estrella nixtamalera* ("Roasted Corn Drink Star or Nixtamalization Star") (Girard 1949, 227). I have not found any other direct relationships among the Ch'orti' between Venus and corn, though a possible connection may exist in other Mesoamerican cultures (see Šprajc 1993, 35).

Venus as the Morning Star has particular relevance to the Ch'orti' as an indicator that it is time to wake up and leave to the cornfield. In addition, according to Girard, the planet Venus also plays a large role in their celestial theogony. The elders observe its morning or evening position and its degree of luminosity. Venus "'opens the way so that The Lord (sun) may pass,' declares Alcario Alonso, elder of Cayur" (Girard 1995, 100). Therefore, the *lucero de la mañana es la esperanza del día* ("Morning Star is the hope of the day") (ALMG 2005, 49). In times past when watches or clocks (*b'isk'in*) were not used in the Ch'orti' area, farmers relied on the Morning Star to know when to begin their sometimes lengthy walk to their cornfields:

> In the past the ancestors didn't have them [watches]. And we also still use it [Venus] a little when we don't have our watches. We look at the sky where the moon is when it gets dark. And we know as it dawns, if we understand it, if it is the middle of the night, or we are in darkness, we look for the moon where it sits. And we say now it is night and we no longer can go, or the night has now set in on us. And if we see the moon sitting in the middle of the sky, or if we see where it sits, we go with the moon, we call it "Our Mother." Now there it sits now and it is now telling us that we are in the middle of the night. And thus in this way indeed, if we know that it is now dawning, we also see the moon or a star. If we see the star when it dawns, we say this star is now low in the sky. Then we are thinking to leave and travel. Or if we want to go out when it is still dark from our houses to leave when the first rays of light are showing on to our work, then we leave when it is still dark, and since we know that the star will indicate to us that it is now dawning. Then we leave to go to our work. And when we leave to our work, or to our cornfields, or where our fields lie, it is starting to become light when we arrive. Therefore the star doesn't deceive us that it is still night. And with this, we measure how much is the path of the length of the night. (Hull n.d.a.; my trans.)[30]

For the field workers, leaving around 3:30 am (4:00 am in the fall) (though some say it comes earlier)[31] when the Morning Star just breaches the mountains in the east ensures they will arrive at their fields just as there is enough light to work[32] while also enabling them to take advantage of the cool of the morning for their labors.

Venus also has darker associations. One pervasive belief throughout Ch'orti' communities is that Venus is related to fighting, wars, and murder. The connection between Venus and war is well established in various Mesoamerican traditions and in precolumbian iconography. While early interpretations of the "star-over-earth" war glyph in Maya hieroglyphic writing were probably overzealous in finding connections to Venus (e.g., Schele and Freidel 1990), there is still an undeniable relationship between war and Venus for the ancient Maya (Taube and Bade 1991). Data from modern-day Ch'orti' support this association.

In an earlier report (Hull 2000), I discussed the connection between a planet and war or killing and surmised that it must be Venus, though some doubt remained. However, since that time I have been able to confirm with several consultants that it is indeed Venus. The tradition states that when Venus rises over the horizon "together"—*a la par con*, or *nut'ur taka*, or more fully, *e katu' nut'ur a'xin taka e ek'* ("the moon goes together with the star")—with the moon, someone will be killed that night in one of the villages. Fought (1972, 427–428) noted this tradition without identifying the planet: "When that star unites its body with the moon, that is a sign that someone is going to kill himself in the villages. If not, a war is going to be made in the towns." A closer analysis of this belief reveals that Venus is said to follow after the moon "begging" her to "sell her children" (*uchoni umaxtak*); that is, allow Venus to kill someone on earth. In other descriptions the moon *hace un pacto con la estrella* ("makes a pact with the star"), the Morning Star, as it follows after the moon. One consultant of mine explained this agreement this way: "The moon is our mother. The star is evil (*malo*). It wants to kill the children of the moon. So it follows closely and keeps asking it to 'sell its children,' and sometimes it wins over and the moon lets the star have some of them. This is when there are killings on the earth." While the tradition seems to hinge of the close proximity of Venus to the moon as they rise, Venus can in fact be above, below, or alongside of the moon, as long as it is in the general proximity. What is more, the dangerous consequences of this "pact" may appear any time during the whole month around this astronomical event, not only on the night the union occurs (although the latter is most common). Another elderly consultant of mine, a traditional healer, spoke of the impending death that results when Venus rises with the moon in the sky:

> Yes, some people are going to be killed. It is coming. Indeed it is coming, just below it, it comes like this. Her children have to be killed; some dead person indeed has to result. The Bad Star is asking for permission with the Queen, with the Moon. It is asking permission for the children of the Moon [i.e., people on the earth]. Yes, it is for the misfortune of that star, maybe also Don Vicente of the Star is coming so that these poor people will be killed—Don Vicente of the Bad Star. (my trans.)[33]

Venus is also closely linked to rites of sorcery. Workers of black magic (*ajb'a'x*) commonly "speak to" (*hablar con*) Venus (Don Vicente) to petition help in placing a curse on someone. A traditional healer explained to me:

> There is also a big star that is called San Pascual (Venus). And there are people who are terrible . . . [who] at times speak to that star. It is called Niño San Pascual. They now place sicknesses on the people, that this [star] causes a sickness such as a headache or dysfunction of the body. And if not, exhaustion of the body, to be void of any desire to do anything. (my trans.)[34]

Venus is also said to be culpable in a number of eye-related diseases. Sorcerers call upon Venus to blind or cause debilitating eye problems for their victims. Other planets are enlisted by sorcerers for causing sicknesses such as fevers or skin rashes.

In short, while the cyclic movements of planets and stars are valuable to the Ch'orti' as indicators of times and seasons, the reality is that they are more often feared or even loathed as bearers of ill fortune.

CONCLUSION

As stated by Antonio Hernández, a consultant of Girard, the Ch'orti' dictum that "whatever is in the heavens is also on the earth" (Girard 1995, 117) could also be reformulated to say "whatever happens in the heavens has an impact on the earth"—meaning that celestial events have traditionally been seen as causative forces with ramifications for humans on earth. Dreaded events such as an eclipse, a comet, or the moon rising with Venus portend unwanted consequences in the lives of the Ch'orti'. A lightning strike from the sky can kill immediately, or even cause a "fright" to someone, thereby weakening their spiritual defenses enough for evil spirits to afflict them. Fortunately, the heavens can also signal rain in a myriad of ways, such as "rainbow" clouds (*cabañuelas*), the position of certain constellations such as the Pleiades or the Southern Cross, the phase of the moon, sun dogs or moon dogs, thunder, the presence of lightning in certain parts of the sky, and so on.

In the Ch'orti' cosmovision, water is the most important element in daily life, while simultaneously one of the most dangerous, for it is the principal facilitator of the movement of malevolent spirits between realms. The Ch'orti' fear any "weakening" of their spirit through embarrassment or fright if near water, since evil spirits can immediately infect them with an illness.

Water penetrates deeply into nearly every aspect of Ch'orti' cosmology. The universe is surrounded by celestial and terrestrial seas of varying colors and characteristics, together informing notions of cause and effect, of sickness and healing, and of divine blessings and evil curses. The quincuncial layout of the Ch'orti' universe, with the native understanding of the five large lakes at each point, are continually recreated in ritual table settings for the rain-bringing angels, at sacred springs in deity appeasement rites, and in field dedication ceremonies (Girard 1995; Hull 2000, 2006; Kufer 2005). Ritual is thus informed by cosmological myth. And, consequently, quotidian routines for the Ch'orti' are in many cases active extensions of their own inherited cosmology.

NOTES

1. *los padrinos, con su fuerte visión localista, no han sido concientizados todavía por el movimiento. Irónicamente, los dirigentes nacionales utilizan los libros de Girard sobre las cofradías para enseñar la cosmovisión maya a los campesinos ch'orti's, mientras que los padrinos que siguen practicando mucho de lo que escribía Girard no son invitados* (Metz 2007, 11).

2. All Ch'orti' data in this chapter are from my fieldwork unless otherwise expressly cited.

3. L. López, one of Girard's consultants, noted that "the clouds full of water are the serpents—or horses—that carry the Angels" (Girard 1995, 159).

4. Fought (1972, 383) contains a reference to *chij tokar* ("horse clouds") that are the *ajwa'k'in* ("lords of the sky"; orthography altered).

5. Compare to the Yukatek Maya belief that the rain gods (Chaacs) are seen as "old men who ride on horses which are seen as clouds" (Redfield, and Villa Rojas 1934, 116).

6. My translation of: *La Gloria es descrita bien como lugar de lindos paisajes y jardines, con bonitas flores y olores fragantes, o bien como una ciudad en el ajetreo de una constante construcción para acoger nuevas almas* (López García and Metz 2002, 234).

7. The watery realms above and below the earth are the basins from which numerous supernatural beings draw water in order to produce rain. One of Girard's consultants said that "the deity of the center is 'Master of the guacal [gourd] of the world,' that is of the '*Pila mayor*' (great basin), or central water reservoir from which the rain gods get the water. In the view of my informants there are five '*pilas*' (basins) or water sources, like gigantic guacals, distributed in the five points of the cosmos" (Girard 1995, 140). Clouds are said to be repositories of water that the rain-bringing angels access to fill their own "containers" in order to pour water onto the earth.

8. The term *príncipos* is a common Ch'orti' variation of the standard Spanish *príncipes* ("princes").

9. *Abundan las leyendas que se refieren a incantaciones; los cerros, los cursos de agua, las pozas, las montañas que se juntan, las lagunas legendarias, las cuevas encantadas, las*

barrancas, cualquier accidente de terreno es un paraje místico poblado de espíritus para los ojos nativos (Girard 1949, 333).

10. The different positions of the sun could be time-based, perhaps implying certain levels, but current ethnographic data only seems to suggest three or four layers: sky, earth, and underworld, which also has its own separate sky.

11. The term *oraj* means something akin to "central" or "time" in a restricted sense, and it may ultimately derive from the Spanish *hora* ("time").

12. Cf. Yukatek *mun luna* ("green, or unripe, moon") (Redfield, and Villa Rojas 1934, 205).

13. There is variation among the few speakers I found who still know this verb between the verbal suffix *-ka* and *-k'a*. Handwritten field notes by John Fought, however, give it as *pechk'a* in *apechk'a tokto* (orthography altered), "it's beginning to come out" (Fought n.d., Fought Ch'orti' Field Notes #300–354 CAA005R005I001, page 0000320; orthography altered; http://www.ailla.utexas.org. See also *pechk'a* in Fought 1972, 382).

14. Metz (2006, 132) notes the Ch'orti' see the moon as "growing" during this stage, just as a gourd fills with water.

15. Wisdom corroborates this belief: "She is worried if an eclipse of the sun or moon occurs during her pregnancy, as this is said to cause great danger that her child will be born without one of its external body parts, such as a part of one of the extremities or a portion of the face or head" (Wisdom 1940, 285).

16. *Akel tiempo uche ani kilisar e luna. Entonce lo que hacen e onya'n jente alok'o'b' ani patir taka yar e kandelirob.' Ucho'b' resar twa' ma'chi ache'na amenesar e luna che'no'b', katu' che'no'b. Entonces e ti'n ayan uch'urkab'ob' kachar unak'ob' tama e kordon twa' ma'chi uwiro'b' ub'an, twa' ma'chi ak'ujxa e ch'urkab'. Ukacho'b' ani twa' ma'chi ache'na kilisar. Entonces ma'chi uwiro'b'. Vaya pwes apaxtak che ani jay uwira e Katata' war ache'na amenesar che'no'b' ani porke' uk'uxo'b' o lok'oy impedido e maxtak che ani e difunto kadoliente. B'an pwes. B'an ani uyajk'o'b ukansejo'b' ani. Ja'xta'ni ukaso'b' onya'n jente kwando ache'na akilisar e katu'. B'an pwes.*

17. Kerr numbers refer to the ceramic vessels in the Maya Vase Database provided by Justin Kerr (http://research.mayavase.com/kerrmaya.html).

18. This notion is pan-Mayan. For the K'iche', see Christenson (2001, 74).

19. *Los ángeles, montados sobre los caballos oscuros, formaron las nubes densas y grises; los hombres, montados sobre los caballos tordillos, formaron nubes blanquecinas. Cada vez que manipulaban su espada, formaban rayos y truenos* (Dary, Elías, and Reyna 1998).

20. Girard also notes the *makchan* is a *culebra de color*, or "colored serpent" (Girard 1949, 101, 323).

21. Traditional healers sometimes refer to rainbows as *Doña Teresa or Santa Teresa*.

22. The Yukatek Maya say rainbows are *u ciz cizinob* ("the flatulence of the demons") (Redfield and Villa Rojas 1934, 206).

23. In Yukatek Mayan *cabañuelas* is *xook k'iin* (lit. "counting-days"), since the first 12 days of January are counted and then compared to the following 12 months for weather prognostication (Navarette 2009, 40). Membreño describes the cabañuelas as: *Cálculo que, observando las variaciones atmosféricas en los diez y ocho primeros días de enero, forma el pueblo para pronostica el tiempo que ha de haber en cada mes del año. Del primero al doce de enero, á cada día corresponde un mes, y del trece al diez y ocho, la mañana corresponde á un mes y la tarde al siguiente* ("Calculation that, observing the atmospheric variations in the first eight days of January, forms the town to forecast the time that must be in each month of the year. From the first to the twelfth of January, each day corresponds to one month, and from thirteen to eighteen, the morning corresponds to one month and the afternoon to the next") (Membreño 1897, 25; my trans.).

24. *Na'tanwa'r tama e makchan tokar xe' ach'a'n tu't e k'in twa' uwirnib' e jaja'r. K'ani ink'ajti ub'an e k'ub'seyaj xe' kana'ta jay k'ani ak'axi e jaja'r wa'kche wa'kchetaka o k'ani e'yxna. I tu't e k'in tara no'n tya' turo'n kawira konde matuk'a me'yra tokar tu't e k'in kawira ach'a'n inte' makchan xe' xoyoyoj. Kawira ach'a'n tu't e k'in cha'te' u't, ach'a'n e makchan ira. Saksak tokar uk'eche. Jay kawira ke' saksak inyajrer o me'yra tokar ach'a'n tu't kawa're ke' a'xin ak'axi e jaja'r e makchan ira uk'eche e jaja'r. Ma'chi'x a'xin e'yxna e jaja'r. I a'xin ak'axi me'yra intaka ke' ma' ak'otoy e ajk'in kawa're. Inton war kawira ke' makchan ira uyajk'o'n kawira ke' e jaja'r a'xin ak'axi. I tya' o konde kawira ke' ma'chi uk'eche e tokar me'yra charantaka yer e makchan ira. Charantaka u't ach'a'n tu't k'in. Kawa're ma'chi a'xin ak'axi wa'kchetaka e jaja'r. A'xin e'yxna. A'xin uche me'yra k'in. E'ra ma'to tuk'a e jaja'r tartaka ke' war kawira ke' e makchan ma'chi uk'eche e jaja'r. Inton b'anixto e'ra kak'ub'seyaj ani no'n i kawira ub'an jay ach'a'r to'r e witzir k'atarb'ir kawa're e jaja'r wati'x.*

25. Søren Wichmann independently recognized this same linguistic and iconographic connection between these "bathing" scenes and the halos around the sun and moon, but among the K'iche' Maya. Wichmann notes the K'iche' expressions *ratin iik'* (Spanish: *halo alrededor de la luna*; English: "halo around the moon") and *ratin q'iij* (Spanish: *halo alrededor del sol*; English: "halo around the sun"), where the term *ratin* literally means "bath," i.e., "bath of the moon" and "bath of the sun," respectively (Wichmann 2004, 82–83).

26. This translation assumes the root *at-* has a transitive reflex in Mayan languages, although there is currently no evidence of this. It is possible, though still problematic, that this a possessed nominal form, "the bathing of."

27. Ixlu Altar 1 (C3-D3) also has a reference to the "bathing" of the Paddler Gods.

28. The Ch'orti' notion of stars as "watchers" is comparable to page 34 in the *Madrid Codex*, where the starry sky is depicted as a circle of disembodied eyes (see Milbrath 1999, 252–533, fig. 7.1a).

29. The Ch'orti' do not seem to distinguish comets from falling stars, although falling stars are actually meteor showers that happen annually at certain times of years when the earth passes through the trail of a comet. Comets are seen at much more infrequent intervals.

30. *O'nix ma'ni tuk'a e onya'n paka'b'ob'. I no'n ub'an kak'ampesto kora konde matuk'a kab'isk'ink'ab'. Kawira tu't e k'in tya' turu e katu' tya' akb'are. I tya' kana'ta kocha asakojpa jay k'ani kana'ta jay yuxin akb'ar wa'ron o e akb'ar wa'ron kalok'oy kach'ujku e katu' tya' turu. I kawa're e'ra e akb'arix ma'chi'x uyub'i ka'xin o e'ra uwajpyo'nix akb'ar. I jay kawira tu'yuxin k'in turu e katu' o kawira tya' turu ka'xin tama e katu' kawa're e katu'. Yaji'x turu e'ra warix uya'ryo'n ke' uyuxin akb'ar wa'ron. I b'an koche'ra b'an jay k'ani kana'ta ke' warix asakojpa kawira ub'an e katu' o ingojr ek'. Jay e ek' kawira tya' asakojpa kawa're e ek' ira t'erpi'x sakojpix e'ra rajxa asakojpa. Ton no'n warix kajb'ijnu twa' kalok'oy kaxana. O jay no'n k'ani kalok'oy akb'arto te' ko'tot twa' kak'otoy war asakch'i' to'r kapatna'r inton no'n kalok'oy akb'arto i kocha kana'ta ke' e ek' war uwirsyo'n ke' warix asakojpa. Inton no'n kalok'oy ka'xin ti kapatna'r. I tya' kak'otoy tama kaptna'r o ti kachor o tya' ch'a'r kajinaj war a'nch'akna kak'otoy. Inton e ek' ma'chi umajreson ke' akb'arto. I taka e'ra no'n kab'isi ko'b'a ub'i'r uwamir e akb'ar.*

31. A publication by the Academia de Lenguas Mayas de Guatemala (ALMG) states Venus as the Morning Star is visible after 2:00 am: *La estrella moledor (de la mañana), que nace a partir de las dos de la mañana* ("The grinder star [of the morning], that comes out after two in the morning") (López de Rosa and Rodas Chactún 2004, 49).

32. The connection between Venus and warfare is well established in parts of Mesoamerica, but the relationship to the timing of specific war events is still debatable (see H. Bricker and V. Bricker 2011, 245–248).

33. *Ja'xto e ti'n e k'ani achamesno'b'. Watar ya'. Watar atz'i, yeb'ar tix, watar koche'ra. Ufamilia k'ani achamesna; algún muerto k'ani ache'na ya'. War uk'ajti e permiso Lucero de Malo. Está pidiendo permiso con la Reina, taka e Katu'. War uk'ajti e permiso kochwa e familia e Katu' Reina. Ja'xto twa' desgracia e lusera yaja'. Ja'xto kaxa watar e don Vicente de Lucero ub'an twa' achamesno'b' e pobre gente—don Vicente de Lucero Malo.*

34. *Hay también una estrella grande que le dicen San Pascual. Y hay gente que son malísimos . . . [que] a veces a esa estrella hablan. Le hablan al Niño San Pascual. Ya ponen enfermedades a las personas que esa [estrella] brota una enfermedad que es el dolor de la cabeza, quebrantamiento de cuerpo. Y si no, el desmayamiento de cuerpo, de sin ganas de no hacer nada.*

REFERENCES

Aldana, Gerardo. 2005. "Agency and the 'Star War' Glyph: A Historical Reassessment of Classic Maya Astrology and Warfare." *Ancient Mesoamerica* 16: 305–320.

ALMG (Academia de Lenguas Mayas de Guatemala). 2005. *Na'tanyaj xe'ayan taka e pak'ab'ob', tzijb'ab'ir tama ojronerob' Ch'orti' yi e castellano*. Guatemala: Academia de Lenguas Mayas de Guatemala (ALMG).

Barrera Vásquez, Alfredo, Juan Ramón Bastarrachea Manzano, William Brito Sansores, Refugio Vermont Salas, David Dzul Góngora, and Domingo Dzul Poot, eds. 1980. *Diccionario Cordemex: Maya Español–Español Maya*. Mérida, Mexico: Ediciones Cordemex.

Bricker, Harvey M., and Victoria R. Bricker. 2011. *Astronomy in the Maya Codices*. Memoirs of the American Philosophical Society, Vol. 265. Philadelphia, PA American Philosophical Society.

Christenson, Allen J. 2001. *Art and Society in a Highland Maya Community: The Altarpiece of Santiago Atitlán*. Austin: University of Texas Press.

Cojti Cuxil, Demetrio. 1992. *Idiomas y Culturas de Guatemala*. Universidad Rafael Landívar, Guatemala: Instituto de Lingüística URLPRODIPMA.

Dary, Claudia, Sível Elías, and Violeta Reyna. 1998. *Estrategias de sobrevivencia campesina en ecosistemas frágiles*. Guatemala: FLASCO.

Fleming, D. M. 2002. "A Perspective from General Practice on Selected Influenza Topics: Near Patient Tests, Influenza in Pregnancy, Influenza in Children." In *Perspectives in Medical Virology*, Vol. 7: *Influenza*, ed. C. W. Potter, 123–144. Amsterdam: Elsevier.

Fought, John G. 1972. *Chortí (Mayan) Texts*, ed. Sarah S. Fought. Philadelphia: University of Pennsylvania Press.

Fought, John. n.d. Fought Ch'orti' Field Notes #300-354 CAA005R005I001, page 0000320. http://www.ailla.utexas.org.

Girard, Rafael. 1947. "The Medicine Chest of the Chorti Indians." *Boletín Indigenista, Instituto Indigenista Interamericano* 7: 346–361.

Girard, Rafael. 1949. *Los chortís ante el problema maya: Historia de las culturas indígenas de América, desde su origen hasta hoy*, vol. 1. Mexico: Antigua Librería Robredo.

Girard, Rafael. 1958. *Indios selváticos de la Amazonia Peruana*. México: Libro Mex-Editores.

Girard, Rafael. 1962. *Los mayas eternos*. Mexico: Antigua Librería Robredo.

Girard, Rafael. 1966. *Los mayas: Su civilización, su historia, sus vinculaciones continentals*. México, Libro Mex Editores.

Girard, Rafael. 1979. *Esotericism of the Popol Vuh*. Pasadena, CA: Theosophical University Press.

Girard, Rafael. 1995. *The People of the Chan*. Richardson, TX: Continuum Foundation.

Houston, Stephen, David Stuart, and Karl Taube. 2006. *The Memory of Bones: Body, Being, and Experience among the Classic Maya*. Austin: University of Texas Press.

Hull, Kerry. n.d.a. "Story 33: How the Ancestors Told Time." Unpublished story from field notes.

Hull, Kerry. n.d.b. "Story 37: Eclipses and Pregnancy." Unpublished story from field notes.

Hull, Kerry. n.d.c. "Story 39: Belief about the Rainbow Cloud." Unpublished story from field notes.

Hull, Kerry. 2000. "Cosmological and Ritual Language in Ch'orti'." A report submitted to FAMSI. URL: http://www.famsi.org/reports/99036/99036Hull01.pdf.

Hull, Kerry. 2003. "Verbal Art and Performance in Ch'orti' and Maya Hieroglyphic Writing." PhD diss., Department of Anthropology, University of Texas at Austin. Printed by UMI.

Hull, Kerry M. 2009. "Dualism and Worldview among the Ch'orti' Maya." In *The Ch'orti' Maya Area: Past and Present*, ed. Brent E. Metz, Cameron L. McNeil, and Kerry M. Hull, 187–197. Gainsville: University Press of Florida.

Hull, Kerry. 2016. *A Dictionary of Ch'orti' Mayan-Spanish-English*. Salt Lake City: University of Utah Press.

Katz, Esther. 1997. "Ritos, representaciones y meteorología en la 'Tierra de la Lluvia' (Mixteca, México)." In *Antropología del clima en el mundo hispanoamericano*, ed. Marina Goloubinoff, Esther Katz, and Annamaria Lammel, 99–134. Colección Biblioteca Abya-Yala, núm. 50. Editorial Quito, Ecuador: Abya-Yala.

Kerr, Justin. n.d. *Maya Vase Database, FAMSI*. Available at: http://research.famsi.org/kerrmaya.html (verified 12-5-2011).

Kufer, Johanna. 2005. "Plants Used as Medicine and Food by the Ch'orti' Maya: Ethnobotanical Studies in Eastern Guatemala." PhD diss., Centre for Pharmacognosy and Phytotherapy. The School of Pharmacy, University of London.

de León Pasquel, María de Lourdes. 2005. *La llegada del alma: Lenguaje, infancia y socialización entre los mayas de Zinacantán*. México: CIESAS-INAH.

López de Rosa, Lucas, and Mario Augusto Rodas Chactún. 2004. *Derecho consuetudinario maya ch'orti': Los ámbitos communal, familiar y medio ambiente = Utwa'chirob' una'tanwa'rob' e ch'ortyob': Tama uturerob' yi xe' ayan to'r e rum*. Guatemala: Consejo Indígena Maya Ch'orti-COIMCH.

López García, Julián. 2010. *Kumix: La lluvia en la mitología maya-ch'orti'*. Guatemala: Ed. Cholsamaj.

López García, Julián, and Brent E. Metz. 2002. *Primero Dios: Etnografía y cambio social entre los mayas ch'orti's del oriente de Guatemala*. FLACSO, Sede Guatemala.

Membreño, Alberto. 1897. *Hondureñismos: Vocabulario de los provincialismos de Honduras*. Honduras: Tipografía Nacional.

Metz, Brent. 1998. "Without Nation, Without Community: The Growth of Maya Nationalism among Ch'orti's of Eastern Guatemala." *Journal of Anthropological Research* 54: 325–349.

Metz, Brent. E. 2001. "Representación colaborativa: Un gringo en el movimiento maya-ch'orti." In *Los Derechos humanos en el Área Maya: Política, Representaciones y Moralidad*, ed. P. Pitarch and J. López, 311–340. Madrid: Sociedad Española de Estudios Mayas.

Metz, Brent E. 2006. *Ch'orti–Maya Survival in Eastern Guatemala: Indigeneity in Transition*, Albuquerque: University of New Mexico Press.

Metz, Brent E. 2007. "De la cosmovision a la herencia: La mayanizacion y los bases cambiantes de la etnia en el area ch'orti'." In *Mayanización y vida cotidiana: La ideología y el discurso cultural en la sociedad guatemalteca*, Volume 2: *Estudios de caso*, ed. S. Bastos and A. Cumes, 445–467. Guatemala: FLASCO.

Metz, B. E., C. L. McNeil, and K. M. Hull. 2009. "Searching for Ch'orti' Maya Indigenousness in Contemporary Guatemala, Honduras, and El Salvador." In *The Ch'orti' Maya Area, Past and Present*, 161–173. Gainesville: University Press of Florida.

Milbrath, Susan. 1999. *Star Gods of the Maya: Astronomy in Art, Folklore, and Calendars*. Austin: University of Texas Press.

Miller, Mary E., and Taube, Karl A. 1993. *An Illustrated Dictionary of the Gods and Symbols of Ancient Mexico and the Maya*. New York: Thames and Hudson.

Murúa, Martín de 1986 [1613]. *Historia general del Perú*. Edición de Manuel Ballesteros. Madrid: Historia 16.

Navarette, Javier Abelardo Gómez. 2009. *Diccionario Introductorio: Español–Maya, Maya–Español*. Chetumal, Quintana Roo: Universidad de Quintana Roo.

Neff Nuixa, Françoise. 1994. *El rayo y el arcoiris: La fiesta indígena en la montaña de Guerrero y el Oeste de Oaxaca; Col. Fiestas de los pueblos indígenas, México*. México: Instituto Nacional Indigenista, Secretaría de Desarrollo Social.

Neff Nuixa, Françoise. 1997. "Los caminos del aire: Las idas y venidas de los meteoros en el estado de Guerrero (México)." In *Antropología del clima en el mundo hispanoamericano*, ed. Marina Goloubinoff, Esther Katz, and Annamaria Lammel, 297–315. Colección Biblioteca Abya-Yala, núm. 50. Editorial Quito, Ecuador: Abya-Yala.

Palma Ramos, Danilo A. 2001. *Así somos y así vivimos los Ch'orti'*. Guatemala: Universidad Rafel Landívar.

Past, Ambar. 2014. *Incantations: Songs, Spells and Images by Mayan Women*. El Paso, TX: Cinco Puntos Press.

Paul, Benjamin D., and Lois Paul. 1952. "The Lifecycle." In *Heritage of Conquest: The Ethnology of Middle America*, ed. Sol Tax, 174–192. Glencoe, IL: The Free Press.

Redfield, Robert, and Alfonso Villa Rojas. 1934. *Chan Kom: A Maya Village*. Chicago, IL: University of Chicago Press.

Roys, Ralph L. ed., 1965. *Ritual of the Bacabs*. Norman: University of Oklahoma Press.

Šprajc, Ivan. 1993. "The Venus–Rain–Maize Complex in the Mesoamerican World View: Part I." *Journal for the History of Astronomy* 24 (1–2): 17–70.

Stuart, D., S. Houston, and J. Robertson. 1999. *The Proceedings of the Maya Hieroglyphic Workshop: Classic Mayan Language and Classic Maya Gods*. Transcribed and ed. Phil Wanyerka. March 13–14, 1999. University of Texas at Austin.

Tarn, Nathaniel, and Martin Prechtel. 1986. "Constant Inconstancy: The Feminine Principle in Atiteco Mythology." In *Symbol and Meaning beyond the Closed Community: Essays in Mesoamerican Ideas*, ed. Gary H. Gossen, 173–184. Studies in Culture and Society, vol. 1. Albany: Institute for Mesoamerican Studies, State University of New York at Albany.

Tavárez, David, and John Justeson. 2008. "Eclipse Records in a Corpus of Colonial Zapotec 260-Day Calendars." *Ancient Mesoamerica* 19(1): 67–81.

Taube, Karl. 1993. *Aztec and Maya Myths*: Austin: University of Texas Press.

Taube, Karl A. 2004. "Flower Mountain: Concepts of Life, Beauty, and Paradise among the Classic Maya." *RES: Anthropology and Aesthetics* 45 (1):69–98.

Taube, Karl A., and B. L. Bade. 1991. "An Appearance of Xiuhtecuhtli in the Dresden Venus Pages." *Research Reports on Ancient Maya Writing* 35: 105–127.

Thompson, J. Eric. S. 1930. *Ethnology of the Mayas of Southern and Central British Honduras*. Ed. Berthold Laufer. Anthropological Series 17 (2): 25–213. Field Museum of Natural History, Publication 274.

Thompson, Eric S. 2001. "The Nature of Ancient Maya Writing." In *The Decipherment of Ancient Maya Writing*, ed. Steven D. Houston, Oswaldo Chinchilla Mazariegos, and David Stuart, 127–138. Norman: University of Oklahoma Press.

Tz'utujiil Tinaamitaal, Comunidad Lingüística Tz'utujiil. 1998. *Nawalin taq tziij: Tradición oral tz'utujiil*. Guatemala: Editorial Cholsamaj.

Vail, Gabrielle, and Christine Hernández. 2013. *Re-creating Primordial Time: Foundation Rituals and Mythology in the Postclassic Maya Codices*. Boulder: University Press of Colorado.

Wichmann, Søren. 2004. "The Names of Some Major Classic Maya Gods." In *Continuity and Change: Mayan Religious Practices in Temporal Perspective, 5th European Maya Conference, University of Bonn, December 2000*. Acta Americana, Vol. 14, ed. Daniel Graña Behrens. Nikolai Grube, Christian M. Prager, Frauke Sachse, Stefanie Teufel, and Elisabeth Wagner, 77–86. Markt Schwaben: Verlag Anton Saurwein.

Wisdom, Charles. 1940. *The Chorti Indians of Guatemala*. Chicago, IL: University of Chicago Press.

Wisdom, Charles. 1950. *Materials on the Ch'orti' Language*. Microfilm Collection of Manuscripts on Middle American Cultural Anthropology, No. 28. Chicago, IL. Transliterated and computerized by Brian Stross. http://www.utexas.edu/courses/stross/chorti/.

Wisdom, Charles. 1952. "The Supernatural World and Curing." In *Heritage of Conquest: The Ethnology of Middle America*, ed. Sol Tax, 119–141. Glencoe, IL: The Free Press.

PART III
Complexity

Breaking Paradigms on Cosmological Conceptions

7

Distance and Power in Classic Maya Texts

ALEXANDRE TOKOVININE

This chapter addresses the relationship between traveling to distant places and claims to political authority in Classic-period inscriptions from the southern Maya lowlands (figure 7.1). Classic Maya lords relied on various narrative strategies involving evocation of and traveling to locations in deep time, historical places beyond the immediate confines of the Classic Maya world, and powerful political centers near the protagonists. Some Maya rulers practiced fundamentally different strategies in relating themselves to the historical and deep-time landscapes. Along with other contributions to this edited volume (in this volume, see Vail, chapter 4; Nielsen and Sellner Reunert, chapter 1), the overview of Maya travel logs indicates that the Classic-period landscape of distant places was neither uniform nor structured along the notion of vertical layers of heavens and underworlds.

An investigation of distant places in Classic Maya narratives should begin with a review of the indigenous notions of place and distance. How is the landscape classified? What qualifies a location as a distant "there" and not "here?" How many kinds of "there" are present in Classic Maya texts? How is the distance between "here" and "there" represented?

Although the topic of Classic Maya landscape categories has been extensively covered in several studies (Stuart and Houston 1994, 7–18; Vogt and Stuart 2005; Martin 2004; Brady and Colas 2005; Helmke and Brady 2014), including contributions by the author of

DOI: 10.5876/9781607329534.c007

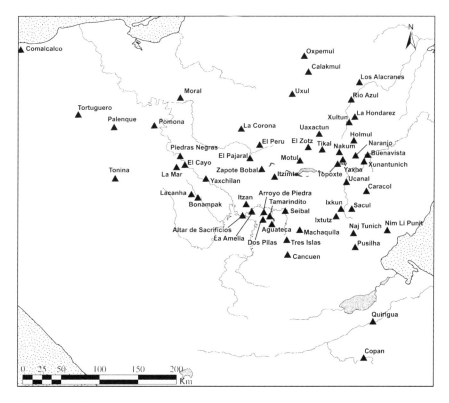

FIGURE 7.1. *Classic-period sites in the southern Maya lowlands.*

this chapter (Tokovinine 2013, 19–55; 2011; 2015), a summary is in order for the present discussion. Nearly all known iconographic representations of places on Classic Maya monuments are binary. The protagonist is depicted on top of the so-called "band" or "register" and another "band" may be placed above the protagonist (Stuart and Houston 1994, 57–68; Schele and Miller 1986, 47). The top register is invariably marked as **CHAN** ("sky") and does not incorporate place name(s) referenced in the textual and visual narrative. The bottom register with the toponym(s) may be marked as **CHAN** ("sky"), **KAB** ("land"), **CH'EEN** ("sacred grounds" or "city," see below), and **HA'** ("water"). Occasional depictions of skulls (von Schwerin 2011) and **WAY** centipede pincers (Taube 2003) are not accompanied by place names and therefore represent specific (supernatural) locations and not general landscape categories.

When it comes to the hieroglyphic inscriptions, all named places are classified as *ch'een*. Its straightforward translation as "cave" does not reflect the full range of contexts and meanings. The narratives from the temples of the Cross group at Palenque and an unprovenanced panel in the collection of the Dumbarton Oaks Research Library provide the best illustration of the significance of the term (Stuart 2006; Houston and Taube 2012; Tokovinine 2013, 29–30). The royal protagonists "ascend" *(t'abaay)* or "step" *(tek')* to a *ch'een* of the local patron god, Hux Bolon Chahk. The very same place is also referred to as a "house" *(naah)*, a "temple" *(waybil)*, a "rock outcrop" *(titz)*, and a "mountain" *(witz)*. This and other contexts strongly indicate that *ch'een* indexes temples of patron deities and deified ancestors and, by extension, the surrounding area. "City" and "sacred grounds" are adequate translations. *Ch'een* may be paired with *chan* ("sky") and *kab* ("land"), which also appear together as in "8000 celestial [and] terrestrial gods" (Stuart 2011b). "The sky [and] the city" *(chan ch'een)* and "the realm [and] the city" *(kab ch'een)* are common combinations. The inscriptions on Stela 31 at Tikal and Altar S at Copan contain a rare triplet of *chan kab ch'een* that may be translated as "the world [and] the city" (Tokovinine 2013, 38–41).

Students of Classic Maya inscriptions would look in vain for explicit references to distance in space between places. Nor would they find much in colonial-period historical and mythical narratives like the Maldonado-Paxbolon papers (Smailus 1975), the books of Chilam Balam (Roys 1973), or the *Popol Vuh* (Christenson 2003). Some places in these later texts are occasionally qualified as "far" or "near," but the only measure of distance is the amount of days it takes to move from one place to another (e.g., Smailus 1975, 80–81; Houston, Stuart, and Taube 2006, 260).

In a world devoid of wheel and beasts of burden, days of travel might well have served as a standard measure of distance. Some Classic-period depictions of travel feature strings of day signs, potentially indicating the length of the trip (Houston, Stuart, and Taube 2006, 260–261). The pictorial maps of Central Mexico also represent distances in terms of days of travel (Yoneda 2007, 168, 178). Classic Maya narratives give an impression of a constant speed of movement. "Running" is reserved for defeated enemies (Beliaev 2006). Everybody else goes, leaves, comes, ascends, and descends. The same verbs describe the coming and going of celestial objects, deities, and even days themselves (Stuart 2004b), merging the notions of space and time altogether.

No location is described as "far" or "near" in Classic Maya inscriptions. However, the use of the locative construction *tahn ch'een* ("in the midst of the ch'een") and *tahn ha'* ("in the midst of the water") with place names instead of the more general *t(a)-u-ch'een* ("in the *ch'een* of") seems to indicate some kind of centrality of the protagonist or event in relation to place. For instance,

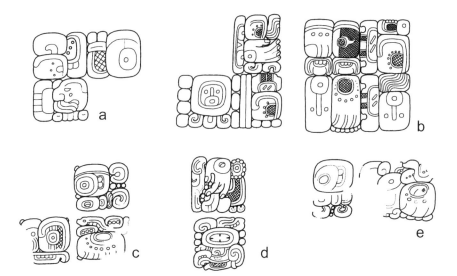

FIGURE 7.2. *Variation in references to the same places: (a) detail of El Cayo Panel 1 (after Chinchilla M. and Houston 1993, fig. 1); (b) detail of El Cayo Altar 4 (after Mathews 1994, fig. 3); (c) detail of Dos Pilas Stela 8 (after photograph by Ian Graham, Peabody Museum of Archaeology and Ethnology, 2004.15.1.519.2); (d) detail of Dos Pilas Stela 1 (after photograph by Ian Graham, Peabody Museum of Archaeology and Ethnology, 2004.15.1.233.4); (e) detail of Dos Pilas Hieroglyphic Panel 19 (after photograph by the author).*

one's return and burial "in the *ch'een* of Yaxniil [the ancient name of El Cayo]" in the narrative on El Cayo Panel 1:C12–C14 (figure 7.2a) may be contrasted to the dedication of an altar during the period-ending ritual "in the midst of the *ch'een*, [in] green/blue turtle water, [in] Yaxniil" (*tahn ch'een yax ahkuul ha' yaxniil*) in the inscription on one of the supports of El Cayo Altar 4 (figure 7.2b) that combines the *tahn ch'een* expression with an elaborate version of the local place name. This accent on the centrality of the period-ending ritual presumably extends to the physical location of the dedicated monument itself. In a similar way, while the inscription on Motul de San Jose Stela 2:A6–B7 locates the dancing ceremony "[in] *Ik'a'*" (Stuart and Houston 1994, fig. 28a), the narrative on Motul de San Jose Stela 1:C5–D8 (Tokovinine and Zender 2012, fig. 2.2) places the period-ending ritual and the dedication of the monument "in the midst of the water, [in] *Ik'a'*."

Classic Maya written accounts also vary in detail when referring to the same locales, sometimes reflecting changes in representation of places just a few days away. For instance, most of the specific locations of period-ending rituals and other ceremonies in the narratives on Tikal Stelae 5, 16, 18, and 22 (Jones and Satterthwaite 1982, fig. 7, 22, 26, 33), and 31 (Stuart and Houston 1994, fig. 58) are not mentioned elsewhere. Inscriptions on monuments at the site of Dos Pilas seem to distinguish between two place names: T369 Ha'al (figure 7.2c), associated with the Main Plaza, and K'in Ha' Nal (figure 7.2d), associated with the El Duende group (Stuart and Houston 1994, 84–88). However, the narrative on Hieroglyphic Panel 19 (figure 7.2e) cites the two place names together without distinguishing between the two parts of the ancient city. Some outside references to events at Dos Pilas also follow the latter pattern. The inscription on the Cancuen Hieroglyphic Stairway 1 (figure 7.8b:pH1–pH2) states that the accession ceremony at Dos Pilas "happened [in] T369 Ha'al K'in Ha' Nal."

Outsiders may replace or conflate place names for cities with names of the regions where they are located. For example, this seems to be the case with the place names of Chi'k Nahb and Huxte' Tuun, which, as the author argued previously (Tokovinine 2007b), correspond to the site of Calakmul and the surrounding area. The inscription on the Naranjo Hieroglyphic Stairway (L2–N3) clarifies the relationship between the two toponyms by stating that the Calakmul ruler is "Kaanul lord in Huxte' Tuun, the man from Chi'k Nahb." However, Dos Pilas (PN 6:B7, ST 9:F4) and La Corona (PN 1:H2, MSC 1) narratives locate distant Calakmul events "in Chi'k Nahb, in Huxte' Tuun," and even "in Huxte' Tuun Chi'k Nahb."

Some texts evoke not only spatial but social, cultural, and possibly ethnic distance. Rulers of Classic Maya polities are sometimes referred to by the names of the geopolitical entities to which they belong, like Seven and Thirteen Divisions, Twenty Eight Lords, and others (Tokovinine 2013, 98–115). The term for foreigners—*tz'ul*—known from colonial-period narratives and dictionaries (Barrera Vásquez et al. 1995, 892) also appears in Classic Maya inscriptions, although only two secure cases are known so far. One example comes from the name of the captive in the inscription on the stingray spine from Comalcalco (figure 7.3a). It is significant that, given Comalcalco's location (see figure 7.1), its rulers might indeed have waged wars against non-Maya groups. The second occurrence of the *tz'ul* term on a recently published monument fragment (figure 7.3b) is particularly important. Its inscription mentions the arrival of a "foreign *kaloomte*" (*kaloomte*' is a title of paramount rulers) in AD 758 followed by an attack against an unknown location four days later (Stuart 2010; Luin and Matteo 2010). Therefore, the term *tz'ul* seems to be reserved for people and things beyond the confines of the Classic Maya world.

FIGURE 7.3. *Inscriptions using the term* tz'ul *("foreigner"): (a) detail of a carved stingray spine from Comalcalco (after Zender 2004, fig. 78a); (b) detail of an unprovenanced monument fragment, possibly from southern Peten, Guatemala (after Luin and Matteo 2010, fig. 1, 2).*

The attribution of directional titles to rulers placed at the boundaries of the southern lowlands conveys the sense that the lords of Classic Maya polities imagined themselves in a larger, unified world (figure 7.4). For example, Copan rulers placed themselves on the southern edge of the world, their Altun Ha counterparts were tied to the east, and a paramount ruler near Ek Balam in Northern Yucatan was apparently in charge of the north (Tokovinine 2013, 94).

In addition to these titles, there are several peculiar quadripartite lists of royal houses previously interpreted as evidence of some kind of macropolitical organization (Marcus 1973), but which most likely reflect Classic Maya ideas about the spatial extent of their civilization (Tokovinine 2013, 91–98). The earliest example at Tonina mentions the *ajawtaak* of Dzibanche

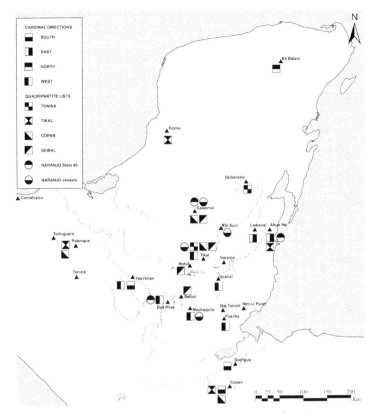

FIGURE 7.4. *Archaeological sites associated with the royal families that appear in quadripartite lists and/or that identify themselves with specific cardinal directions.*

and Tikal along with two unidentified royal families. Eighth-century Tikal lists include the dynasties of Edzna, Palenque, Altun Ha, and Copan. The contemporaneous Copan version mentions the kings of Tikal, Calakmul, Copan, and Palenque. The Terminal Classic list at Seibal (Ceibal) features the royalty from Seibal, Tikal/Dos Pilas, Calakmul/Dzibanche, and Motul de San Jose. Naranjo Stela 46 possibly contains an eighth-century version with Tikal/Dos Pilas, Calakmul/Dzibanche, Altun Ha, and an otherwise unknown place of Yobe' (Martin et al. 2017). Nearly a century later, depictions of maize deities on Naranjo pottery suggest that the list changed to Tikal, Calakmul/Dzibanche, Rio Azul, and Machaquila (Tokovinine 2013, 115–121). The implication of this regional and temporal variation is that the emic perspective strongly depended on the location of the viewer and the historical context. Later variants also attest to a substantial shrinking of the *oikumene* as seen from Naranjo and Seibal.

Finally, there are places in the otherworld, those in deep time and "beyond the sea" (Sachse and Christenson 2005; Sachse 2008) but no less important because they provide the setting of creation narratives and dwellings of deities. However, these extreme spatial and/or temporal distances did not prevent Maya lords from interacting with and sometimes even visiting these highly important locations, at least according to the narratives that they commissioned.

DEEP-TIME ORIGINS

The term "deep time" is preferable to "mythical" in describing a class of distant places in Classic Maya written and visual narratives. Maya inscriptions offer no clear distinction between history and myth. However, a whole range of places are effectively separated from the historical landscape by immense spans of time—from thousands to millions of years (Stuart 2011a, 229–251).

Some of these locations, including the Jo' Chan Naah Witz ("Five Sky House Mountain"), are placed in the north—a cardinal direction that is seemingly reserved for deep-time events in the texts of the southern lowlands (table 7.1). Yet such attributions are not always consistent. For example, the so-called White Shell Mountain may be located in the north, according to one text from Naranjo (table 7.1). Yet the same mountain is repeatedly mentioned in the painted cosmograms on the tomb walls at Río Azul, where it is twice linked to the north (Tombs 2 and 25), once to the south (Tomb 6), and once to the east (Tomb 5) (Acuña 2015, 179). Nor there is any indication of vertical layering beyond the vague above–below dichotomy discussed above. The mostly northern "Five Sky" abode of the Paddler Gods is a mountain (Stuart and Houston 1994, 71). The "Six Sky" dwelling of Itzamna/Itzam Kokaaj/God D (Boot 2008) is also a mountain. It is unlikely that the "Six Sky" name implied a vertical arrangement of heavens, because it was also associated with one of the four Maize God manifestations, along with the mountains of "Six Monkeys," "Six Dogs," and "Six Jaguars" (Tokovinine 2013, 115–122).

Maya rulers chose two strategies with respect to deep-time places. Matwiil lords of Palenque or Kaanul lords of Calakmul traced their royal lines to places thousands of years away from the present (Stuart 2005, 22, 169; Martin 1997). One of the codex-style sherds recently discovered near Structure 20 at Calakmul (figure 7.5a) mentions "entering the *ch'een* [in] Kaanul" (*och-ch'een kaanul*). The surviving image on the sherd suggests that the caption

TABLE 7.1. North in Classic Maya inscriptions in the southern lowlands

Context	Object
siyaj k'uh uhtiiy jo' chan naah witz xaman "God was born. It happened [at] the Jo' Chan Naah Mountain [in] the north."	Vase (Kerr n.d., K688)
[. . .] huxlajuun . . .-pikjiiy huk ajaw hux k'anjalaw yiljiiy ahkuul k'an . . . sabaak naahal ajaw "Thirteen 160,000 years since [the day] 7 *Ajaw*, [day] three [of] *K'anjalaw*; Ahkuul K'an . . . Sabaak north lord had seen it."	Quirigua Stela D: C20–C21
t'abaay wak chan . . . naahal waxak . . . naah uk'aba' yotoot xaman "His dwelling in the north named Wak Chan . . . Eight 'G1' House was dedicated"	Palenque Temple of the Cross Tablet: C9–C13
bolon ipnaj sak baak naah chapaht uway k'awiil ukabjiiy bolon ookte' k'uh uhtiiy xaman "K'awiil's demon Sak Baak Naah Chapaht was strengthened nine times; Bolon [Y]ook Te' god(s) had ordered it. It happened in the north."	Palenque Temple XIV Panel: G2–H4
[. . .] huliiy sak . . . witz nal xaman [. . .] ". . . since he came to the place of the 'White Shell' mountain place [in the] north."	Vase (Kerr n.d., K1398)

accompanied a representation of the mythical confrontation between Chak Xib Chahk and the wind deity, a scene known from several unprovenanced codex-style vessels (Garcia Barrios and Carrasco Vargas 2006; Garcia Barrios 2006). The divine protagonists of the confrontation are usually waist deep in water (figure 7.5b). *Matwiil* (figure 7.5c) is certainly a deep-time place in Palenque narratives and is referred to as "precious shell waters" and depicted likewise (Stuart and Houston 1994, 73–77; Stuart 2005, 169). Just like Tulan in the *Popol Vuh* (Sachse and Christenson 2005), Kaanul and Matwiil are represented as aquatic locations, maybe even origin places within or beyond the sea. Nevertheless, a number of recently discovered monuments indicate that Kaanul was also the ancient name for the archaeological site of Dzibanche, where the Kaanul dynasty resided during the Early Classic period (Helmke and Awe 2016; Martin and Velásquez García 2016).

Several Classic Maya dynasties across the southern lowlands (figure 7.6a) evoked the same deep-time place of Chih Ka' variously known to epigraphers as "Maguey Throne," "Maguey Metate," "Chi Witz," or "Chi-Bent Cauac" (Grube 2004; Schele 1992; Stuart 2004a, 2014). In the case of several narratives,

FIGURE 7.5. Deep-time places: (a) ceramic Fragment 7, Structure 20, Calakmul (after Garcia Barrios and Carrasco Vargas 2006, fig. 10); (b) detail of a codex-style vessel (after Kerr n.d., K2011); (c) detail of a tablet in the Temple of the Foliated Cross, Palenque (after Robertson 1991, fig.153); (d) detail of Altar 1, Naranjo (after photographs by the author); (e) detail of an unprovenanced Late Classic vessel (after Kerr n.d., K7750); (f) detail of an unprovenanced Middle Classic vessel (after Kerr n.d., K1288).

events involving this location are dated to a few centuries before the Classic period and serve as a kind of bridge between the historical canvas and the deep-time background. Fewer dynasties trace themselves to Chih Ka'. Only Tikal and Dzibanche combine references to specific events at Chih Ka' and claims that the dynastic founders originated from Chih Ka'. The connection seems prestigious and yet somehow ambiguous, because not a single royal dynasty of the Classic period asserted a direct descent from the kings of Chih

Ka'. For example, the text on Yaxchilan Lintel 21 (Graham and Von Euw 1977, 49) claims that the local dynastic founder was a war captain (*yajawte'*) from Chih Ka'. The rulers of Dzibanche and Arroyo de Piedra evoke the same military office in relation to their links to Chih Ka'.

A different strategy is adopted by Naranjo rulers who claim that their present location at Maxam and Sa'aal (Tokovinine and Fialko 2007) is also a deep-time place where their divine ancestors ruled before the latest creation. The first "Sa'aal lord" is a deity mentioned at other Classic Maya sites, including Palenque and Copan. According to the inscription on Altar 1: A2–A9 at Naranjo (Graham and Von Euw 1978, 103–104), he acceded to the kingship 22,000 years ago (figure 7.5d). Later narrative on Naranjo Stela 1: F7–E14 (Graham and Von Euw 1975, 12) places his accession even further back in time, 896,000 years ago. The abovementioned Altar 1 (figure 7.5d) also refers to an event at Maxam in 257 BC (in blocks D7–C8) and then goes on to report its dedication at Maxam in the reign of the sixth-century king (in blocks F11–G11). This insistence on the indigenous nature of the local dynasty is consistent with a remarkable continuity in ritual activity centered on the key temple complex at the site, Triadic Acropolis C-9 associated with the Sa'aal place name (Tokovinine and Fialko 2007).

Other locations near Naranjo also seem to be part of a deep-time landscape populated by gods. Two painted stucco friezes of the sixth-century temple of the Naranjo hummingbird patron god Lem Aat evoke a specific mountainous location of Sak/Xak Witz [Nal] ("Seed/Bud Mountain [place]"), where the deity hailed from (Tokovinine and Fialko 2019). However, a lengthy war narrative on a painted vessel from the site of Baking Pot in Belize to the east of Naranjo mentions Sak/Xak Witz as one of the actual places affected by a protracted conflict between Yaxha and Naranjo at the end of the eighth century (Helmke et al. 2018, 60).

Nearby adversaries of Naranjo rulers also occupied deep-time places. The scenes on two unprovenanced vases show six deity figures before Itzam Aat/God L (Kerr n.d., K2796, K7750). The imagery is usually interpreted as a gathering of gods in the underworld palace of God L (Martin 2015, 200). The glyphic commentary to one of the two vessels, also known as the Vase of Seven Gods (Coe 1973, 107–109), states that the deities "were brought together" (*tz'a[h]kaj*) on the creation date of 4 *Ajaw* 8 *Kumk'u*. Stuart (2011a, 224) recently suggested that the list and the figures stood for collectivities and not individual gods. An even likelier interpretation is that the text is not a straightforward list at all, but a set of parallelisms, as is often the case in Classic Maya literature (Lacadena García-Gallo 2009). The sequence features a reference to the totality of "celestial god[s] [and] terrestrial god[s]," also cited in other narratives as "8000 celestial god[s] [and] terrestrial god[s]" (Stuart 2011b) or simply as "8000 god[s]" (Tokovinine 2013, 39–40). Yet it includes the so-called Paddler Gods, two poorly understood deity names or epithets, and a term for the gods of the Palenque Triad (Stuart 2006). Finally, the text mentions an entity or a group of deities called Bolon [Y]ook Te' (Eberl and Prager 2005), who is (are) simultaneously present at another gathering of gods on the same day, this time upon the Six Sky Mountain of Itzamna/God D, according to a scene on a painted vase from Motul de San Jose (Boot 2008).

There is no place name in the scene on K2796, although the iconography indicates that the event took place inside a mountain. However, a nearly identical representation on the other vessel (K7750) features a longer text (figure 7.5e), which specifies that the event "happened in K'inchil" (*uhtiiy k'inchil*). The dedicatory inscription on this vase states that it belongs to the Naranjo ruler K'ahk' Ukalaw Chan Chahk and the style of this and the other vessel is consistent with the other vases attributed to the Naranjo region. Therefore, it is highly likely that K'inchil is the same place as "K'inchil land" (*k'inchil kab*) in the vicinity of Naranjo. The conquest of K'inchil is celebrated in the narrative on Naranjo Stela 22 ("downfall" in blocks F12–E14 and "burning" in blocks G11–H12) and in the scene on Naranjo Stela 24 showing a captive from "K'inchil land" treaded by the Naranjo queen (Graham and Von Euw 1975, 56, 63). Consequently, at least from the Naranjo viewpoint, God L resided in a specific mountain near a settlement not far from Naranjo, much like the present-day cave-dwelling mountain deities (Vogt 1981; Watanabe 1990) and not in a generic underworld. God L's role as a patron deity of a place conquered by Naranjo also explains some rather extreme depictions of his defeat and humiliation on Naranjo royal vessels from the same period (Beliaev and Davletshin 2007, 188–189; Tokovinine and Beliaev 2013).

A final example of a place with deep-time roots near Naranjo is the hilltop city of Bahlam Jol (the archaeological site of Witzna), which was burned by K'ahk' Tiliw Chan Chahk according to the narrative on Naranjo Stela 22 (Wahl et al. 2019). An elaborate hieroglyph of Bahlam Jol is depicted as a seat of Itzamna/Itzam Kokaaj (God D) in a mythical scene on a vessel that belonged to a sixth-century Naranjo ruler (figure 7.5f). Needless to say, this vision of the sacred landscape directly contradicts the models, which assign deities to layered heavens and underworlds.

Naranjo narratives reflect a distinct concept of the origins of people and their landscape: places that never changed, populated by those who never moved. Not only did the Naranjo kings claim to have originated locally according to the narratives on their monuments, they never made pilgrimages to or received regalia from faraway places. Even in the case of the famously powerful queen "Lady Six Sky" from Dos Pilas (Martin and Grube 2000), the narratives at Naranjo, like the inscription on Stela 29 (Graham and Von Euw 1978, 78), center exclusively on what she did after her arrival at Sa'aal and not on her time at the court of her father, Dos Pilas ruler and holy Mutal lord Bahlaj Chan K'awiil. It is potentially significant that local deep-time origins claims are seemingly restricted to members of a distinct geopolitical group known as *huk tzuk* or the "Seven Divisions" (Beliaev 2000). Unfortunately, the corpus of inscribed monuments from other Seven Divisions sites like Yaxha, Topoxte, Buenavista del Cayo, and possibly Holmul is too small to ascertain whether such insistence on local deep-time origins is a trait shared by all dynasties of Seven Divisions, but the available evidence suggests that it was a regional tradition.

MAKING A PILGRIMAGE

Just as Aztec emperors could send messengers to Aztlan (Durán 1994, 214–222), Classic Maya lords were able to travel to some faraway deep-time places. The narrative on the unprovenanced panel that likely came from the site of Cancuen (it is commonly designated as Cancuen Panel 1) offers a possible example of such pilgrimage (Guenter 2002; Kistler 2004). This inscription commissioned by the king Tajal Chan Ahk presents a retrospective account on the history of four generations of Cancuen rulers. The most interesting section of the text deals with the accession of the king K'iib Ajaw at Calakmul, his subsequent pilgrimage to a faraway place, and the establishment of the royal court at Cancuen (figure 7.8a).[1]

The story begins in Chi'k Nahb (Calakmul) and ends in Haluum (Cancuen)—a place of pools, islands, and springs. A visit to the "Makan Mountain" (*makan witz*) is inserted between the accession in the presence of Calakmul lords and gods and the (re)foundation of the royal court at Cancuen. The travel distances to the Makan Mountain are 102 days from Calakmul and 257 days from Cancuen. The latter distance is particularly impressive and possibly carries a symbolic significance as it approximates a full *tzolk'in* round of 260 days. This is the longest known travel distance in Classic Maya inscriptions and it places Makan Mountain virtually beyond all historical locations in the landscape. Interestingly enough, Calakmul occupies a sort of intermediary position between Makan Mountain and Cancuen.[2]

In addition to deep-time places, at least one historical location beyond the confines of the Classic Maya world plays a prominent role in a set of narratives evoking essential, formative events in the histories of Classic Maya royal

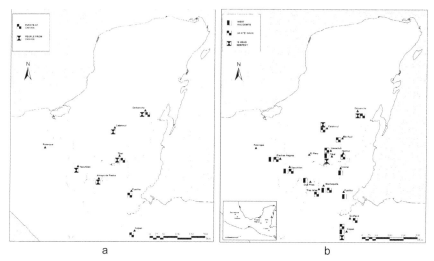

FIGURE 7.6. *Shared origins: (a) Classic Maya sites associated with the Chih Ka' place name; (b) Classic Maya sites associated with Teotihuacan.*

courts. In one way or another, the underlying context of these narratives is the interaction between Classic Maya polities and Teotihuacan during the Early Classic period (figure 7.6b). Key events in this interaction happened in the last quarter of the fourth century, when Ochk'in K'awiil Siyaj K'ahk' became the overlord of many Maya kings, including the rulers of Tikal, one of the most ancient and powerful Classic Maya dynasties (Stuart 2000; W. Fash and B. Fash 2000; Estrada-Belli et al. 2009).

The key place in these narratives is Wiinte' Naah[3] associated with Teotihuacan imagery as well as with the themes of firemaking, calendar rituals, and the foundation of the new political order. It has been even suggested that the Adosada of the Sun Pyramid in Teotihuacan was in fact known to Classic Maya as Wiinte' Naah (W. Fash, Tokovinine, and B. Fash 2009; Estrada-Belli and Tokovinine 2016). Wiinte' Naah was the location of pre-accession rituals and of the final accession ceremony of a Tikal ruler installed by Siyaj K'ahk' (Stuart 2000). Tikal Stelae 4 and 31 depict that new king, Yax Nuun Ahiin, as a young Teotihuacan warrior. According to the text on Tikal Stela 31 (Jones and Satterthwaite 1982, fig. v52), 283 days after his predecessor died, Yax Nuun Ahiin "ascended to Wiinte' Naah." A much later inscription on the incised bone from Tikal Burial 116 (4P-113(61e)/2, MT. 35; Moholy-Nagy and Coe 2008, fig. 195e, 205a) reports that Yax Nuun Ahiin "descended [from?] Wiinte'

Naah" in 61 days. However, in the narrative on Stela 31, he was still at Wiinte' Naah some 261 day later when he acceded to the kingship and "took twenty-eight provinces" by the order of Siyaj K'ahk'.

Even though Tikal became the center of its own hegemony in the reign of Yax Nuun Ahiin's son Siyaj Chan K'awiil, the connection between Tikal rulers and Teotihuacan remained one of the defining narratives in the history of the dynasty during the rest of the Classic period (Haviland 1992; Stuart 2000, 489–490). The very last Tikal inscribed stela (Stela 11:C14) evokes Wiinte' Naah (Jones and Satterthwaite 1982, fig. 16).

Wiinte' Naah also plays a key role in the foundation story that deals with the accession of the first king of the Copan dynasty, K'inich Yax K'uk' Mo'. According to later Copan inscriptions, particularly those on Altar Q, Stela J, and the Hieroglyphic Stairway, the first king of Copan traveled to Wiinte' Naah, where he received the lightning god *K'awiil*, the key insignia of rulership, and only then arrived at Copan in AD 426 (Stuart 2004a, 226–240; 2004c, 376–379). The inscription on Copan Altar Q states that K'inich Yax K'uk' Mo' "takes K'awiil [at] Wiinte' Naah" and "departs Wiinte' Naah" in three days. After traveling for 153 days, K'inich Yax K'uk' Mo' arrives at the Copan Valley, where "K'awiil rests his feet."

Evocations of Teotihuacan imagery and references to K'inich Yax K'uk' Mo' in association with Wiinte' Naah are common in Late Classic Copan monuments and architecture, particularly after its resurgence from a hiatus caused by a defeat in a war against Quirigua (Stuart 2004c; W. Fash and B. Fash 2000; Stuart 2000, 2004a; W. Fash 2002; B. Fash 1992). Quirigua inscriptions also mention the visit of K'inich Yax K'uk' Mo' to Wiinte' Naah as an event related to the foundation of the Quirigua dynasty (Looper 2003). In summary, pilgrimage to the Wiinte' Naah and accession of the ancestor at Wiinte' Naah become a kind of master narrative in the history of the Copan dynasty.

The third example of the Wiinte' Naah–related narrative comes from the inscriptions at Piedras Negras. The main text on Piedras Negras Panel 2 (Stuart 2000, fig. 15.25) informs us that the Early Classic ruler Yat Ahk received a *ko'haw* helmet from "*kaloomte'* of the West Tajoom Uk'ab Tuun" in AD 510. The text also refers to a reenactment of the event, when Mo' Xook Chahk Itzam K'an Ahk (also known as Ruler 2) "took 5 [10?] *ko'haw*" in the presence of his gods in AD 658. The scene on the panel depicts a standing Piedras Negras king accompanied by a prince facing six kneeling lords from various locations including Yaxchilan and Lacanha. As Stuart (2000, 498–499) points out, all are dressed as Teotihuacan warriors and wearing "trapeze-and-ray" helmets identified as *ko'haw* in the text.

A wooden box, currently at Álvaro Obregón Museum in Tabasco, sheds further light on the Early Classic event and suggests that it probably did not take place at Piedras Negras (Anaya Hernández, Guenter, and Mathews 2002; Skidmore 2002; Zender 2007). The inscription on the box mentions the same Early Classic event preceded by a distance number of 155 days. Tajoom Uk'ab Tuun is referred to as "Wiinte' Naah lord." Therefore, the Piedras Negras narrative deals with the pilgrimage to and accession at Wiinte' Naah, much like the narratives at Copan and Tikal.

Early Classic Maya–Teotihuacan interactions and pilgrimages to Wiinte' Naah are illuminated further by two surviving contemporaneous representations of landscapes and traveling. One of them is the mural in Structure 1 at La Sufricaya that was commissioned no later than just a year after Siyaj K'ahk's arrival at Tikal (Estrada-Belli et al. 2009). This mural shows two places represented as temple-like structures connected by the road with a group of characters apparently meeting in the middle (figure 7.7a). The vertical arrangement of features on this *lienzo*-like scene possibly evokes the west–east movement (see Marcus 1992, 153–189). Retrospective accounts highlight a connection of the local Chak Tok Wayaab family to Wiinte' Naah (Estrada-Belli and Tokovinine 2016). Similar murals apparently adorned one of the earliest structures of the Copan royal court, although only small fragments of these murals have survived (figure 7.7b). Either landscape features rows of footprints, which indicate travel and direction of movement in all Mesoamerican visual traditions (Mundy 1998; Berlo 1989, 37–39). The footprints at La Sufricaya mark the road and head from each of the two places to meet in the middle, same as the characters in the scene. The footprints at Copan are not framed by a path and seemingly head in either direction and overlap as if the painter wanted to show a round trip of some sort. It is also important to note that footsteps, especially when not framed by a road or a path, may indicate not only travelling but also social and genealogical connections (Asselbergs 2007, 130–133, 144).

The travel distances between Wiinte' Naah and Classic Maya cities are roughly comparable—145, 153, and 155 days—suggesting that they may in fact reflect travel to the same historical location. With the exception of the Cancuen Panel 1 narrative discussed above, these distances are unmatched in other Classic Maya texts. During the Late Classic period, while there were no more travels to the Wiinte' Naah or other new events involving this place as a real geographical location, the list of royal families associating themselves with the Wiinte' Naah became longer. For instance, the Kaanul lords from Calakmul and Dzibanche acquired the title of "those of Wiinte' Naah" (Prager 2004). However, by the beginning of the seventh century AD, the original

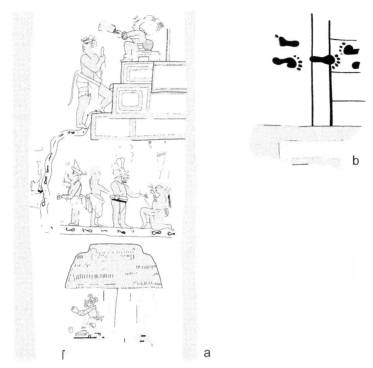

FIGURE 7.7. *Early Classic representations of traveling: (a) detail of Mural 6N, La Sufricaya (after Hurst 2009, fig. 48 and 104 and photographs by the author); (b) mural fragment from K'uk' or Mo' structure, Copan (after Bell 2007, fig. 5.14).*

Wiinte' Naah was a place in the past: the archaeological evidence indicates that Teotihuacan burned to the ground in AD 550 (López Luján et al. 2006).

The *ko'haw* events at Piedras Negras are of additional interest, because they implicate the participation in or the awareness of the *ko'haw* and Wiinte' Naah narratives by other royal courts in the region. This awareness or a shared narrative is confirmed by references in imagery and inscriptions commissioned by members of these dynasties. Late Classic Yaxchilan king Itzamnaah Bahlam appears dressed as a Teotihuacan warrior on Yaxchilan Lintel 25 while his spouse Lady *K'abal Xook* is impersonating a Wiinte' Naah goddess, according to the caption on the same monument (Graham and Von Euw 1977, 55). Middle Classic Lacanha Stela 7 (AD 593) reveals that Teotihuacan associations were likely instrumental in the construction of the political identity of Lacanha and Bonampak rulers (O'Neil 2001). If we take the narrative on Piedras Negras Panel 2 at face value, then the travel of Piedras Negras ruler to Wiinte' Naah and his return home with a new token of greater political authority was followed by visits from his vassals who traveled to Piedras Negras and possibly even received their own *ko'haw* helmets. These travels likely produced new local narratives, now centered on Piedras Negras as the source of Wiinte' Naah relics.

Inscriptions on Cancuen monuments offer one of the best examples of similarly local pilgrimage narratives. As mentioned above, the text on Cancuen Panel 1 (figure 7.8a) claims that Tajal Chan Ahk's ancestors traveled to Calakmul to receive royal crowns and maybe even made pilgrimages to deep-time places. Tajal Chan Ahk's own travel story detailed on the Cancuen Hieroglyphic Stairway (Fahsen, Demarest, and Luin 2003) is less ambitious. According to this text, he "took the *Xot' Te' K'ahk' Hu'n* ('Piece of Wood [?] Fire Headband')" at Dos Pilas, "[in] the sky, [in] the *ch'een*," before the eyes of its lord K'awiil Chan K'inich, and arrived at Cancuen ("the midst of the *ch'een* [at] Haluum") "in four days" (figure 7.8b).[4] The event is also described on the recently discovered Cancuen Panel 2 (figure 7.8c), which clarifies that the object in question was a kind of *piit* palanquin (Fahsen, Demarest, and Luin 2003, 710). Apparently, the same *Xot' Te' K'ahk' Hu'n* palanquin had been previously taken upon accession by the Dos Pilas ruler Itsamnaah K'awiil, according to Dos Pilas Stela 8 (figure 7.8d). It seems unlikely that Tajal Chan Ahk left Dos Pilas with or upon a royal relic. A more plausible explanation would be that he undertook a ritual aimed to boost his political status and to demonstrate his loyalty to Dos Pilas kings. At best, he might have gotten a "copy" of the object, just like those Usumacinta lords who received their *ko'haw* at Piedras Negras.

While Tajal Chan Ahk sent his ancestors on long journeys to Calakmul and Makan Witz in the narrative on Cancuen Panel 1, his own much celebrated trip to hold a Dos Pilas royal headband four days away from home might reflect Cancuen's actual position in the geopolitical landscape. However, these Cancuen monuments were commissioned after the collapse of the Dos Pilas hegemony and after Tajal Chan Ahk's victories over Machaquila rulers (Martin and Grube 2000; Kistler 2004; Fahsen, Demarest, and Luin 2003). Therefore, the narratives on the Cancuen Hieroglyphic Stairway and Panel 1 may reflect two distinct strategies: stressing a direct continuity between Dos Pilas and Cancuen dynasties in terms of claims of regional hegemony and placing Cancuen in a larger ideational landscape in order to strengthen these claims.

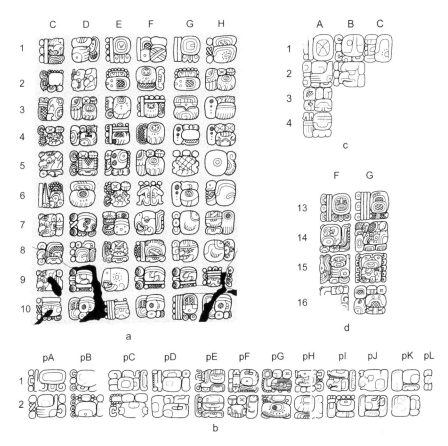

FIGURE 7.8. *Traveling in Cancuen narratives: (a) detail of Cancuen Panel 1 (drawing by Yuriy Polyukhovych); (b) detail of Cancuen Hieroglyphic Stairway 1 (after photographs by the author); (c) detail of Cancuen Panel 2 (after photograph by Dmitri Beliaev); (d) detail of Stela 8, Dos Pilas (after photograph by Ian Graham, Peabody Museum of Archaeology and Ethnology, 2004.15.1.516.1).*

If we consider the narratives commissioned by subroyal elites like biographies on tombstones of *sajal* courtly officials (Tokovinine 2007a), travel distances get even smaller. In such cases, the travel of one's lifetime is commonly composed solely of a particularly successful visit to the royal court. One of the best examples of this genre is the Dumbarton Oaks panel PC.B.539 (Miller 1984; Houston 2012), which highlights the *sajal*'s visit to Piedras Negras, where he danced together with the king at the royal palace (Tokovinine 2007a, 4,

15–16, fig. 3). This text leaves an impression that it was the only important thing that happened to the protagonist besides the fact that he was born, acceded to *sajal*-ship, and lived for over 80 years. Another monument from the same area, El Cayo Panel 1 (Chinchilla M. and Houston 1993), details how its *sajal* "ascended" to Piedras Negras. Interestingly enough, the main character of the El Cayo narrative travels to the court while his father accedes to *sajal*-ship after the death of the predecessor. The protagonist of the narrative on the Dumbarton Oaks panel also travels before the appointment and accedes to the *sajal* office several years after the visit. His age at the time of the visit, however, suggests that it was hardly his first time at the court and that the dance constituted some kind of acceptance ceremony, possibly an introduction to the new monarch, and that it was an important step prior to the actual appointment. The narrative on El Cayo Panel 1 highlights a visit to the capital well before the protagonist's accession but it also implies some kind of important transition that the visit brought about.

YOUR FRIENDLY NEIGHBORHOOD SACRED MOUNTAIN

Distant places play an important role in Classic Maya narratives about the origins of royal power and about changes in the political fortunes of individual kings or even entire dynasties. Classic Maya rulers chose between two different strategies of weaving the histories of their families into a larger historical and mythical canvas. While some royal dynasties apparently traced themselves to distant ancient places, others claimed to have occupied the same landscape for thousands of years. This latter strategy is known primarily from the inscriptions at Naranjo and it is potentially significant that Naranjo's rulers never sought to travel to boost their political standing, or at least never mentioned such travels on public monuments.

Other powerful dynasties looked back at the political landscape in the fourth century AD and the pilgrimages to the city of Teotihuacan as another set of foundational events supporting their claims to greater geopolitical standing. The number of royal families evoking their ancestral links to Teotihuacan peaked well after the city was burned and abandoned. One may even wonder if Teotihuacan ceased to be a distant and yet potentially accessible location and became yet another deep-time place of origins.

Lesser lords like Cancuen kings relied on narratives that highlighted visits to the capitals of other Maya rulers. Cancuen inscriptions also offer an excellent example of how the narratives on Classic Maya monuments could reflect the changing fortunes of the lords who commissioned them. The narrative strategies adopted by Tajal Chan Ahk seem to combine an emphasis on the continuity or even ascendancy between the former regional hegemony of Dos Pilas and Cancuen with evocations of larger historical and deep-time landscape in the accounts of the ancestral travels to Calakmul and distant Makan Mountain.

On the even smaller political scale of subroyal elites, distances become shorter and the list of powerful places is reduced to the court of the king. However, the underlying trope is not unlike the royal pilgrimage narratives. The emphasis seems to be on earlier pre-accession visits, which were presumably of some transformative nature and prepared the protagonist for the future office.

At any given scale, Classic Maya landscapes defy uniform models and definitions. Perceptions changed depending on the spatial and temporal location of the narrative. Same mythological places could be assigned different cardinal directions and even the perceived confines of the Classic Maya world varied through time and space. Moreover, the sacred geography was linked to actual places, local history, and politics, much like the ethnographically documented Maya landscapes.

ACKNOWLEDGMENTS

This book chapter grew out of a presentation for the session "Discourses on distance among the Maya" at the 107th Annual Meeting of the American Anthropological Association in San Francisco in 2008. I would like to thank the session chairs, Cameron McNeil and Timothy Pugh, and other participants for many insightful comments and suggestions, which led to a substantial improvement of the original manuscript. The Classic Maya Place Name Database Project was supported by grants from FAMSI and Harvard University. My mentors and colleagues Stephen Houston and Dmitry Beliaev have contributed enormously to the success of that project. I appreciate the invitation from Ana Díaz to participate in the present volume. My thanks also go to Yuriy Polyukhovych, who kindly agreed to use part of his drawing of the Cancuen panel. Of course, I remain solely responsible for any errors and omissions.

NOTES

1. (B10) **11-AJAW** (C1) **8-MUWAAN-ni** (D1) **K'AL-ja** (C2) **K'AHK'-?-wa** (D2) **XOOK-HU'N** (C3) **tu-BAAH** (D3) **k'i-ba-AJAW** (C4) **K'UH-ya-?-AHK-AJAW** (D4) **yi-chi-NAL** (C5) **KAL-TE'** (D5) **ya-AJAW-MAN-na** (C6)

5-KOHKAN-K'UH (D6) YAX-HA' (C7) CHAHK-ki (D7) u-KAB-ji-ya (C8) yu-ku-CH'EEN-na (D8) 3-TE'-TUUN-ni (C9) KAL-TE' (D9) u-TZ'AK-AJ (C10) 2-10-WINIK-ji-ya (D10) i-u-ti (E1) 9-IK' (F1) 5-IHK'-a-AT (E2) HUL-li (F2) ma-ka-na (E3) wi-tzi (F3) 7-9-?-ni (E4) 6-12-<u>pa-k'a</u> (F4) ju-? (E5) a-AK'-no-ma (F5) k'i-ba-AJAW (E6) K'UH-ya-?-AHK-AJAW (F6) 4-? (E7) CHAK-ka-ja-li-bi (F7) a-ni (E8) ?-<u>CHAN-AHK</u>-na (F8) AJ-?-na (E9) MO' (F9) u-TZ'AK-AJ (E10) 17-12-WINIK (F10) i-u-ti (G1) 10-KAWAK (H1) 2-YAX-SIHOOM (G2) HUL-li (H2) k'i-ba-AJAW (G3) KAJ-yi (H3) ?WAL-AKAN (G4) ?WAL-tz'e-ka (H4) ?WAL-NAHB (G5) 3-AHK (H5) PET-ne (G6) ?WAL-yo-OHL (H6) a-ku (G7) yu-lu (H7) CHAN-na-HA' (G8) ha (H8) lu-mi

[ta] buluk ajaw waxak[te'] muwaan k'a[h]laj k'ahk' . . . xook hu'n tu baah k'iib ajaw k'uh[ul] . . . ahk ajaw yichnal kaloomte' yajawman jo' kohkan k'uh yax ha'[al] chahk ukabjiiy yuk[noom] ch'een huxte' tuun kaloomte' utz'a[h]kaj cha'[hew] jo' winikjiiy i-u[h]ti bolon ik'jo'ihk'at huli makan witz huk bolon . . . wak lajcha' pak' . . . ak'noom k'iib ajaw k'uh[ul] . . . ahk ajaw chan . . . chakajlib a[h]ni . . . chan ahk aj- . . . mo' utz'a[h]kaj huklajuun[hew] lahca' winik[jiiy] i-u[h]ti [ta] lajuun kawak cha'[te'] yax sihoom huli k'iib ajaw kajaay wal akan wal tz'eek wal nahb hux ahk peten wal yohl ahk yul chan ha' haluum

"[On the day] eleven *Ajaw*, eight *Muwaan*, Fire . . . Shark headband was bound on the head of the holy . . . Ahk lord K'iib Ajaw in front of *kaloomte'*, *yajawmaan*, five stingray spine god, Yax Ha'al Chahk. Huxte' Tuun *kaloomte'* Yuknoom Ch'een had ordered it. Two [days and] five months were put together and then it happened [on the day] nine *Ik'*, five *Ihk'at*; holy . . . Ahk lord K'iib Ajaw, seven-nine . . . , six-twelve *pak'*, the giver, the reddener, came to Makan Witz. . . . Chan Ahk of . . . Mo' ran. Seventeen [days and] twelve months were put together and then it happened [on the day] ten *Kawak*, two *Yax Sihoom*, *K'iib Ajaw* came [here]. He settled in the grass, on the shore, at the pool, three turtle island, in the heart of the turtle, polished sky water, Haluum."

2. Houston (Finamore and Houston 2010) argues that Haluum corresponds to the final destination of the pilgrimage, possibly a place on the seashore because of all its aquatic connotations. However, this hypothesis is effectively refuted by the inscription on the Cancuen Hieroglyphic Stairway 1 (figure 7.8b, see note 4 below) that places Haluum within less than four days of travel from Dos Pilas. Given the available evidence, Haluum should be the ancient name of Cancuen.

3. The translation of Wiinte' Naah remains uncertain, although, according to Albert Davletshin (personal communication 2011) its etymology may have something to do with the term *wi(l)te'* for *tapanco* ("loft") attested in Huastec and Chuj languages. *Wi(l)te'* seems to contain the gloss *wil* that means "to tie" and "to knot" in several Mayan languages including Huastec, Q'anjobal, and Yucatec and a common gloss *te'* for "tree" and "wood."

4. (pA1) 7-[ETZ'NAB] (pB1) 1-UUN-wa (pA2) u-CH'AM-wa (pB2) xo-t'o-TE' (pC1) K'AHK'-HU'N-na (pD1) AJ-CHAK-?JU'-TE' (pC2) K'UH-ya-?-AHK-AJAW (pD2) yi-chi-NAL (pE1) K'AWIIL-la-CHAN-na (pF1) K'INICH-chi-ni (pE2) u-CHAN-na (pF2) AHK-lu-AJAW (pG1) K'UH-MUT-la-AJAW (pH1) u-ti-ya (pG2) T369-HA'-la (pH2) K'IN-HA'-ni-NAL? (pI1) CHAN-na (pJ1) CH'EEN (pI2) tu-4-la-ta (pJ2) i-HUL-li (pK1) TAHN-CH'EEN-na (pL1) ha (pK2) lu-mi

[ta] huk etz'nab juun[te'] uuniiw uch'amaw xot' te' k'ahk' hu'n aj-chak ju' te' k'uh[ul] . . . ahk ajaw yichnal k'awiil chan k'inich k'uh[ul] mutal ajaw ucha'n ahkul ajaw u[h]ti . . . ha'al k'in ha' nal chan ch'een tu-chan-lat i-huli tahn ch'een haluum

"[on the day] seven *Etz'nab*, one *Uuniiw*, man of red ju' tree, holy . . . Ahk lord took the *xot' te'* fire headband in front of holy Mutal lord K'awiil Chan K'inich. It happened [in] . . . Ha'al, K'in Ha' Nal, [in] the sky, [in] the *ch'een*. In four days then he arrived [here] in the midst of the *ch'een*, [at] Haluum."

REFERENCES

Acuña, Mary Jane. 2015. "Royal Death Tombs and Cosmic Landscapes: Early Classic Maya Tomb Murals from Río Azul, Guatemala." In *Maya Archaeology 3*, edited by Charles Golden, Stephen Houston, and Joel Skidmore, 168–185. San Francisco: Precolumbia Mesoweb Press.

Anaya Hernández, Armando, Stanley Guenter, and Peter Mathews. 2002. "An Inscribed Wooden Box from Tabasco, Mexico. Mesoweb Report." http://www.mesoweb.com/reports/box/index.html.

Asselbergs, Florine G. L. 2007. "A Claim to Rulership: Presentation Strategies in the Mapa de Cuauhtinchan No. 2." In *Cave, City, and Eagle's Nest: An Interpretive Journey through the Mapa de Cuauhtinchan No. 2*, ed. Davíd Carrasco and Scott Sessions, 121–146. Albuquerque: University of New Mexico Press.

Barrera Vásquez, Alfredo, Juan Ramón Bastarrachea Manzano, William Brito Sansores, Refugio Vermont Salas, David Dzul Góngora, and Domingo Dzul Poot. 1995. *Diccionario maya: Maya–español, español–maya*. 3rd ed. México: Porrúa.

Beliaev, Dmitri D. 2000. "Wuk Tsuk and Oxlahun Tsuk: Naranjo and Tikal in the late Classic." In *Sacred and the Profane: Architecture and Identity in the Maya Lowlands*, ed. Pierre Robert-Colas, 63–81. Berlin: Verlag von Flemming.

Beliaev, Dmitri D. 2006. "'Verbs of Motion' and Ideal Landscape in the Maya Hieroglyphic Inscription: Paper presented at the 11th European Maya Conference." In *Ecology, Power, and Religion in Maya Landscapes*. Malmö, Sweden.

Beliaev, Dmitri D., and Albert Davletshin. 2007. "Los sujetos novelísticos y las palabras obscenas: Los mitos, los cuentos y las anécdotas en los textos Mayas sobre la cerámica del periodo Clásico." In *Sacred Books, Sacred Languages: Two Thousand Years of Ritual and Religious Maya Literature*, ed. Rogelio Valencia Rivera and Geneviève Le Fort, 21–44. Markt Schwaben: Verlag Anton Saurwein.

Bell, Ellen E. 2007. "Early Classic Ritual Deposits within the Copan Acropolis: The Material Foundations of Political Power at a Classic Period Maya Center." PhD diss., Department of Anthropology, University of Pennsylvania.

Berlo, Janet C. 1989. "Early Writing in Central Mexico: In Tlilli, In Tlapaplli before AD 1000." In *Mesoamerica after the Decline of Teotihuacán, AD 700–900*, ed. Richard A. Diehl and Janet C. Berlo, 19–48. Washington, DC: Dumbarton Oaks Research Library and Collection.

Boot, Erik. 2008. "At the Court of Itzam Nah Yax Kokaj Mut: Preliminary Iconographic and Epigraphic Analysis of a Late Classic Vessel." http://www.mayavase.com/God-D-Court-Vessel.pdf.

Brady, James E., and Pierre R. Colas. 2005. "Nikte Mo' Scattered Fire in the Cave of K'ab Chante: Epigraphic and Archaeological Evidence for Cave Desecration in Ancient Maya Warfare." In *Stone Houses and Earth Lords: Maya Religion in the Cave Context*, ed. Keith M. Prufer and James E. Brady, 149–166. Boulder: University Press of Colorado.

Chinchilla M., Oswaldo, and Stephen D. Houston. 1993. "Historia política de la zona de Piedras Negras: Las inscripciones de El Cayo." In *VI Simposio de Investigaciones Arqueológicas en Guatemala*, ed. Juan Pedro Laporte, Héctor L. Escobedo A., and Sandra Villagrán de Brady, 63–70. Guatemala: Museo Nacional de Arqueología y Etnología.

Christenson, Allen J., ed. 2003. *Popol Vuh: The Sacred Book of the Maya*. Winchester: O Books.

Coe, Michael D. 1973. *The Maya Scribe and His World*. New York: Grolier Club.

Durán, Diego. 1994. *The History of the Indies of New Spain*. Norman: University of Oklahoma Press.

Eberl, Markus, and Christian Prager. 2005. "B'olon Yokte' K'uh: Maya Conceptions of War, Conflict, and the Underworld." In *Wars and Conflicts in Prehispanic Mesoamerica and the Andes, Selected Proceedings of the Conference Organized by the Société des Américanistes de Belgique with the Collaboration of Wayeb (European Association of Mayanists): Brussels, 16–17 November 2002*, ed. Peter Eeckhout and Geneviève Le Fort, 28–36. Oxford: John and Erica Hedges.

Estrada-Belli, Francisco, and Alexandre Tokovinine. 2016. "A King's Apotheosis: Iconography, Text, and Politics from a Classic Maya Temple at Holmul." *Latin American Antiquity* 27 (2): 149–168.

Estrada-Belli, Francisco, Alexandre Tokovinine, Jennifer Foley, Heather Hurst, Gene A. Ware, David Stuart, and Nikolai Grube. 2009. "A Maya Palace at Holmul,

Peten, Guatemala and the Teotihuacan 'Entrada': Evidence from Murals 7 and 9." *Latin American Antiquity* 20 (1): 228–529.

Fahsen, Federico, Arthur A. Demarest, and Fernando Luin. 2003. "Sesenta años de historia en la escalinata jeroglífica de Cancuen." In *XVI Simposio de Investigaciones Arqueológicas en Guatemala, 2002*, ed. Juan Pedro Laporte, Barbara Arroyo, Héctor L. Escobedo A., and Hector E. Mejía, 703–713. Guatemala: Museo Nacional de Arqueología y Etnología.

Fash, Barbara. 1992. "Late Classic Architectural Sculpture Themes in Copán." *Ancient Mesoamerica* 3 (1): 89–104.

Fash, William L. 2002. "Religion and Human Agency in Ancient Maya History: Tales from the Hieroglyphic Stairway." *Cambridge Archaeological Journal* 12 (1): 5–19.

Fash, William L., and Barbara Fash. 2000. "Teotihuacan and the Maya: A Classic Heritage." In *Mesoamerica's Classic Heritage: Teotihuacán to the Aztecs*, ed. Davíd Carrasco, Lindsay Jones, and Scott Sessions, 465–513. Niwot: University Press of Colorado.

Fash, William L., Alexandre Tokovinine, and Barbara Fash. 2009. "The House of New Fire at Teotihuacan and Its Legacy in Mesoamerica." In *Art of Urbanism: How Mesoamerican Kingdoms Represented Themselves in Architecture and Imagery*, ed. William L. Fash and Leonardo López Luján, 201–229. Washington, DC: Dumbarton Oaks Research Library and Collection.

Finamore, Daniel, and Stephen D. Houston, eds. 2010. *Fiery Pool: The Maya and the Mythic Sea*. Salem, MA, and New Haven, CT: Peabody Essex Museum in association with Yale University Press.

Garcia Barrios, Ana. 2006. "Confrontation Scenes on Codex-Style Pottery: An Iconographic Review." *Latin American Indian Literatures Journal* 22 (2): 129–152.

Garcia Barrios, Ana, and Ramon Carrasco Vargas. 2006. "Algunos fragmentos cerámicos de estilo códice procedentes de Calakmul." *Los Investigadores de la Cultura Maya* 14 (1): 126–136.

Graham, Ian, and Eric Von Euw. 1975. *Corpus of Maya Hieroglyphic Inscriptions*, Vol. 2, Part 1: *Naranjo*. Cambridge, MA: Peabody Museum of Archaeology and Ethnology, Harvard University.

Graham, Ian, and Eric Von Euw. 1977. *Corpus of Maya Hieroglyphic Inscriptions*, Vol. 3, Part 1: *Yaxchilan*. Cambridge, MA: Peabody Museum of Archaeology and Ethnology, Harvard University.

Graham, Ian, and Eric Von Euw. 1978. *Corpus of Maya Hieroglyphic Inscriptions*, Vol. 2, Part 2: *Naranjo, Chunhuitz, Xunantunich*. Cambridge, MA.: Peabody Museum of Archaeology and Ethnology, Harvard University.

Grube, Nikolai. 2004. "El Origen de la Dinastía Kaan." In *Los Cautivos de Dzibanché*, ed. Enrique Nalda, 117–131. Mexico: Instituto Nacional de Antropología e Historia.

Guenter, Stanley. 2002. "A Reading of the Cancuén Looted Panel." *Mesoweb*. www.mesoweb.com/features/cancuen/Panel.pdf.

Haviland, William A. 1992. "From Double Bird to Ah Cacaw: Dynastic Trouble and the Cycle of Katuns at Tikal, Guatemala." In *New Theories on the Ancient Maya*, ed. Elin C. Danien and Robert J. Sharer, 71–80. Philadelphia: University of Pennsylvania.

Helmke, Christophe, and Jaime J. Awe. 2016. "Sharper Than a Serpent's Tooth: A Tale of the Snake-Head Dynasty as Recounted on Xunantunich Panel 4." *The PARI Journal* 17 (2): 1–22.

Helmke, Christophe, and James E. Brady. 2014. "Epigraphic and Archaeological Evidence for Cave Desecration in Ancient Maya Warfare." In *A Celebration of the Life and Works of Pierre Robert Colas*, ed. Christophe Helmke and Frauke Sachse, 195–228. Munich: Verlag Anton Saurwein.

Helmke, Christophe, Julie A. Hoggarth, and Jaime Awe. 2018. *A Reading of the Komkom Vase Discovered at Baking Pot, Belize*, Vol. Monograph 3. San Francisco: Precolumbia Mesoweb Press.

Houston, Stephen. 2012. "Carved Panel." In *Ancient Maya Art at Dumbarton Oaks*, ed. Joanne Pillsbury, Miriam Agnes Doutriaux, Reiko Ishihara, and Alexandre Tokovinine, 49–57. Washington, DC: Dumbarton Oaks Research Library and Collection.

Houston, Stephen D., David Stuart, and Karl A. Taube. 2006. *The Memory of Bones: Body, Being, and Experience among the Classic Maya*. Austin: University of Texas Press.

Houston, Stephen D., and Karl A. Taube. 2012. "Carved Panel." In *Ancient Maya Art at Dumbarton Oaks*, ed. Joanne Pillsbury, Miriam Agnes Doutriaux, Reiko Ishihara and Alexandre Tokovinine, 38–47. Washington, DC: Dumbarton Oaks Research Library and Collection.

Hurst, Heather. 2009. "Murals and the Ancient Maya Artist: A Study of Art Production in the Guatemalan Lowlands." PhD diss, Yale University, New Havenb, CT.

Jones, Christopher, and Linton Satterthwaite. 1982. *The Monuments and Inscriptions of Tikal: The Carved Monuments*. University Museum Monograph 44. Philadelphia: University Museum, University of Pennsylvania.

Kerr, Justin. n.d. *Maya Vase Data Base: An Archive of Rollout Photographs*. URL: http://www.mayavase.com/.

Kistler, Ashley S. 2004. "The Search for Five-Flower Mountain: Re-Evaluating the Cancuen Panel." *Mesoweb*. www.mesoweb.com/features/kistler/Cancuen.pdf.

Lacadena García-Gallo, Alfonso. 2009. "Apuntes para un estudio sobre literatura maya antigua." In *Text and Context: Yucatec Maya Literature in a Diachronic Perspective*, ed. Antje Gunsenheimer, Tsubasa Okoshi Harada, and John F. Chuchiak, 31–52. Aachen: Shaker Verlag.

Looper, Matthew George. 2003. *Lightning Warrior: Maya Art and Kingship at Quirigua*. Linda Schele Series in Maya and Pre-Columbian Studies. Austin: University of Texas Press.

López Luján, Leonardo, Laura Filloy, William L. Fash, Barbara Fash, and Pilar Fernandez. 2006. "The Destruction of Images in Teotihuacan." *RES* 49/50: 13–39.

Luin, Camilo, and Sebastian Matteo. 2010. "Notas sobre algunas textos jeroglíficos en colecciones privadas." In *XXIII Simposio de Investigaciones Arqueológicas en Guatemala*, 1235–1250. Guatemala: Museo Nacional de Anrqueologia y Etnologia.

Marcus, Joyce. 1973. "Territorial Organization of the Lowland Classic Maya." *Science*, New Series 180 (4089): 911–916.

Marcus, Joyce. 1992. *Mesoamerican Writing Systems: Propaganda, Myth, and History in Four Ancient Civilizations*. Princeton, NJ: Princeton University Press.

Martin, Simon. 1997. "Painted King List: A Commentary on Codex-Style Dynastic Vases." In *Maya Vase Book 5: A Corpus of Rollout Photographs of Maya Vases*, 847–867. New York: Kerr Associates.

Martin, Simon. 2004. "Preguntas epigráficas acerca de los escalones de Dzibanché." In *Los cautivos de Dzibanché*, ed. Enrique Nalda, 105–115. Mexico: Instituto Nacional de Antropología e Historia.

Martin, Simon. 2015. "The Old Man of the Maya Universe: A Unitary Dimension to Ancient Maya Religion." In *Maya Archaeology 3*, ed. Charles Golden, Stephen Houston, and Joel Skidmore, 186–227. San Francisco, CA: Precolumbia Mesoweb Press.

Martin, Simon, and Nikolai Grube. 2000. *Chronicle of the Maya Kings and Queens: Deciphering the Dynasties of the Ancient Maya*. New York: Thames and Hudson.

Martin, Simon, Alexandre Tokovinine, Elodie Treffel, and Vilma Fialko. 2017. "Estela 46 de Naranjo Sa'al, Petén: Hallazgo y texto jeroglífico." In *XXX Simposio de Investigaciones Arqueológicas en Guatemala*, ed. Bárbara Arroyo, Luis Méndez Salinas, and Gloria Ajú Álvarez, 669–684. Guatemala: Ministerio de Cultura y Deportes, Instituto de Antropología e Historia.

Martin, Simon, and Erik Velásquez García. 2016. "Polities and Places: Tracing the Toponyms of the Snake Dynasty." *The PARI Journal* 17 (2):23–33.

Mathews, Peter. 1994. "Una lectura de un nuevo monumento de El Cayo, Chiapas, y sus implicaciones politicas." In *Modelos de entidades politicos mayas; Primer Seminario de Mesas Redondas de Palenque*, ed. Silvia Trejo, 113–130. Mexico: Instituto Nacional de Antropologia e Historia.

Miller, Mary Ellen. 1984. "Four Maya Reliefs." *Apollo* CXIX (266): 17–20.

Moholy-Nagy, Hattula, and William R. Coe. 2008. *The Artifacts of Tikal: Ornamental and Ceremonial Artifacts and Unworked Material*. University Museum Monograph 118. Philadelphia: University of Pennsylvania Museum of Archaeology and Anthropology.

Mundy, Barbara E. 1998. "Mesoamerican Cartography." In *Cartography in the Traditional African, American, Arctic, Australian, and East Asian Socieities*, ed. David Woodward and G. Malcolm Lewis, 183–256. Chicago: University of Chicago Press.

O'Neil, Megan E. 2001. "Lacanja Stela 7: The Juxtaposition of Past and Present for an Image of the Future." Paper presented at the 6th European Maya Conference "Recapturing Mayan Memories: Historical Representation in Text, Image and Ritual," Hamburg, December 5–9.

Prager, Christian. 2004. "A Classic Maya Ceramic Vessel from the Calakmul Region in the Museum Zu Allerheiligen, Schaffhausen, Switzerland." *Human Mosaic* 35 (1): 31–40.

Robertson, Merle Greene. 1991. *The Sculpture of Palenque*. Vol. 4. Princeton, NJ: Princeton University Press.

Roys, Ralph L. 1973. *The Book of Chilam Balam of Chumayel*. Norman: University of Oklahoma Press.

Sachse, Frauke. 2008. "Over Distant Waters: Places of Origin and Creation in Colonial K'iche'an Sources." In *Pre-Columbian Landscapes of Creation and Origin*, ed. John E. Staller, 123–160. New York: Springer.

Sachse, Frauke, and Allen J. Christenson. 2005. "Tulan and the Other Side of the Sea: Unraveling a Metaphorical Concept from Colonial Guatemalan Highland Sources." *Mesoweb*. www.mesoweb.com/articles/tulan/Tulan.pdf.

Schele, Linda. 1992. "Founders of Lineages at Copán and Other Maya Sites." *Ancient Mesoamerica* 3 (1): 135–144.

Schele, Linda, and Mary Ellen Miller. 1986. *The Blood of Kings: Dynasty and Ritual in Maya Art*. 2nd pbk., repr. with corrections. ed. New York: Braziller; Fort Worth: Kimbell Art Museum.

Skidmore, Joel. 2002. "New Piece or Precolumbian New Box." Mesoweb Report. http://www.mesoweb.com/reports/box/piece.html.

Smailus, Ortwin. 1975. *El maya–chontal de Acalán: Análisis lingüístico de un documento de los años 1610–12. Centro de Estudios Mayas, Cuaderno 9*. México: UNAM Coordinación de Humanidades.

Stuart, David. 2000. "The Arrival of Strangers: Teotihuacan and Tollan in Classic Maya History." In *Mesoamerica's Classic Heritage: Teotihuacán to the Aztecs*, ed. Davíd Carrasco, Lindsay Jones, and Scott Sessions, 465—513. Niwot: University Press of Colorado.

Stuart, David. 2004a. "The Beginnings of the Copan Dynasty: A Review of the Hieroglyphic and Historical Evidence." In *Understanding Early Classic Copan*, ed. Ellen E. Bell, Marcello A. Canuto, and Robert J. Sharer, 215–248. Philadelphia: University of Pennsylvania Museum of Archaeology and Anthropology.

Stuart, David. 2004b. "The Entering of the Day: An Unusual Date from Northern Campeche." *Mesoweb*:www.mesoweb.com/stuart/notes/EnteringDay.pdf.

Stuart, David. 2004c. "A Foreign past: The Writing and Representation of History on a Royal Ancestral Shrine at Copan." In *Copan: The History of an Ancient Maya Kingdom*, ed. E. Wyllys Andrews and William L. Fash, 373–394. Santa Fe, NM: School of American Research Press.

Stuart, David. 2005. *The Inscriptions from Temple Xix at Palenque: A Commentary*. San Francisco, CA: Pre-Columbian Art Research Institute.

Stuart, David. 2006. "The Palenque Mythology: Inscriptions and Interpretations of the Cross Group." In *Sourcebook for the 30th Maya meeting, March 14–19, 2006*, 86–194. Austin: University of Texas at Austin.

Stuart, David. 2010. "Notes on an Inscription Fragment from the Southern Peten." In *Maya Decipherment: A Weblog on the Ancient Maya Script*. http://decipherment.wordpress.com/2010/10/03/notes-on-an-inscription-fragment-from-the-southern-peten/.

Stuart, David. 2011a. *The Order of Days: The Maya World and the Truth about 2012*. New York: Harmony.

Stuart, David. 2011b. "Some Working Notes on the Text of Tikal Stela 31." *Mesoweb*. www.mesoweb.com/stuart/notes/Tikal.pdf.

Stuart, David. 2014. "A Possible Sign for Metate." *Maya Decipherment*. https://decipherment.wordpress.com/2014/02/04/a-possible-sign-for-metate/.

Stuart, David, and Stephen D. Houston. 1994. *Classic Maya Place Names*. Studies in Pre-Columbian Art and Archaeology, No. 33. Washington, DC: Dumbarton Oaks Research Library and Collection.

Taube, Karl A. 2003. "Maws of Heaven and Hell: The Symbolism of the Centipede and Serpent in Classic Maya Religion." In *Antropología de la eternidad: La muerte en la cultura maya*, 405–442. Madrid: Sociedad Española de Estudios Mayas.

Tokovinine, Alexandre. 2007a. "Art of the Maya Epitaph: The Genre of Posthumous Biographies in the Late Classic Maya Inscriptions." In *Sacred Books, Sacred Languages: Two Thousand Years of Religious and Ritual Mayan Literature*, ed. Rogelio Valencia Rivera and Geneviève Le Fort, 1–20. Markt Schwaben: Verlag Anton Saurwein.

Tokovinine, Alexandre. 2007b. "Of Snake Kings and Cannibals: A Fresh Look at the Naranjo Hieroglyphic Stairway." *The PARI Journal* 7 (4): 15–22.

Tokovinine, Alexandre. 2011. "People from a Place: Re-Interpreting Classic Maya Emblem Glyphs." In *Ecology, Power, and Religion in Maya Landscapes*, ed. Christian Isendahl and Bodil Liljefors-Persson, 81–96. Möckmuhl: Verlag Saurwein.

Tokovinine, Alexandre. 2013. *Place and Identity in Classic Maya Narratives*. Studies in Pre-Columbian Art and Archaeology 37. Washington, DC: Dumbarton Oaks Research Library and Collection.

Tokovinine, Alexandre. 2015. "Holy Lords and Holy Lands: Territory in Classic Maya Inscriptions." 80th Annual Meeting of the Society for American Archaeology, San Francisco, CA, April 17, 2015, April 17, 2015.

Tokovinine, Alexandre, and Dmitri D. Beliaev. 2013. "People of the Road: Traders and Travelers in Ancient Maya Words and Images." In *Merchants, Markets, and Exchange in the Pre-Columbian World*, ed. Kenneth G. Hirth and Joanne Pillsbury, 169–200. Washington, DC: Dumbarton Oaks Research Library and Collection.

Tokovinine, Alexandre, and Vilma Fialko. 2007. "Stela 45 of Naranjo and the Early Classic Lords of Sa'aal." *The PARI Journal* 7 (4): 1–14.

Tokovinine, Alexandre, and Vilma Fialko. 2019. "En el cerro de los colibris: El patrón divino y el pasiaje sagrado de la ciudad de Naranjo." In *XXXII Simposio de Investigaciones Arqueológicas en Guatemala, 2018*, ed. Bárbara Arroyo, Luis Méndez Salinas, and Gloria Ajú Álvarez, 825–838. Guatemala City: Ministerio de Cultura y Deportes; Instituto de Antropología e Historia.

Tokovinine, Alexandre, and Marc Zender. 2012. "Lords of Windy Water: The Royal Court of Motul de San José in Classic Maya Inscriptions." In *Politics, History, and Economy at the Classic Maya Center of Motul de San José, Guatemala*, ed. Antonia Foias and Kitty Emery, 30–66. Gainesville: University Press of Florida.

Vogt, Evon Z. 1981. "Some Aspects of the Sacred Geography of Highland Chiapas." In *Mesoamerican Sites and World-Views*, ed. Elizabeth P. Benson, 119–142. Washington, DC: Dumbarton Oaks Research Library and Collections.

Vogt, Evon Zartman, and David Stuart. 2005. "Some Notes on Ritual Caves among the Ancient and Modern Maya." In *In the Maw of the Earth Monster: Mesoamerican Ritual Cave Use*, ed. James E. Brady and Keith M. Prufer, 155–185. Austin: University of Texas Press.

von Schwerin, Jennifer. 2011. "The Sacred Mountain in Social Context: Symbolism and History in Maya Architecture: Temple 22 at Copan, Honduras." *Ancient Mesoamerica* 22 (2): 271–300.

Wahl, David, Lysanna Anderson, Francisco Estrada-Belli, and Alexandre Tokovinine. 2019. "Palaeoenvironmental, Epigraphic, and Archaeological Evidence of Total War among the Classic Maya." *Nature Human Behavior* 3 (10): 1049–1054.

Watanabe, John M. 1990. "From Saints to Shibboleths: Image, Structure, and Identity in Maya Religious Syncretism." *American Ethnologist* 17 (1): 131–150.

Yoneda, Keiko. 2007. "Glyphs and Messages in the Mapa de Cuauhtinchan No. 2: Chicomoztoc, Itzpapalotl, and 13 Flint." In *Cave, City, and Eagle's Nest: An*

Interpretive Journey through the Mapa de Cuauhtinchan no. 2, ed. Davíd Carrasco and Scott Sessions, 161–203. Albuquerque: University of New Mexico Press.

Zender, Marc. 2004. "A Study of Classic Maya Priesthood." PhD diss., Department of Anthropology, University of Calgary.

Zender, Marc. 2007. "Mexican Associations of the Early Classic Dynasty of Turtle Tooth I." The 31st Annual Maya Meetings, Austin, TX, March 11.

8

The Sky, the Night, and the Number Nine

Considerations of the Nahua Vision of the Universe[1]

KATARZYNA MIKULSKA

The aim of this chapter is to contribute to what in the Western tradition is called cosmography, *imago mundi* ("image of the world"), or its German equivalent, *Weltbild*,[2] as conceived of by the ancient inhabitants of central Mexico. Undeterred by the generally accepted model of 13 heavens and nine underworld levels vertically placed one upon another, Jesper Nielsen and Toke Sellner (Nielsen and Sellner Reunert 2009) and Ana Díaz (2009) have recently demonstrated the lack of data confirming that such colonial reconstructions reflect precolumbian beliefs. The authors identified a strong influence of late medieval ideas in these descriptions and images[3] and even a reworking of models belonging to both cultures (Díaz 2009). Here I suggest that the Mesoamerican otherworld is not, as in the Western tradition, something composed of a heaven above and an underworld below, with the human world located in between. Instead, I argue that precolumbian imagery distinguishes between a dark and opaque primordial world, nonhuman, full of dangerous and creative divine forces, and another mundane, human world ordered by the framework or limits of time and the sun and hence more predictable and secure.

THE SKY AND THE "UNDERWORLD"

I begin by briefly summarizing the representations of heaven and the supposed underworld in screenfold books or codices and on artifacts from central Mexico,

DOI: 10.5876/9781607329534.c008

including the codices of what is often called the Borgia Group (primarily the *Borgia* and *Vaticanus B*), and then turn to their descriptions in postconquest alphabetic sources. These observations will allow me to propose a reconsideration of what "above" and "below" imply in precolumbian imagery.

In central Mexican codices the most characteristic representation of the sky is the night sky, with dark gray being the dominant color, with circles and black dots, or with little details in the form of horseshoes, or with stars, conventionally represented as "eyes of the night" (see Mikulska 2008a: 238–241; 2008b). Sometimes the image also contains the design of flint knives (*Codex Borgia* 1976, 49–52 sup.; see also figure 8.1a). The night sky can be recognized by the presence of the moon (*Codex Borgia* 1976, 71; see also figure 8.1b), as is confirmed by examples from the *Codex Mendoza*. In one of them appears the gloss, "this painting signifies the night" (*Essential Codex Mendoza* 1997, fol. 60r; see figure 8.1c, henceforth *Codex Mendoza*), and in other examples the place names Yoallan or "Place of Night" (*Codex Mendoza* 1997, fol. 37r) and Yoaltepec or "Hill of Night" (*Codex Mendoza* 1997, fol. 8r; see also figure 8.1d) are recorded alphabetically, along with very similar graphic forms (see Mikulska 2008a, 238–241; 2008b, 161–164). In codices from the Basin of Mexico, a variant representation of the night sky can be observed, in which the depictions of stars are completely white, and do not match the conventional form of stars known as "night eyes." There is no doubt, however, that this also refers to stars because the toponym of Citlaltepec, "Hill of Stars" (present-day Pico de Orizaba), contains the same pattern (*Codex Mendoza* 1997, fol. 17v; see also figure 8.2b). As suggested by Élodie Dupey, these circles probably represent meteors in the night or cloud-covered sky (Dupey 2010, I, 404), and I will return to the similarity between representations of these two later. At times these circles are accompanied by conventional star signs, as on leaf 16 of the *Codex Borbonicus* (n.d.; see also figure 8.2a). In the same image, another variant of the star motif can be observed: below the black band with stars drawn as white circles, there is another with stars represented as "bulging eyes," so-called Venus signs and flint knives. These combinations are also found in some three-dimensional artifacts such as three pieces currently exhibited at the Museo Nacional de Antropología in Mexico (MNA): a stone *cuauhxicalli*, or altar-like receptacle for human hearts (Seler 1990–2000, 3, 135; see also figure 8.2c), the so-called Altar of Venus (figure 8.2d), and the engravings on two bone musical instruments or *omichicahuaztli* (figure 8.3). The presence of the sun in the engravings on the *omichicahuaztli*, by the side of the star signs and of the meteors, indicates that the sun can appear equally in the dark sky, whether it is a night sky or the day sky covered by clouds.

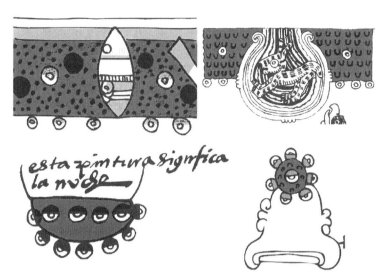

FIGURE 8.1. *Night sky: (a) eyes, black dots, and flint knife (*Codex Borgia, *49); (b) eyes, "horseshoes," and the moon (*Codex Borgia, *71); (c) "Esta pintura significa la noche" (*Codex Mendoza, *folio 60r); (d) place name* Yoaltepec *("Hill of Night"), (*Codex Mendoza, *f. 60r). (Drawings by Nadezda Kryvda).*

More problematic is the identification of the representation of the day or clear sky. In leaves 57–60 of the *Codex Borgia*, where images of a night sky are present, in some of them an image of the sun appears by its side, representing the daytime sky (*Codex Borgia* 1976, 57–60; see also figure 8.4a). In the same codex, in a section in which four weather prognostications are found, and two of rain and two of heat and draught, under the control of the god Tlaloc (*Codex Borgia* 1976, 27), the image of the heated or warm sky is similar to the sun, as if the latter were "unrolled" horizontally (figure 8.4b), since the same ochre, red, and yellow bands appear as well as a line of white dots outlined in black along with red rays. In turn, the representations of the sky in the rainy periods are in reality images of clouds, with the same curved volutes that appear in toponyms associated with "clouds" (figure 8.5a–b), for example, Mixtlan, "Place of Clouds" (*Codex Mendoza* 1997 fol. 46r; see also figure 8.5c). What is noteworthy, and I return to this later, is that although a sky covered with clouds laden with rain is shown, stars can still be seen (in the form of "night eyes"). This provides us with an important indication: at the representational or conventionalized iconographic level stars are drawn to give the notion of "sky," but without following faithfully their appearance in

FIGURE 8.2. *Starry sky: (a) eyes and white dots* (Codex Borbonicus, *16*; drawing by Nadezda Kryvda); *(b) Place name* Citlaltepec *("Hill of Stars")* (Codex Mendoza, *f. 17v*; drawing by Ana Díaz); *(c) a* cuauhxicalli *with bulging eyes, or Venus signs (National Museum of Anthropology, Mexico City); (d) Altar of Venus, also with bulging eyes (National Museum of Anthropology, Mexico City; digital reconstruction by Ana Díaz).*

reality. These representations of the sky covered with clouds, taken together with the center of the following leaf (*Codex Borgia* 1976, 28; see figure 8.5b) are extremely similar to those of the night sky, with its signs of stars, little points, and black circles, and in fact were it not for the curves and volutes, it would probably have been interpreted this way. This, in my opinion, leads to the conclusion that painting the sky with a dark color (black, gray, or dark blue), along with dots, black circles, and stars, primarily indicates darkness, whether resulting from night or thick clouds covering the sun.

An image in the *Codex Borgia* represents the interior of the earth or the World of the Dead (*Codex Borgia* 1976, 42; see also figure 8.6a). A dead man passes through the mouth of the earth monster to find himself before the Lord of the World of the Dead. The latter is characterized by the presence of owls or owl-like creatures, which in other sources are mentioned as messengers of the underworld (e. g., *Popol Vuh* f. 13v, in Craveri 2013, 59; Sahagún 1950–1982, V, 161, henceforth *FC*). Human body parts, which in the *Manuscrito*

FIGURE 8.3. *Detail of the starry band at the top of a musical instrument or omichicahuaztli. (Drawing by Nadezda Kryvda.)*

de la Real Academia de la Historia (*RAH*)—published as the *Primeros Memoriales* (Sahagún 1997, henceforth *PM*)[4]—are listed as food of the lords of the World of the Dead (*RAH* 1997, fol. 84r; *PM* 1997, 177; Mikulska 2008a, 289). What interests me more, however, is that the background of this enclosure is painted in the same way as the night sky in this same codex, and is similar to the images of the sky obscured by clouds. Other depictions in which the interior of the earth is painted in the same way are found in the *Codex Borgia* (1976; see also figure 8.6b) and in the *Codex Laud* (Mikulska 2008a, 241–242; 2008b, 164–165). These representations of the "night sky" painted in a form so similar to those of the cloud-covered sky in the *Codex Borgia* (1976, 27, 28), and to those of the interior of the earth or precinct of the World of the Dead, may indicate that there was some conceptual similarity between these spaces. This recalls what Michel Graulich called "consubstantiality" between the night sky and the earth, giving as examples images of the body of the earth covered with stars (*Codex Borgia* 1976, 43) or the findings of Konrad Theodor Preuss that among the Cora (Náayarite), the earth is the starry sky and the moon, and at sunset, "black water and stars emerge from the innards of the earth and darken the sky" (Graulich 1990, 76). This also brings to mind what Johannes Neurath has pointed out concerning the concept of the sky among the modern Huichol (Wixáritari). To them, the night sky or the sky of the rainy season is not directly opposed to the day or dry-season sky because they

Figure 8.4. *Clear daytime sky: (a) the sun, representing the daytime sky (Codex Borgia, 58); (b) the heated sky with "unrolled" sun (Codex Borgia, 27) (Anders, Jansen, and Reyes 1993, redrawing by Ana Diaz).*

are ontologically different. The "underworld" is the same as the dark world, the rainy season, and the night, and is something that has always existed. It is an eternally existing time of rain and fertility, source of creation, which comes to dominate the world when the day sky is destroyed (see Neurath, chapter 9, this volume).

Considering the expressions encountered in the written texts, we find that the word *ilhuicatl* ("sky") is used in reference not only to the "sky" which from the Western point of view is above the earth, but also to various places which are at the same time destinations of the dead (Schwaller 2006, 397–399; Mikulska 2008a, 231–232; 2008b, 152–154).[5] Thus, this place can be "the House of the Sun, the Sky" (*tonatiuh ichan ilhuicatl*) (*FC*, III, 49; VI, 162, 38), where warriors killed in combat go, where women who die in first childbirth go as well as children offered to Tlaloc.[6] It can be the "Tlalocan Heaven" (*in ilhuicac Tlallocan*), the place of Tlaloc, where all those who die from some aquatic cause go, such as those who drown or are struck by lightning (*FC*, XI, 68). But the majority of the dead go, as colonial sources maintain, to Mictlan, "Place of the Dead" (*FC*, III, 41), which, like the two places mentioned previously, also appear qualified by the noun *ilhuicatl* ("sky"), as if this term were used to designate any realm outside of the world of men (*FC*, VI, 31).

In some cases, Mictlan occurs in conjunction with the locative *topan* ("above" or "above us") (*FC*, VI, 2, 5, 11, 33, 48), despite the fact that the

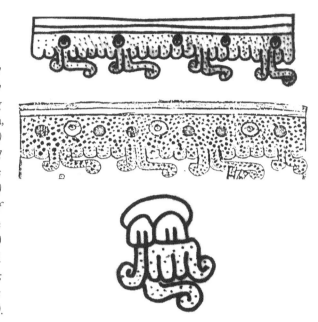

FIGURE 8.5. *Cloudy sky: (a, b) sky with volutes representing clouds (*Codex Borgia, *27 and 28, respectively) (Anders, Jansen, and Reyes 1993; redrawings by Ana Diaz); (c) Mixtlan, "place of clouds" (*Códice Mendoza, f. 46r) *(after* The Essential Codex Mendoza *1997; redrawing by Ana Diaz).*

supposed "underworld" ought to be located "below." The problem, I believe, is that in utilizing the English words "underworld" or "netherworld" (or the Spanish equivalent "inframundo"), we are led by implicit assumptions to imagine that this place is always found "below." In a place that is *topan* we should rather expect to find Omeyocan, home of Ometecuhtli and Omecihuatl (considered to be the primordial pair of deities). Philological analyses of the texts of Book VI of the *Florentine Codex* show that Mictlan and Omeyocan exhibit more similarities than differences. Here I present only the conclusions drawn from other works (Mikulska 2008a, 225–235; 2008b). Both are places of residence of a creator deity, described with the words "he who creates people," "he who engenders people," "he who gives beginning to people," "progenitor, he who engenders people," "he who makes or composes people" (*FC*, VI, 4, 141, 195). Other common titles are "father, mother," a diphrasism (see below) that indicates the quality of parental and protective beings, and Mictlantecuhtli and Mictecacihuatl are referred to interchangeably with this title (*FC*, VI, 21, 27, 31, 48, 59), as well as Ometecuhtli and Omecihuatl (*FC*, VI, 168, 183, 187). As much in Mictlan as in Omeyocan, the creation and birth of human beings are accomplished through acts expressed with the verbs "to say" (*itoa*) and "to create" or "to make" (*iocoya*) (*FC*, VI, 2,

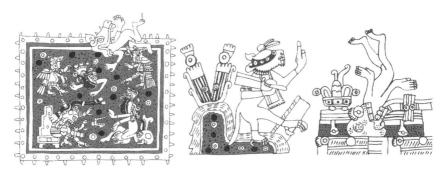

Figure 8.6. *The interior of the earth, or the World of the Dead, with backgrounds similar to the night sky. (Codex Borgia: (a) 42, (b) 13, and (c) 57; drawings by Nadezda Kryvda.)*

31, 49). An examples that shows the proximity, if not identity, of both places, is a fragment of a speech given to a woman recently made pregnant. We are thus told that the place where the decision is made regarding the creation of a child is "above us, in Mictlan, in the place/time of night" (*in topan in mictlan, in iooaian*; *FC*, VI, 141), and then it is stated that Ometecuhtli and Omecihuatl were the ones who decided it. After the birth, in the ritual oration, the arrival of the child from Mictlan or from Omeyocan is described, and using the same verbs, such as "to be sent" (*mioallia*), "to come" (*huitz*), and "to descend" (*temo*). Therefore, I suggest that both places, Mictlan as well as Omeyocan, were understood as places of creation, resident places of gods who were creators and protectors of human beings. The only difference between them noted in these ritual speeches is that the dead went only to Mictlan, while Omeyocan is not mentioned in this context, just as the latter place is not associated with the dead. Thus, some of the *huehuetlatolli* of Book VI of the *Florentine Codex* state:

In oquinpolo, in oquintlati totecujo,	[destroyed them, hid them], our lord,
in vevetque, in ilamatque, . . . ca oquinmihoali	[to the old men, to the old women ("the ancestors")] . . . sent them
in atlan in oztoc in Mictlan (*FC*, VI, 195)	[in the water, to the cave] to Mictlan
ca ie quicevitoque	already they are resting/ becoming cold
in itloc, in inaoac	[near there, by its side]
in tonan, in tota in mictlan tecuhtli (*FC*, VI, 152; see also VI, 136, 27)[7]	of [our mother, our father] Mictlan tecuhtli

But still, is it actually the presence of the dead that distinguishes Mictlan from Omeyocan? We have just seen that the three places mentioned most frequently as those where the dead go are called *ilhuicatl* apart from their "proper names" of *Mictlan, Tlalocan,* or *Tonatiuh ichan*. Moreover, various Mesoamerican myths recount histories of gods who after their death are converted into stars and appear as such in the sky. This happens with various groups of "four hundred": the four hundred *mimixcoa* (plural of Mixcoatl, the "Cloud Snake") of the myth of the creation of war (*Leyenda de los Soles* 2002, 184–187; henceforth *LS*) closely related to the creation of pulque (*Historia de los mexicanos por sus pinturas* 2002, 40–43; henceforth *HMP*), and who are inveterate drunkards, the *huitznahua* of the myth of the birth of Huitzilopochtli; and the four hundred youths of the *Popol Vuh*, drunkards like the *mimixcoa*, and slain by Sipakna (*Popol Vuh* f. 9v, in Craveri 2013, 43–44). The close relationship between these groups has been noted by some investigators (e.g., Graulich 1990, 172–177; Taube 1993). Other categories of deceased who had as their destination the celestial realm, the *tonatiuh ichan*, were warriors killed in battle, women who died in first childbirth, and children sacrificed to Tlaloc. Of the first it is known that after four years they were converted into different types of birds, such as hummingbirds and orioles, or into various types of butterflies (*FC*, III, 49). Those who died in childbirth were converted into clouds, according to Jacinto de la Serna (1953, 306; see also Klein 2000, 28; Ragot 2000, 140; Contel n.d.). This fact is partly confirmed by Fray Gerónimo de Mendieta, who provides a repertoire of the beings into which the dead were converted according to the beliefs of the Tlaxcaltecans: mists, clouds, birds with rich feathers, and precious stones; although those of low class were converted into ground-dwelling and unpleasant animals (Mendieta 1993, 97). Similarly, in a report from 1645, cited by Arturo Gómez (see Pérez Castro 2012, 205–206), there is mention of dead children being converted into clouds, and other persons being changed into winds, whirlwinds, birds, or flies. These beliefs survive to the present day among various indigenous groups across Mesoamerica.[8] It is noteworthy that several of these beings, if not all, appear to be residing in the celestial realm, which in turn seems to have been quite a populated place.

Moreover, the dreaded *tzitzimime* ("the bony ones"; see below),[9] aside from threatening humankind with destruction during eclipses, and at the end of fifty-two year cycles, also appear to have taken on the function of supplying the world with "rain, water, thunder, lightning bolts and lightning" (Alvarado 2001, 262). They were considered stars (*Codex Telleriano-Remensis* 1995, fol. 4v; Graulich 1980, 265; Taube 1993, 5; Boone 1999, 191; Klein 2000, 17; Mikulska 2008b, 248), but at the same time they were skeletal beings. The *Historia de los*

FIGURE 8.7. *The Venus "star."* *(Codex Nuttall, 10; Anders, Jansen, and Pérez 1992; redrawing by Ana Díaz.)*

mexicanos por sus pinturas (2002, 81) describes them as follows: "In the second (heaven) they say that there are some women who do not have flesh, only bones; and they are called tetzauhcihua ['*tetzahuitl* women'],[10] and by another name tzitzimime." Their proper name, *tzitzimitl* in the singular, derives from the ancient Proto-Uto-Aztecan root for "bone" (**øuH*), which by virtue of reduplication follows the pattern for names of smaller objects or imitations (Dakin 1996, 315), indicating therefore that they were "imitations" of skeletons (see also Mikulska 2008a, 219).[11] In fact, in their only surely identified graphic representations (indicated by the name in an accompanying gloss)—that is, the *çiçimime* of the *Codex Tudela* (*Códice Tudela* n. d., fol. 46r) and the *Codex Magliabechiano* (1983, fol. 76r)—are shown without flesh. This tells us not only that there are dead people in the celestial realm, but that they are skeletons, as can be appreciated from the aforementioned *omichicahuaztli* bone rasps (see figure 8.3). There is also a noteworthy depiction in the *Códice Nuttall* (2007, 10) in which the Venus "star" has a fleshless mandible (see figure 8.7), similar to those of beings from the world of the dead.

THE NUMBER NINE: WHAT IS ABOVE, IS BELOW

Beginning with the examples cited above, where similarities between the graphic representations typical of the celestial realm and underworld reinforce each other, it is possible to propose that the concept[12] of Mictlan and that of the night are equivalent (López Austin 1996, I, 428; Graulich 1980, 260; 1990, 282; Siarkiewicz 1982, III, 245; Mikulska 2008a, 235ff.; 2008b). Consequently, the following question arises: Are we speaking of an above or of a below? The night sky is located above, but as already mentioned, in English the terms "underworld" and "netherworld" are inextricably associated with a lower space. Thus, the domain of Mictlantecuhtli, is usually conceptualized intuitively as a specific location, which should perhaps be called "the world beyond," as for example, Elżbieta Siarkiewicz (1982, 111, 245) does, or perhaps "otherworld" or "the beyond." My objective is not to propose an alternative translation or understanding of the term Mictlan, but to use some descriptive term that does not imply any *a priori* projection of "below" and that may possibly be less burdened with connotations belonging to the Western tradition.

The conceptual proximity between Mictlan and night, for example, can be partially explained thanks to ideas recorded in the *Codex Telleriano-Remensis* (1995, fol. 20r): "they say that when the sun sets it goes to give light to the dead"; in the *Florentine Codex* (*FC*, VI, 163; see Mikulska 2008a, 235); or among the ancient Maya (see Schele and Freidel 1990, 76; Milbrath 1999, 251), and the contemporary K'iche' (Milbrath 1999, 40–41) and Lacandon (Ramírez 2002, 17), as well as the Otomi (Hñähñu) (Galinier 1990, 491). This still does not explain, however, why Mictlan, at least in the texts of the ritual speeches, is always located *topan*, "up" or "above us," just like Omeyocan, as can be seen from the following passages:

Cuix ic itolo in topan in Mictlan in ilhuicatl (*FC* 1950–1982, VI, 141)	"perhaps thus it was said [above us, in Mictlan]"
In topan in mictlan in ilvicac (*FC* 1950–1982, VI, 5)	"above us, [in Mictlan, in the sky]."
in topan in omeiocan in chicunauhnepaniuhcan (*FC* 1950–1982, VI, 206)[13]	"above us, in Omeyocan, in the place of the nine unions"

As can be seen, there are two sets of words that appear in reference to Mictlan and Omeyocan. One is *in topan in mictlan in ilhuicac* ("above us, in Mictlan, in the sky") (*FC*, VI, 3, 5, 37), and the other is *in topan in omeiocan in chicunauhnepaniuhcan* ("above us, in Omeyocan, in the 'place of the nine unions'") (*FC*, VI, 175, 176, 183).

Given that these two expressions, augmented by the toponym *topan*, are diphrasisms, it is essential to give some notion of what this term encompasses, although, in spite of the frequency with which it has been used, researchers still do not agree concerning its definition. This is not the place to review all the definitions that have been proposed since Ángel María Garibay K. (2007, 19).[14] Instead, I will concentrate on some of the most recent. Within the last few years, the research of Mercedes Montes de Oca merits special attention. She defines diphrasisms as "the juxtaposition of two terms which are associated to construct a unity of meaning which may or may not be distinct from what each lexeme states," indicating that the relation between them is of distinct types, arising from opposition via "synonymy, complementarity, interdependence, and [a] generic-specific [relationship]" (Montes de Oca 2013). As to the order of the lexemes, this depends on the type of diphrasism: probably in those found in the process of lexicalization, their order is stable (with the second term as final), while in others it is free (ibid.). With regard to diphrasisms, Montes de Oca considers elements that are intrinsic to Mesoamerican languages, that comply with the linguistic criterion of frequency, and that refer to an expected and not an aleatory referent.

In contrast, Danièle Dehouve defines diphrasisms at the discursive level, and her objective has been to show the logic of the creation of diphrasisms in oral discourse while also going beyond the purely linguistic level, because the results of the same cognitive processes are perceptible at the material or ritual level (Dehouve 2009, 21, 23–24; 2011, 157–158). Dehouve's hypothesis is that the diphrasisms (and all their variants: tri-, quadri-, or multiphrasisms)[15] are the result of the definition of a referent by extension or enumeration, from which are created what she calls metonymic or metaphoric series. These operate under these two principles (metaphoric and metonymic), understood according to the definition of Lakoff and Johnson (2007, 39–42ff., 73–78). In other words, the metonymic series with which a referent is defined, are created with a basis in a metaphoric mental concept.[16] These series can be formed by different numbers of components, although binomials will be the preferred form.

As to the particular case of the expression *topan in mictlan in ilhuicac*, Montes de Oca considers that it involves a generic-specific relation between the lexemes that comprise it: *ilhuicatl* being the generic term and *mictlan* a part of that space (Montes de Oca 2000, 213–214; 2013). Meanwhile, Alfredo López Austin argues that it concerns a set of opposed complementaries, basing his explanation on a fragment of a *huehuetlatolli* in a manuscript from the Biblioteca Nacional de México (Ms. 1477 fol. 233 *apud* López Austin 2003, 155), that relates the following:

Ihuan onca mani in coyahuac tezcatl in necocxapon	and there is a great mirror perforated from both sides;
in mictlan ontlaneci,	here appears the region of the dead;
inic oncan tontlachixtica; in quenamican,	with it you are looking at the place which is in some way,
in mictlan	the region of the dead,
in ilhuicac	the sky:
inic tonitztica in nohuian in cemanahua	with it you are seeing all parts of the world[17]

López Austin thus suggests that "it is explicit here that the diphrasism 'the region of the dead, the sky'" refers to "all the parts of the world" (López Austin 2003, 155). In my opinion, although such meaning is clear from the above passage, the combinations of *ilhuicatl* with other components that designate other places where some of the dead go (*tonatiuh ichan ilhuicatl*, "the House of the Sun, the Sky"; *in ilhuicac Tlallocan*, "the Sky, Tlalocan"; Montes de Oca 2000, 214; 2013) show *ilhuicatl* as a more general space (and not necessarily opposed to Mictlan). I therefore find the proposal of Montes de Oca more convincing: that Mictlan is a specific part of the more generic *ilhuicatl*. In any event, if Omeyocan is in any sense opposed (or complementary) to Mictlan, *topan* always appears in the discourse, and there is no indication of any "below" (which would be expected at least in cases when Mictlan is mentioned emphasizing its particularity, that is to say, as a place of the dead, to mark its opposition). Nor is there anything to suggest that *in mictlan in ilhuicatl* is, as a whole, any more above than below.

Nevertheless, it is not my intention to demonstrate that Mictlan, the World of the Dead, is above. It would appear thus if we based ourselves only on the citations I have presented so far. However, taking into account other kinds of data, it appears to be a place that more than anything is quite dynamic, or perhaps even all-encompassing. In graphic expressions, images of the "night sky" are situated below, as for example in an image of the *Codex Borgia* (1976, 57; see also figure 8.6c), in which "death" or the earth is represented in the form of a skull that is devouring a dead man, and that is placed on the night-sky background. Additionally, similar conceptual and symbolic overlap between "above" and "below" appears to be present in two versions of the same myth recorded by Konrad Theodor Preuss (among the Nahua of Durango) that tells the story of a man who goes to the land of the dead to look for his wife. According to one informant, the man flew below, and according to another, the man was carried by a buzzard "up until they reached the sky" (Preuss 1982, 474–477, 482–483; see also Mikulska 2008a, 233–234).

This appears to be the result of the same phenomenon on which Richard Haly comments, regarding the same expression *in topan in mictlan*. According to his interpretation, the polysemic characteristic of Nahuatl comes into play here, or it may be that *topan* refers to the two poles of the dialectic relation "below/above," given that "typical of the dynamics of an oral tradition, while the names change, the relationship encoded between them remains constant" (Haly 1992, 289). In other words, it is a matter here not so much of a category that deals with opposition as with reciprocity. In the World of the Dead, Haly suggests, things are "created, formed, arranged determined, and imagined (*oyocoloc*) as they are *topan* 'over us'" (Haly 1992, 288–289). Here the accession of a ruler to the throne or the birth of a child is decided (Haly 1992, 288), and again we may note an ambiguity regarding a possible association "fixed" in a particular space or cosmological region. In a ritual speech that mentions the arrival of a child from the realm of creation to the human world, we read:

ca çan in otonvitza,	from where did you come?
ca çan in otonquiçaco,	from where did you leave?
in otimoquetzaco	raise yourself up,
ca mictlan	from Mictlan,
ca yluicac	from the sky
in otontemoc	you descended[18]
(*FC*, VI, 33)	

It is not clear if the human being "raises itself" or "descends" but it is evident that the words used form opposing pairs. Returning to the question whether the diphrasisms *in mictlan in ilhuicac* ("in Mictlan, in the sky") and *in omeiocan in chicunauhnepaniuhcan* ("in Omeyocan, in the 'place of the nine unions'") are formed by means of lexemes that are in relations of synonymy, complementarity, or interdependence, but taking into account the prior conclusion that Mictlan is part of the more general *ilhuicatl* (the type of relation that Montes de Oca describe as "generic-specific"), more light on the identity of Mictlan and Omeyocan may provide a deeper understanding of what *Chicunauhnepaniuhcan* refers to. It is exactly this question I turn to next.

WHERE IS THE "NINE-PLACE"?

Chicunauhnepaniuhcan is described in various sources as the place of residence of the supreme pair. This couple is designated sometimes with a pair of names, sometimes with more, with the most frequent being Tonacatecuhtli and Tonacacihuatl, Ometecuhtli and Omecihuatl, and Citlallatonac with Citlalicue. It is, however, noteworthy that, first, they are not necessarily in all the sources located in the supposed thirteenth heaven, and second, that when they appear as separate pairs, they can be located in another supposed "level." Thus, according to the *Historia de los Mexicanos por sus Pinturas* (*HMP*), Tonacatecuhtli and Tonacacihuatl reside in what is called the thirteenth heaven (*HMP* 2002, 25). The couple are the creators of other deities (*HMP* 2002, 80–81), and according to the *Leyenda de los Soles (LS)*, it was they who received Nanahuatl when he was transformed into the sun (*LS* 2002, 183). In this same source, in the form of Citlalicue and Citlallatonac, they are the ones who become angry with Tata and Nene (*LS* 2002, 176–177), and according to the *HMP* the pair are situated in "the first heaven" (*HMP* 2002, 80–81). Finally, a pair named Ometecuhtli and Omecihuatl is situated above the supposed 12 heavens according to two different texts (*FC*, X, 169; Ponce de León 1985, 121) and in the thirteenth heaven according to another (Thevet n.d., fol. 83r; henceforth *HM*).[19] In the later work of Fray Juan de Torquemada, the couple is mentioned as: "Ometecuhtli and Umecihuatl, called by another name Citlallatonac and Citlalicue" and is situated on top of the 11 heavens (Torquemada 1976, III, 66–67). In the *Anales de Cuauhtitlan* (*AC*) it is reported that the ruler of Tollan, Quetzalcoatl, "was praying inside heaven," named Omeyocan, and was not invoking Ometecuhtli-Omecihuatl, but instead *çitlali ycue, çitlallatonac tonacaçihuatl tonacateuctli, tecolliquèqui, yeztlaquenqui, tlallamanac tlallichcatl* (*Annals of Cuauhtitlan* 1992, 8, *side* 4) who were in "the Place of Duality, to the Nine Layers" or *ommeyocan chiucnauhnepaniuhcan* ("to the Place of Duality, to the Nine Layers") (Bierhorst 1992, 30; cf. trans. by Primo Feliciano Velázquez in *Anales de Cuauhtitlán* 1975, 8). A new pair of names is given to the couple in the already-mentioned text of Sahagún, where Ometecuhtli and Omecihuatl are called Ilhuicatecuhtli and Ilhuicacihuatl, or "Lord of the Sky, Woman of the Sky" (*FC*, X, 169).

As can be seen, in only two of the mentioned primary sources (*HMP* and *HM*) is the residence of the pair supposed to be the "thirteenth heaven," but neither of the two sources is in Nahuatl and both are quite truncated. In contrast, we have mentions of the "highest heaven" of "nine heavens," "11 heavens," "12 heavens," and including "first heaven" for Citlalicue and Citlallatonac. It is also striking that in the *Florentine Codex, matlactlanepanolli vmome*, or 12 *tlanepanolli*, is found, the latter word undoubtedly formed from the root *-nepan-*, the translation of which I will explore more fully in what follows.[20] According to the *AC*, the place where Quetzalcoatl was praying was *Ommeyocan Chiucnauhnepaniuhcan*, which is again a place name formed by "nine" and by the root *-nepan-*. Nor do 13 "heavens" or strata appear in the *Codex Vaticanus A* (*Códice Vaticano A* 1996), since we only find nine colored layers in folio IV, the figure of the supreme gods on top of them, and then the two "celestial" levels lower in the following folio (fol. 2r) and the surface of the earth below. Thus, a good deal of imagination or manipulation is required in order to join the nine layers of one folio, two of the other, then add the terrestrial level and finally consider that the presence of Tonacatecuhtli is evidence of yet another level, and thus reach a total count of 13. Nevertheless, as Díaz (2009, 38) showed, in the processing of the image of the universe contained in the *Codex Vaticanus A*, the indigenous idea of something "nine" has been adopted, joining it with two celestial layers to form a model corresponding to the 11 levels of the Renaissance model. This is confirmed by the Italian commentary, which says "the Otomi call this place where he is *hive narichnepaniucha*, which means 'on the nine structures or compositions', and by another name Homoyocan, or 'the Place of the Triune Lord'" (*Códice Vaticano A* 1996, fol. IV; italics added).[21] That is to say, the place called *Chicnauhnepaniuhcan* appears again (Anders, Jansen, and Reyes 1996, 45), which should correspond to the *ommeyocan chiucnauhnepaniuhcan* of the *AC* and to *in topan in omeiocan in chicunauhnepaniuhcan*, known from the *huehuetlatolli* of the *FC* (see above).

The word *chicnauhnepaniuhcan*, although it is written in various manners, consists of the numeral "9" (*chicnauh / chiconauh / chicunauh*), the verbal root *-nepaniuh-*, and the locative suffix *-can*. Understanding the verbal is both essential and problematic (Mikulska 2008a, 231; Díaz 2009, 20–22; Nielsen and Sellner Reunert 2009, 403–404). According to López Austin (1995, 91), *nepaniuh* comes from the transitive verb *nepanoa.nitla* that Alonso de Molina translates as "to join or put together one thing with another," and as "to join one thing with another, or throw one thing on top of another" (Molina 2005, [1] fol. 73v, [2] fol. 68v).[22] The word can equally refer to binding on a horizontal level, for example *monepanoa yn atl* is "the confluence of rivers" (Molina 2005, [1] fol. 7v). Based on the word "cross vault," *vitoliuhcanepaniuhqui*, which designates the architectural element that unites arches in Christian churches, Díaz proposes that *chicnauhnepaniuhcan* can also refer to horizontal composition or union, as a "crossing of roads, ways or whatever other concept, and not precisely the superposition of floors, as up to now has been proposed" (Díaz 2009, 21–22; Nielsen and Sellner Reunert 2009, 405).[23] Moreover, López Austin, although he explains the word *chicnauhnepaniuhcan* as a "succession of nine cross-shaped figures, nine superpositions," also provides arguments that allow it to be understood as a point of union, confluences, crosses, and "crossroads" (López Austin 1995, 91–93). As Nielsen and Sellner Reunert (2009, 402; chapter 1, this volume) have observed, the representation of the horizontal universe based on the *Codex Fejérváry-Mayer* (1971, 1) in fact shows a union of nine spaces: that is, four directions, four intercardinal directions between them, and the center.

In the conjuries collected by Hernando Ruiz de Alarcón, *chicnauhnepaniuhcan* appears written with a small modification, as *chicnauhtlanepaniuhcan*, that the curate himself translates as "the nine junctions or pairings" (Ruiz de Alarcón 1953, 140). This word occurs in the conjury used to recuperate the lost *tonalli* of a child (VI-3).[24] Specifically, the curer asks where the *tonalli* of the child is detained, if it is in the "Nine-above, Nine-confluences" (*chicnautopa, chicnauhtlanepaniuhcan*). However, as Richard Andrews and Ross Hassig also note in their commentary (Andrews and Hassig 1984, 166), there is no indication here that this concerns a strictly celestial domain. For comparison, among the contemporary Nahua of San Miguel Tzinacapan (Puebla), the lost *tonalli* usually travels to Talocan, which today is imagined as existing under the *tlalticpac* or the surface of the earth (Knab 1991, 34–35). Obviously, we do not know to what degree this current conceptualization is influenced by coexistence with Euro-Christian ideas, because we are dealing with concepts recorded five hundred years after the Spanish conquest. Nevertheless, what is important is that the lost *tonalli*, according to this community's conceptualization, is not found in the celestial domain and "above." The issue becomes more complex when making a comparison with the some of Maya groups of Chiapas. Thus, according to the Tzeltal, the places in which the *ch'ulel* (an animistic entity, or soul-force comparable to the *tonalli* of the Nahua), can abandon the body during life and later return to it. While lost and away from the human body it can be within the earth, in one of the hearthstones, in the buds in the crown of a tree, in the interior of a rock, in the rattle of a serpent, in a niche of a church, and so forth (Pitarch 2006, 46–47). Equally, in a prayer to recover a lost *ch'ulel* recorded by Ulrich Köhler, the places where it can be found are equally varied: "outside the community," "in the four pillars of the sky, in the four pillars of the earth," "in the dark blue hill, in the dark blue sierra," "on the seashore, on the shore of the ocean," or "in a Ladino town, in an Indian town" (Köhler 1995, 33–37). Moreover, among the Totonacs, the soul that leaves the body has to be sought at the place where it was lost, or where the sudden fright or *espanto* was produced (Ichon 1973, 176–177). Although these examples make for only a limited comparison, it shows that the places where a *tonalli* or other animistic entity susceptible to loss can be found are quite varied. Thus, it is difficult to establish if Ruiz de Alarcón's *chicnautopa, chicnauhtlanepaniuhcan* of the Nahua conjuries refers definitively to a place "above," taking into account that some fifty years before, in the *huehuetlatolli* collected by Bernardino de Sahagún, Omeyocan as much as Mictlan is described as being *topan*.

Furthermore, *chicnauhnepaniuhcan* or "place of the nine confluences" is not the only place that appears to be named with this numeral. *Mictlan, topan, chicnauhmictlan* or "Mictlan, above, Nine-Mictlan" appears in another conjury documented by Ruiz de Alarcón, and is mentioned as the location to where it is necessary to go to obtain a new bone, that is, a bone to cure a broken bone. Obviously, the conjury reflects the well-known origin myth, according to which the bones of the dead ancestors, retrieved from Mictlan, served for the creation of new humans (*LS* 2002, 179–179; Andrews and Hassig 1984, 371n8). The curer says: "I go to Mictlan, I go above us, I go to Nine-Mictlan" (*niani mictlan, niani topan, niani chicnauhmictlan*; VI-22). In this case it is more probable that the name of the otherworldly realm is provided merely by synonymic words, and in the myth the dead ancestors, Mictlan and the domain of Mictlantecuhtli are explicitly mentioned. In addition, *Chicnauhmictlan* or "Nine-Mictlan" appears in another Nahua conjury written down by Ruiz de Alarcón, and which serves "to induce sleep" (II-2). Here I reproduce the fragments of interest:

Nòmatca nèhuatl ninoyoalitoatzin,	I myself, I am He who Speaks of Night,
inic nèhuatl,	because I am
inic chicnauhtopa,	because in nine-above us
iniquac	when [. . .]
tlaxihualhuin in temicxoçh,	Please, come, flower of dream
iniquac in nicanato, in nohueltitih	When I came to seize my elder sister
chicnauhtopa.	Nine-above us
Nitlamacazqui	I am *tlamacazqui*,
in nohueltiuh xochiquetzal	My elder sister is Xochiquetzal
inic çenca quipiaya in tlamacazque,	Because the *tlamacazque* guarded her much
in mochintin in quahuili in occelome [ocēlōmeh]	all the eagles and ocelots
in ayhehuel [ayac huel] *calaquia*;	none was able to enter
inic nictzàtzili in cochiztli,	so I called the dream
inic chicnauhmictlan yàque,	thus they went to nine-Mictlan
inic nèhuatl nixolotl, nicapani tli [nicapanilli]	because I am Xolotl, I am Capanilli
in çan tlalhuiz notivan nitzàtzi . . .	he who without care shouts everywhere . . .
inic ye nichuicaz in chicnauhmictlan,	So that I will carry her to nine-Mictlan
in oncan nic-huicaz tlalli innepantla	So that I will carry her there, to the middle of the earth

Original text from Ruiz de Alarcón 1953, 63; with corrections from Andrews and Hassig 1984, 79. Translation from Nahuatl to Spanish by Katarzyna Mikulska and from Spanish to English by Jerome Offner.

The conjurer first sends the guardians of the woman and then the woman herself to the place named "Nine-Mictlan" (*chicnauhmictlan*), but it means that she is falling asleep "Nine-Mictlan" must be, therefore, a time-space of dream and of unconsciousness. It is a place—but also time—"in the middle of the earth" (*tlalli inepantla*); it is then similar to the interior of the earth that we saw before drawn in graphic form in the same way as the night sky. This interpretation can be confirmed through the view of the contemporary Nahua, whose *tonalli* wanders around during the sleep (Sandstrom 2010, 343), and if by any chance the *tonalli* loses itself in this other world, or any other health problem happens, the curer sends his/her *tonalli* there, also during the dream, to look for the lost *tonalli* or to consult with the beings living there about the origin of the illness (Knab 1991, 34–37). As for the region named at the beginning of the conjury, "Nine-above us" (*chicnauhtopan*), Andrews and Hassig interpret it as "state of consciousness" perhaps influenced by Ruiz de Alarcón himself, who in another conjury (against scorpion stings, see VI-32) identifies it with the heavens. The priest comments: "With this the goddess Xochiquetzal climbed up, and covering him with her *huipil*, caused him to fail in his purpose; and the cause of this fall was the said Xochiquetzal's being a stranger and Goddess who came from the heavens, that they called *chicnauhtopan*, which means of the nine places" (Ruiz de Alarcón 1953, 177). However, Andrews and Hassig themselves observe a recurrent association of the locative *topan* with Mictlan, supposing that the number "nine" serves to refer to the otherworldly realm in its totality (Andrews and Hassig 1984, 21–22). Given this consideration, it is perhaps not strange that the border between the human and supernatural worlds would be called *chicunavatenco*, "Nine-at the Water's Edge." This place is mentioned in the context of the death of a dog, after which its owner asks it to wait for him at *chicunavatenco*, "Nine-at the Water's Edge," until he himself dies. In addition, according to the text of the *Manuscrito de la Real Academia de la Historia*, this same dog is called *chicunavizcuintli*, "Nine-Dog" (*RAH* 1997, fol. 84r; *PM* 1997, 178). *Chicunauhmictlan* or "Nine-Mictlan" is also the name of the final destination of the dead where they arrive accompanied by their dogs (*FC*, III, 44; Haly 1992, 289).[25] It would thus seem that in all the examples mentioned here, that the key term is the number "nine," which possesses its own semantic value.[26] What interests me is to try understand how such value is created: is it as a reference to a number *per se* present in the universe, or is it rather a matter of a qualitative principle, which consists in a conventional symbolism assigned to the number (see Dehouve 2011, 153). Dehouve lists the names that contain the numeral "nine"—the days "Nine-Death," "Nine-Grass," and "Nine-Dog," which are all related to individuals with supernatural powers, and "Nine-*Cipactli*," which is the name of Mictecacihuatl, goddess of the dead (Dehouve 2011, 169). In addition, the commentary of the *Codex Vaticanus A* (*Códice Vaticano A* 1996, fol. 30v) explicitly states: "He who was born in the day of the nine [in a day with the number nine] was bad, because it was a day dedicated to sorcerers, necromancers, who transformed themselves into other animals" (see Anders, Jansen, and Reyes 1996, 183).[27]

Moreover, not only was the dog who accompanied a person to the otherworld called "Nine-Dog," as already mentioned, it was also the calendrical name of *Chantico*, goddess of the domestic fire (*Codex Telleriano-Remensis* 1995, fol. 21v). In the same way, the god of fire could be called *Chicnauhyotecuhtli*, "Lord Quality of Nine" (Dehouve 2011, 169). These deities of fire are associated with the center where, as we have seen, the nine horizontal regions or compartments are possibly joined (*Codex Fejérváry-Mayer* 1971, 1; Díaz 2009, 21–22; Nielsen and Sellner Reunert 2009, 405). Another common denominator of all the beings designated with the number "nine" is the capacity to penetrate into the otherworld; additionally, "nine" refers to places of death and of darkness (Dehouve 2011, 169).

In sum, "nine" is associated with regions outside of the human world. Indeed, the conjurer who called himself "knower of Mictlan, knower of above" (V-2), ending up calling himself Mictlantecuhtli (V-1), along with other deity names. Thus, Mictlan ought to be this nonhuman space, the world of the ancestors, where it is possible to know your destiny, and it is worth pointing out that the conjurer practically always emphasizes his nonhuman quality and in the same way, his quality of being equal (or even more powerful) than the gods (Contel and Mikulska 2011, 41–55). Another expression that confirms this value of Mictlan is the "ladder of Mictlan," which is the name given to the arm of the patient measured with the conjurer's hand in order to understand the origin of his disease. This procedure is called "to climb the precious ladder" (*tlatlecahuia in chalchiuhecahuaztli*, V-1),[28] but Ruiz de Alarcón himself (1953, 121) indicates that "our precious ladder" (*tochalchiuhecahuaz*) may also be called *tomictlan ecahuaz*, "our ladder [to] Mictlan" (V-1). It is well known that *ecahuaztli* is "ladder of wood to reach something" as Molina explains (Molina 2005, 2, fol. 28r); in other words, it does not mean steps as in a building, but instead a tool. Another "knower of Mictlan, knower of above" (*mictlan mati, topan mati*), is tobacco, invoked in the majority of the conjuries as an indispensable substance for contact with the supernatural realm (II-3). Obviously, we could again use here the interpretation of "opposites" in which Mictlan merely refers to the world below, and *topan* to the realm above, and so in combination, by means of opposition, they designate all the otherworld space. However, it is noteworthy that tobacco is also always and everywhere named in the conjuries with the number "nine." Usually, the conjurer joins two or three names referring to the processing of the tobacco: "Nine-crushed with stone" (*chiucnauhtlatecapanilli*),

"Nine-pounded with stone" (*chiucnauhtlatetzotzonalli*), and "Nine-shredded between the hands" (*chiucnauhtlatlamatellolli*[29]). I want to emphasize that in each of these names, a cardinal number is intended, in no case are repetitions (things to be done nine times) mentioned, as these names are frequently translated, even by Ruiz de Alarcón. The number "nine" appears in still other names: those referring to places of origin (e.g., Haly 1992, 289). These involve the mythical toponyms *Chiucnauhtilihuican*, or "Nine-Mountain," and *Chiucnauhixtlahuatl*, or "Nine-Plain," (*AC*; see Bierhorst 1992, 23)[30] that designates the place from which the Chichimecs emerged to begin their migration.[31] In Mesoamerican creation mythology, the places of origin *par excellence* are caves and it would be difficult to list all the authors who have written on this point (e.g., Heyden 1975; Taube 1986; López Austin 1995, 167–229; Nielsen and Brady 2006, 206–208; Mikulska 2008a, 143–150; Martínez and Núñez 2009, 43 ff.). The *Histoire du Mechique* reports that the first creation took place in a cave at Tamoanchan (*HM* 2002, 148–149). This is the residence of the ancestors, who in archetypal form are presented as a couple (*Códice Vaticano A* 1996, fols. 4v, 6r–6v), known among the Nahua as Oxomoco and Cipactonal (*Codex Borbonicus* n.d., 21; relief of Coatlan, Morelos; painting on the Templo Calendárico de Tlatelolco; Boone 2007, 24–25), and that frequently can be represented in the sexual act as a creative act (Nielsen and Brady 2006). The cave is what specifies this complex at the graphic level as well as the conceptual—in a *huehuetlatolli* a woman says: "we, who are women . . . in us is the cave, the cavern" (*FC*, VI, 118)—which conceptualization is much more universal.[32]

The word "cave" also forms part of another reference to Mictlan: *in atlan in oztoc* ("in the water, in the cave"). This diphrasism is used to name precisely the place where the dead go:

in oquinpolo,	he destroyed them,
in oquintlati totecujo . . .	he hid them, our lord . . .
ca oquinmotlatili in totecujo,	he put them, our lord,
à ca oquinmotoptemili,	he filled the box with them
ca oquinmopetlacaltemili,	he filled the flask with them
ca oquinmihoali in atlan	he sent them into the water,
in oztoc	to the cave
in mictlan	to Mictlan.[33]
(FC, VI, 195)	

It is here that the dead remain, deprived of their bodies and their individualities, left in the form of bones (e.g., López Austin 1995, 222; Mikulska 2008a,

227; Martínez and Núñez 2009, 45), the same as the ruling couple of the same place, Mictlantecuhtli and Mictecacihuatl. In their case, however, the bones do not indicate only their quality as dead, but to the contrary, also their capacity to create or recreate. This is precisely the concept of bone in Mesoamerica, as many authors have indicated. "Bone" thus refers simultaneously to death and to subsequent re-creation and rebirth (Furst 1982; Haly 1992, 278–290; Mikulska 2008a, 217–225; Martínez and Núñez 2009, 45–46) and the bones "generate life" as López Austin formulates it (López Austin 1995, 173–174). For the same reason, creator deities are often bony in appearance—for example, "Lord One-Deer" and "Lady One-Deer" of the Mixtecs (see *Codex Vindobonensis* 1974, 51)—just as earth and terrestrial goddesses, responsible for reproduction, have fleshless mandibles and, frequently, skirts adorned with bones (*Codex Borgia* 1976, 49–53; see also Mikulska 2008a, 256–260). Likewise, the platforms made to celebrate the completion of the 52-year cycles are adorned with skulls and crossed bones (Mikulska 2008a, 254–256); Huitzilopochtli is a "bony" deity, "made of bones," because he himself is the one charged with dying and being reborn (Mikulska 2008a, 263–274); the clothes of the new *tlatoani* in the accession ceremony are adorned with bones, because the ruler suffers a ritual death (Townsend 1987; Olivier 2008; Mikulska 2008a, 262–263), just to mention some of the most telling examples.

The creative powers of the couple Mictlantecuhtli and Mictecacihuatl are revealed as they are addressed as *motechiuhcauh* ("your progenitor") and *in tlacachioani, in tlacaiolitiani, in tlacapeoaltiani* ("he who creates, engenders, gives beginning to men") (*FC*, VI, 4, 195), just as their activities are described with expressions that include the verbs *itoa* and *iocoya*, "to say" and "to make" (*FC*, VI, 2, 31, 49; Mikulska 2008a, 227–229; 2008b, 154–155; see also above).[34] Summing up these seemingly ambiguous and, from a Western perspective, conflicting associations of Mictlantecuhtli and Mictecacihuatl, Richard Haly (1992, 286) remarks:

> I am trying to overcome our resistance to seeing the deity who created life as the "Lord of the Land of the Dead." This distinction is easier for us if we disassociate the skeletal figure of Omiteuctli/Mictlanteuctli from that of the Grim Reaper. The "Lord of the Land of the Dead" is not "Death" itself. Or if he is, he is also "Life." In Mesoamerica, death precedes life.

I previously pointed out that Mictlan and Omeyocan present more similarities than differences, both being places of creation (see also Haly 1992, 288–289). When speaking of the dead, only Mictlan is mentioned, which in this context can also be called *Chicunamjctlan*, or "Nine-Mictlan" (*FC*, III, 44;), we have seen that it is not clear to which of the two places or realms, Mictlan or Omeyocan, the name *Chiuhnepaniuhcan* refers, since it refers more generally to a realm beyond the ordinary human world. If one of the most important places of origin is Chicomoztoc, translated always as "Seven Caves," this calls attention to Michel Oudijk's observation that, in Zapotec sources, such places are named *Billegaa* and *Billehegache* ("Cave Nine" and "Cave Seven"); Oudijk emphasizes that we are probably not dealing with words that mean "Seven Caves" or "Nine Caves" (Oudijk 2011, 158, 169, no. 9). It should also be mentioned that no plural is indicated in the word *Chicomoztoc* either, because *oztotl* ("cave") is an inanimate noun and therefore cannot be pluralized, and because *Chicomoztoc* is a locative as well. What is relevant, in my opinion, is that numbers incorporated in names function more to transmit a certain semantic value than to indicate the quantity of something, and for this reason I consider that pluralizing nouns that form composite names changes their original idea.

Without going into a more comprehensive analysis of the interplay of the numbers seven and nine, I consider it clear that *in atlan in oztoc* ("in the water, in the cave") is another name for the World of the Dead, the domain of Mictlantecuhtli and Mictecacihuatl, a place beyond the human world, and as such it can be as much a place of death as of origin and creation. Although the "the water, the cave" could indicate the World of the Dead as being situated "below" (Mikulska 2008a, 234), but, being a place of ancestors and fleshlessness, it also appears to be the same space-time as the interior of the earth and night. Referring to it as "the water, the cave" underlines its principal characteristics, namely that it is a place of moisture and of darkness, as is typical of most caves. The informants of Sahagún described "cave" in this manner: "The cave: . . . is a frightening place, a fearful place, place of the dead; thus it is called the Place of the Dead, because there is death there, there the night entered, it is night" (*FC*, XI, 276). Caves are always dark and humid places, and in them are found the sources of springs and rivers, and at a conceptual level, of life and fertility. These expressions again coincide with what Neurath remarks concerning the ontological identity of the night sky with the place of eternal fertility, or eternal place of fertility, that is at the same time darkness, night, and the rainy season (Neurath, chapter 9, this volume).

CONCLUSIONS

Without wanting to propose definite conclusions, I think that a number of viable suggestions and hypotheses have been developed in this contribution.

First, the differences between the supposed celestial world or the heaven and the world of "below," that is, Mictlan or the World of the Dead, which we are accustomed to call the "underworld," have been shown to be fewer than the similarities. Mictlan and Omeyocan are both places of creation, although the first is also the place of the dead, confirming its creative capacity. There appears to be a temporal distinction between them, since, while the day sky is in the celestial dome during the day, by night the nocturnal sky occupies its place, full of the dead: stars or mythical ancestors. This is most probably a matter of ontologically different spaces or realms: the day sky being the space organized by time, and the night being an eternal source of creation, fertility, or the enigmatic, disordered, and dangerous primordial "chaos." Perhaps this realm can best be understood as an "otherworld" existing *par excellence* outside of the human world. However, access to this realm could be obtained by those specialized in contacting the divine world through dreams or through altered states of consciousness.

This otherworld is frequently qualified by the number "nine." It forms names that designate the final destination of the dead, the place that is the source of life-giving bone matter; *Chicunauhmictlan* ("Nine-Mictlan"), the place of creation and the residence of the creator couple (*Chicnauhnepaniuhcan* or "Place of Nine-Confluences"); the border between the human and supernatural worlds (*Chicunavatenco*, "Nine-At the Edge of the Water"); and the origin place of humans (*Chiucnauhtilihuican* or "Nine-Mountain"[35] and *Chiucnauhixtlahuatl* or "Nine-Plain"). It is possible that *Chicnauhnepaniuhcan* or "Place of the Nine-Confluences" also refers to the totality of the cardinal directions (as can be appreciated in the *Codex Fejérváry-Mayer* 1971), and it could be precisely these that form the principal division of time-space in ancient indigenous thought.

Therefore, I suggest that the number "nine" refers to the divine or supernatural realm generally inaccessible to humans and that can only be reached at night, in dreams, and by ritual specialists. In closing, it is worth mentioning that my conclusions recall the information collected by Laureano Reyes Gómez about the Zoque vision of the "underworld," according to which the three "spaces" or realms outside of the human world are at times marked by the path of the sun, within which time does not advance at all, or does so slowly (Reyes Gómez 2008, 100–106). Although in the graphic model elaborated by Reyes Gómez the path of the sun at night is seen as "below," in terms of the ancient versions, as has been seen, this space can appear located as much within mountains, caves, or the interior of the earth as above or below the night sky, because it is a nocturnal time-space made of the same "substance" as the interior of the earth, to use Michel Graulich's expression (Graulich 1990, 76), or, as Neurath suggests, because it is ontologically different from the day sky (Neurath, chapter 9, this volume). Therefore, Mictlan or the World of the Dead, being at the same time the night sky, is more than anything a realm beyond, which during the night would be found above, owing to its dynamic nature, or perhaps it was imagined that with the temporary disappearance of the sun, the dark, primordial sky with its immense creative potential reclaimed the entire cosmos until the sun was born anew.

ACKNOWLEDGMENTS

I would like to express my profound gratitude to Ana Díaz, Jesper Nielsen, and Toke Sellner Reunert for the thought-provoking and inspiring discussions that resulted in this chapter, and in particular to Ana for the invitation to contribute to the present volume. Additionally, I owe thanks to various colleagues who read the manuscript and helped me with their invaluable observations and comments: José Contel, Danièle Dehouve, Elodié Dupey, Johannes Neurath, Jerry Offner, and two anonymous reviewers who read the Spanish version of this chapter. To all these as well as to Mercedes Montes de Oca, I also express my appreciation for providing me with texts still in the course of publication, which were of invaluable assistance. Last but not least, my infinite *gracias* to Jerry Offner, for his very careful translation of this text.

NOTES

1. Translated from to English from Spanish by Jerome Offner. Original text published in Mikulska (2015a).

2. *Weltbild* is a concept from the German philosophical tradition that includes the "image of the origin (of the genesis) and of the structure of the world" (Griffioen 1989, cited in Iwaniszewski 2009, 21) and encompasses "the presentations of the world, the concepts of time and of space, the origins of the world, the basic similarities and differences between things, plants, animals and human beings" (Iwaniszewski 2009, 21). The term *cosmography* comes from the Greek tradition and, according to Stanisław Iwaniszewski, "defines the study and the description (or the mapping) of the whole world or of the cosmos such as it appears in a given moment, with respect to its structure, its possible divisions into levels, regions, etc." (Iwaniszewski 2009, 22). Iwaniszewski recently presented a history of these concepts applied to Mesoamerican studies, adding appropriate critical comments concerning the use of the words

cosmovision and *cosmology* (Iwaniszewski 2009, 11–36). I thank Johannes Neurath for calling my attention to the subtleties of these terms and concepts.

3. In 2000 José Contel observed that the early colonial descriptions of the otherworld of the Nahua echoed the medieval vision of the Old World of the three destinations (heaven, hell, and purgatory) of the souls, beginning with the number of these places referred to among the Nahua (Mictlan, and Tonatiuh ichan, without mentioning at the same time Chichihuacuauhco, Cihuatlampa or Cincalco), and ending with the association of Tlallocan with the Earthly Paradise located at the mountain near the Heaven of the Moon, just as this place was located in the *Divina Commedia* of Dante Alighieri (Contel 2000, 617–630).

4. Taking into account the preliminary results of the investigation in progress by Miguel Ángel Ruz Barrio of what are called the *Códices Matritenses*, I choose to indicate from which of the manuscripts (*Manuscrito de la Real Academia de la Historia* [henceforth *RAH*] or *Manuscrito del Real Palacio* [MRP, not cited in this chapter]) the information comes, although I will also provide the reference to the published text, *Primeros Memoriales* [henceforth *PM*].

5. Again it is worth noting that the colonial sources in some manner reproduce the Western model, mentioning the three destinations (Mictlan, Tlallocan, and Tonatiuh ichan) without mentioning the others, which were Chichihuacuauhco, Cincalco or Cihuatlampa (Contel 2000, 617–630; Ragot 2000, 85–179).

6. The text of a prayer to Tlaloc is as follows: *Auh in piltzintli, in conetzintli in oc tototl, in oc atzintli: in aia qujmomachitia, ca chalchiuhtitiaz, ca teuxiuhtitiaz in ilvicac in tonatiuh ichan* (*FC*, VI, fol. 38): "And the child, the little child, still a bird, still a little water, he who still knows nothing, will be made *chalchihuitl*, will be made turquoise, in the sky, in the house of the sun" (translation from Nahuatl to Spanish by Katarzyna Mikulska, and from Spanish to English by Jerome Offner). It would thus seem that the children sacrificed to Tlaloc had as their destination the same glorious space as those warriors killed in battle and women who died in childbirth, in contrast to those who died in early childhood, whose destination was *Chichihualcuauhco*.

7. Translation from Nahuatl to Spanish by Katarzyna Mikulska, and from Spanish to English by Jerome Offner.

8. Regarding the persistence of these concepts among the Huastec (Téenek), see Pérez Castro (2012, 206–207, 225); among the P'urhépecha there appears to have been the conviction that the dead returned to the sky in the form of birds, while they called the stars "our ancestors" (Zantwijk 1967, cited in Martínez 2013, 235); Johannes Neurath reports similar beliefs it among the Huichol (Wixáritari) (Neurath, personal communication, 2012). For his part, Graulich recalls the information collected by Thompson and Preuss, concerning the beliefs of the Tz'utujil Maya of Panajachel and among the

Cora (Náayarite), respectively, according to which the dead were transformed into stars (Graulich 1990, 275).

9. Originally, the beings called *tzitzimime* appear to have been more feminine; in the colonial era, in accordance with being associated more with the Devil and Lucifer, their gender changed to masculine (cf. Klein 2000, 18–21).

10. *Tetzahuitl* is commonly translated as "bad omen," due to its translation by Molina as *cosa escandalosa, o espantosa, o cosa de aguero* ("something terrifying, or something like an omen"; Molina 2005, [2] fol. 111r). However, other translations, like *marauilla* ("miracle"; Molina 2005, [1] fol. 82r), together with the fact that deities are called *tetzahuitl* (e.g., Huitzilopochtli; Castillo 2001, 92; Chimalpahin 1998 I, 178; see Mikulska 2008a, 376), suggest that it means something rather "prodigious, extraordinary, miraculous," and because of these characteristics, can cause horror.

11. Although it has frequently been thought that the *tzitzimime* beings are recognizable because they "come down" from the sky, based on a commentary in the *Codex Telleriano-Remensis* that mentions the "fall of the demons that they say were stars" and calls them "*tzitzimitli*, a person who says a monstrous or frightful thing" (*Codex Telleriano-Remensis* 1995, fol. 4v), in my opinion their most characteristic feature is precisely that they are skeletons (Mikulska 2008a, 219, 245–254) that threaten human beings with being devoured (Boone 1999, 198). This is confirmed by a Nahua myth from the south of Durango, recorded by Preuss (1960, 7), concerning the goddess of the Moon Tepusilam, who was murdered and eaten by the star-men (see Mikulska 2008a, 219).

12. In other works (Mikulska 2008a, 52–53, 76–77; 2008b, 151–152) I have resorted to semiotic terminology to talk about a mental *concept* or *idea* that allows elaboration of different *expressions*, or exterior manifestations of the same mental concept in different media (linguistic, ritual, graphic, and others). Based on the research of Lakoff and Johnson (2007), Danièle Dehouve calls "macrometaphor" what I called *concept* or *metaphoric concept* (basing myself on the same study of Lakoff and Johnson), and considering in the same way that this mental idea is the basis for the formation of different manifestations (Dehouve 2009, 26; Mikulska 2008a, 50–52).

13. Translation from Nahuatl to Spanish by Katarzyna Mikulska, and from Spanish to English by Jerome Offner.

14. I refer the reader to the works by Montes de Oca (2000, 5–15; 2013, 10–14) and by Dehouve (2009, 20–21).

15. In a similar manner, Michela Craveri (2004, 63, 2012, 78), who works with Maya diphrasisms, also considers that diphrasisms are created at the discursive level, and that the di-, tri-, cuadri-, and other multiphrasisms are variants of the same process, although the binomial is the preferable and most frequent form. Moreover, she observes that in some texts preference is given to binomials, while in others to multiphrasisms (Craveri 2012, 77, 89; personal communication, 2013).

16. Dehouve considers that it is not only the metaphoric process that basically forms this cognitive procedure, as Lakoff and Johnson proposed, but also the metonymic process (Dehouve 2013).

17. Translation from Nahuatl to Spanish by Alfredo López Austin (2003, 155), and from Spanish to English by Jerome Offner.

18. Translation from Nahuatl to Spanish by Katarzyna Mikulska, and from Spanish to English by Jerome Offner.

19. Although Ometeotl and Omecihuatl appear as inhabitants of the thirteenth heaven in Rafael Tena's edition of the *Histoire du Mechique* (2002, 145), in my opinion, after having consulted the original of the work, what is written is *Ometecuhtli* and *Omecihuatl*, although with a different spelling; see Mikulska (2015b, 70–74).

20. According to Alonso de Molina, *tlanepanoliztli* is "joined in this way" (Molina 2005, [1] fol. 73v), and *tlanepanolli* is "thing collated or verified" (Molina 2005, [2] fol. 128r), while in Cortés and Zedeño, in GDN we find "[j]oined with another joined thing" (GDN).

21. Translation from Italian to Spanish by Ferdinand Anders and Maarten Jansen (Anders, Jansen, and Reyes 1996, 45), and from Spanish to English by Jerome Offner.

22. In Spanish: *juntar o ayuntar vna cosa con otra* and *juntar vna cosa con otra, o echar vna cosa sobre otra* (Molina 2005, [1] fol. 73v, [2] fol. 68v). However, another possibility is that -*nepaniuh*- is the agentive of the impersonal form of the verb, *nepanihui*, which Molina translates as "to agree or harmonize in what they say" (Molina 2005, [1] fol. 28v), and Wimmer as "to be joined, to be united, to accord with each other, to mingle" (GDN). See also Andrews and Hassig (1984, 224).

23. Regarding errors in the translation of *chicnauhnepaniuhcan*, see Nielsen and Sellner Reunert (2009, 403–404), and the fuller discussion by Nielsen and Sellner Reunert (chapter 1, this volume).

24. Following the proposal of Andrews and Hassig (1984), and given the variety of editions of Ruiz de Alarcón's *Tratado*, including the electronic ones, I indicate the reference to the conjuries of Ruiz de Alarcón specifying the number of the treatise in which it appears in Roman numerals and the chapter in Arabic numerals, without providing the pages of any one specific edition.

25. Anders, Jansen, and Reyes (1996, 42n5) present information from the Chatinos and Nahua of Atla who still associate the number "nine" with the otherworld. The Chatinos say: "'the road of the dead' is divided into nine stages. Each one is a real site and, together, they mark a road, which goes towards the north, leaving from the pueblo . . . In the ninth level, in *kiche ko*, 'town-nine', a town with nine houses, the spirit abandons the geography of its world and leaves it behind" (Greenberg 1987, 159–160, as cited in Anders, Jansen, and Reyes 1996, 42n5). The Nahua of Atla speak of a journey through nine places "four rivers (of water, of blood, of pus, of embers), two

plains (without grass but with fat sheep, with grass but with thin sheep), a place where mules fight, a place where two hills crash together and a house where the Virgin Mary is" (Montoya Briones 1964, 167–171, as cited in Anders, Jansen and Reyes 1996, 42n 5; see also Nielsen and Sellner Reunert, chapter 1, this volume).

26. By "semantic value" of numbers, I understand the symbolism of the same, which has recently been studied thoroughly by Dehouve. The French researcher discusses the numeric symbolism constructed by metaphoric and metonymic procedures (Dehouve 2011, 151–153). For my part, following the idea of Paul Ricoeur (1995, 58–66), I consider that the metaphoric connotations are also cognitive, forming the semantic field of a given term; thus we can speak inclusively of the "semantic value" of numbers.

27. Translation from Italian into Spanish by Anders, Jansen, and Reyes (1996, 183), and from Spanish to English by Jerome Offner.

28. This is the reconstruction of the expression that Ruiz de Alarcón erroneously recorded in personal form as *tla totoconecahuican tochalchiuhecahuaz* (V-1), but that Andrews and Hassig correct as *tlā tocontlehcahuīcān tochālchiuhehcahuāz* (Andrews and Hassig 1984, 145, 355n14).

29. Compare the translation by Tavárez (2003: 7).

30. My translations differ from those of Bierhorst, who translates "Nine Hills" and "Nine Fields" (Bierhorst 1992, 23); however, words in Nahuatl that do not refer to something alive necessarily cannot specify singular or plural.

31. Another mention of *Chiucnauhnepaniuhqui* appears in the *Leyenda de los Soles*, in the myth of the creation of the fifth sun, where it is reported that when Tlahuizcalpantecuhtli shot at the sun with his flame-like arrows so that the sun would move, the sun's face was covered by "Nine-Confluence," "Nine-Joined," or "Nine-Crossed" (*chiucnauhnepaniuhqui*) (*LS* 2002, fol. 78; Bierhorst 1992, 91).

32. For example, Martínez and Núñez (2009, 43) mention a piece of information collected by Romero López that indicates that the Nahua of the Sierra Negra call the uterus "cave."

33. Translation from Nahuatl to Spanish by Katarzyna Mikulska, and from Spanish to English by Jerome Offner.

34. It also should be significant that Mictlantecuhtli, according to the description of the *Histoire du Mexique*, lives in the "sixth heaven" (*HM* 2002, 143; Mikulska 2008a, 236).

35. Apparently from *chiucnahui* ("nueve") and *tiliuhcân* ("hill") (fr. *colline*; Wimmer 2004 in GDN by Thouvenot).

REFERENCES

Bibliographic information for primary documents cited by abbreviations in the text is to be found at the following Reference entries:

AC: *Anales de Cuauhtitlán* (*Anales de Cuauhtitlán* 1975)
FC: *Florentine Codex* (Sahagún 1950–1982)
GDN: *Gran diccionario náhuatl* (*Gran diccionario náhuatl*)
HM: *Histoire du Mechique* (*Histoire du Mechique* 2002)
HMP: *Historia de los mexicanos por sus pinturas* (*Historia de los mexicanos por sus pinturas* 2002)
LS: *Leyenda de los Soles* (*Leyenda de los Soles* 2002)
PM: *Primeros Memoriales* (Sahagún 1997)
RAH: *Manuscrito de la Real Academia de la Historia* (Sahagún 1997)

Alvarado Tezozómoc, Hernando de. 2001. *Crónica mexicana*. Ed. Gonzalo Díaz Migoyo, and Germán Vázquez Chamorro. Madrid: Dastin (Crónicas de América).

Anales de Cuauhtitlán (AC). 1975. In *Códice Chimalpopoca*. Trans. Primo Feliciano Velázquez, 3–118. Mexico: Universidad Nacional Autónoma de México, Instituto de Investigaciones Históricas.

Anders, Ferdinand, Maarten Jansen, and Aurora Pérez Jiménez. 1992. *Crónica mixteca: Libro explicativo del llamado* Códice Zouche-Nuttall. Introduction and commentary by Ferdinand Anders and Maarten Jansen. Mexico: Akademische Druck- und Verlagsanstalt, Sociedad Estatal Quinto Centenario, Fondo de Cultura Económica.

Anders, Ferdinand, Maarten Jansen, and Luis Reyes García. 1993. *Los templos del cielo y de la oscuridad; oráculos y liturgia: Libro explicativo del llamado Códice Borgia*. Introduction and commentary by Ferdinand Anders and Maarten Jansen. Mexico: Akademische Druck- und Verlagsanstalt, Sociedad Estatal Quinto Centenario, Fondo de Cultura Económica.

Anders, Ferdinand, Maarten Jansen, and Luis Reyes García. 1996. *Religión, costumbres e historia de los antiguos mexicanos: Libro explicativo del llamado* Códice Vaticano A. Introduction and commentary by Ferdinand Anders and Maarten Jansen. Mexico: Akademische Druck- und Verlagsanstalt, Sociedad Estatal Quinto Centenario, Fondo de Cultura Económica.

Andrews, J. Richard, and Ross Hassig, trans. 1984. *Treatise on the Heathen Superstitions That Today Live among the Indians Native to This New Spain, 1629*, by Hernando Ruiz de Alarcón. Norman: University of Oklahoma Press.

Annals of Cuauhtitlan. 1992. In *Codex Chimalpopoca: The Text in Nahuatl with a Glossary and Grammatical Notes*, ed. John Bierhorst, 1–84. Tucson: University of Arizona Press.

Bierhorst, John, trans. 1992. *History and Mythology of the Aztecs: The* Codex Chimalpopoca. Tucson: University of Arizona Press.

Boone, Elizabeth. 1999. "The 'Coatlicues' at the Templo Mayor." *Ancient Mesoamerica* 10 (2): 189–206.

Boone, Elizabeth H. 2007. *Cycles of Time and Meaning in the Mexican Books of Fate.* Austin: University of Texas Press.

Castillo, Cristóbal del. 2001. *Historia de la venida de los mexicanos y de otros pueblos e historia de la conquista.* Translation and introductory study by Federico Navarrete Linares. Mexico: Consejo Nacional para la Cultura y las Artes (Cien de México).

Chimalpahin Cuauhtlehuanitzin, Domingo. 1998. *Las ocho relaciones y el Memorial de Colhuacan*, 2 vols. Paleography and translation by Rafael Tena. Mexico: Consejo Nacional para la Cultura y las Artes (Cien de México).

Codex Borbonicus. n.d. *Codex borbonicus. Tonalamatl, ou calendrier religieux et divinatoire mexicain.* Ms. Y 120. Paris: Bibliothèque de l'Assemblée Nationale.

Codex Borgia, Der. 1976. Graz: Akademische Druck- und Verlagsanstalt.

Codex Fejérváry-Mayer. 1971. Graz: Akademische Druck- und Verlagsanstalt.

Codex Laud. 1966. Graz: Akademische Druck- und Verlagsanstalt.

Codex Magliabechiano.1983. The Codex Magliabechiano and the Lost Prototype of the Magliabechiano Group, ed. Elizabeth Hill Boone. Berkeley, Los Angeles, London: University of California Press.

Codex Telleriano-Remensis. 1995. *Codex Telleriano-Remensis: Ritual, Divination and History in a Pictorial Aztec Manuscript*, ed. Eloise Quiñones Keber. Austin: University of Texas Press.

Codex Tro-Cortesianus (Codex Madrid). 1967. Graz: Akademische Druck- und Verlagsanstalt.

Codex Vaticanus 3773 (Codex Vaticanus B). 1972. Commentary by Ferdinand Anders. Graz: Akademische Druck- und Verlagsanstalt.

Codex Vindobonensis Mexicanus 1. 1974. Graz: Akademische Druck- und Verlagsanstalt.

Códice Nuttall, lado 1: La vida de 8 Venado. 2006. Introductory study and interpretation by Manuel A. Hermann Lejarazu. *Arqueología Mexicana*, special num. 23.

Códice Nuttall, lado 2: La historia de Tilantongo y Teozacoalco. 2007. Introductory study and interpretation by Manuel A. Hermann Lejarazu. *Arqueología Mexicana*, special num. 29.

Codice Tudela o Códice del Museo de América. n. d. Document 70 400. Madrid: Museo de América.

Códice Vaticano A. 1996. Ed. Ferdinand Anders, Maarten Jansen, and Luis Reyes García. Graz, Mexico: Akademische Druck- und Verlagsanstalt, Fondo de Cultura Económica.

Contel, José. n.d. "De la nube sigmoidea de Chalcatzingo al xonecuilli. Signos selectos e intemporales de la iconografía mesoamericana" (unpublished manuscript).

Contel, José. 2000. "Visiones paradisiacas: Extrañas analogías entre *Tlallocan* y 'Paraíso Terrenal' en la *Historia general de las cosas de Nueva España* de Fray Bernardino de Sahagún." In *Fray Bernardino de Sahagún y su tiempo*, ed. Jesús Paniagua Pérez, and María Isabel Viforcos Marinas, 617–630. León: Universidad de León/Instituto Leonés de Cultura.

Contel, José, and Katarzyna Mikulska. 2011. "'Mas nosotros que somos dioses nunca morimos': Ensayo sobre tlamacazqui: ¿dios, sacerdote o qué otro demonio?" In *De dioses y hombres: Creencias y rituales mesoamericanos y sus supervivencias*, ed. Katarzyna Mikulska and José Contel, 23–65. University of Warsaw, Institute of Iberian and Iberoamerican Studies.

Craveri Slaviero, Michela Elisa. 2004. *El arte verbal k'iche': Las funciones poéticas de los textos rituales mayas contemporáneos*. Mexico: Praxis.

Craveri Slaviero, Michela Elisa. 2012. *El lenguaje del mito: Voces, formas y estructura del* Popol Vuh. Mexico: Universidad Nacional Autónoma de México, Instituto de Investigaciones Filológicas, Centro de Estudios Mayas.

Craveri Slaviero, Michela Elisa, trans. and notes. 2013. Popol Vuh. *Herramientas para una lectura crítica del texto k'iche'*. Mexico: Universidad Nacional Autónoma de México, Instituto de Investigaciones Filológicas, Centro de Estudios Mayas.

Dakin, Karen. 1996. "'Huesos' en el náhuatl: etimologías yutoaztecas." *Estudios de Cultura Náhuatl* 26: 309–325.

Dehouve, Danièle. 2009. "El lenguaje ritual de los mexicas: hacia un método de análisis." In *Image and Ritual in the Aztec World: Selected Papers of the "Ritual Americas" Conferences*, ed. Sylvie Peperstraete, 19–33. Oxford: Archaeopress.

Dehouve, Danièle. 2011. *L'imaginaire des nombres chez les anciens Mexicains*. Rennes: Presses Universitaires de Rennes (Histoire des religions).

Dehouve, Danièle. 2013. "Les métaphores comestibles dans les rituels mexicaines." *Amérique Latine Histoire et Mémoire. Les Cahiers ALHIM* 25.

Díaz, Ana. 2009. "La primera lámina del *Códice Vaticano A*: ¿Un modelo para justificar la topografía celestial de la antigüedad pagana indígena?" *Anales del Instituto de Investigaciones Estéticas* 31 (95): 5–44.

Dupey, Élodie. 2010. "Les couleurs dans les pratiques et les representations des nahuas du Mexique central (XIVe–XVIe siecles)." PhD diss., École Pratique des Hautes Etudes, Paris.

Essential Codex Mendoza, The. 1997. Ed. Frances F. Berdan and Patricia Rieff Anawalt. Berkeley: University of California Press.

Furst, Leslie. 1982. "Skeletonization in Mixtec Art: A Re-Evaluation." In *The Art and Iconography of Late Post-Classic Central Mexico*, ed. Elizabeth Benson, 207–225. Washington, DC: Dumbarton Oaks Research Library and Collections.

Galinier, Jacques. 1990. *La mitad del mundo: Cuerpo y cosmos en los rituales otomíes*. Mexico: Universidad Nacional Autónoma de México, Centro de Estudios Mexicanos y Centroamericanos, Instituto Nacional Indigenista.

Garibay K., Ángel María. 2007 [1953–1954]. *Historia de la literatura náhuatl*. Prologue by Miguel León-Portilla. Mexico: Porrúa.

Gran diccionario náhuatl (*GDN*). Program by Marc Thouvenot. http://www.sup-infor.com.

Graulich, Michel. 1980. "L'au-delà cyclique des anciens Mexicains." In *La antropología americanista en la actualidad: Homenaje a Raphael Girard*, vol. I, 253–270. Mexico: Editores Mexicanos Unidos.

Graulich, Michel. 1990 [1987]. *Mitos y rituales del México antiguo*. Madrid: Istmo.

Haly, Richard. 1992. "Bare Bones: Rethinking Mesoamerican Divinity." *History of Religions* 31 (3): 269–304.

Heyden, Doris. 1975. "An Interpretation of the Cave underneath the Pyramid of the Sun in Teotihuacan, Mexico." *American Antiquity* 40 (2): 131–147.

Histoire du Mechique (*HM*). 2002. In *Mitos e historias de los antiguos nahuas*, trans. Rafael Tena, 115–167. Mexico: Consejo Nacional para la Cultura y las Artes (Cien de México).

Historia de los mexicanos por sus pinturas (*HPM*). 2002. In *Mitos e historias de los antiguos nahuas*, trans. Rafael Tena: 15–114. Mexico: Consejo Nacional para la Cultura y las Artes (Cien de México).

Ichon, Alain. 1973. *La religión de los totonacas de la sierra*. Mexico: Instituto Nacional Indigenista.

Iwaniszewski, Stanisław. 2009. "Algunos rasgos de las cosmografías y cosmologías chamánicas." In *Construyendo cosmologías: consciencia y práctica*, ed. Gustavo Aviña Cerecer y Walburga Wiesheu. Mexico: Programa de Mejoramiento del Profesorado, Escuela Nacional de Antropología e Historia, Instituto Nacional de Antropología e Historia, Consejo Nacional para la Cultura y las Artes.

Klein, Cecilia. 2000. "The Devil and the Skirt: An Iconographic Inquiry into the Prehispanic Nature of the Tzitzimime." *Estudios de Cultura Náhuatl* 31: 16–71.

Knab, Tim. 1991. "Geografía del inframundo." *Estudios de Cultura Náhuatl* 21: 31–57.

Köhler, Ulrich. 1995. *Chonbilal Ch'ulelal–Alma Vendida: Elementos fundamentales de la cosmología y religión mesoamericanas en una oración en maya–tzotzil*. Mexico: Universidad Nacional Autónoma de México, Instituto de Investigaciones Antropológicas.

Lakoff, George, and Mark Johnson. 2007 [1980]. *Metáforas de la vida cotidiana.* Madrid: Cátedra.

Leyenda de los Soles (LS). 2002. In *Mitos e historias de los antiguos nahuas,* ed. Rafael Tena, 173–206. Mexico: Consejo Nacional para la Cultura y las Artes.

López Austin, Alfredo. 1995. *Tamoanchan y Tlalocan.* Mexico: Fondo de Cultura Económica.

López Austin, Alfredo. 1996 [1980]. *Cuerpo humano e ideología: Las concepciones de los antiguos nahuas,* 2 vols. Mexico: Universidad Nacional Autónoma de México–Instituto de Investigaciones Antropológicas.

López Austin, Alfredo. 2003. "Difrasismos, cosmovisión e iconografía." *Revista Española de Antropología Americana,* extraordinary vol.: 143–160.

Martínez González, Roberto. 2013. *Cuiripu: cuerpo y persona entre los antiguos p'urhépecha de Michoacán.* Mexico: Universidad Nacional Autónoma de México, Instituto de Investigaciones Históricas.

Martínez González, Roberto, and Luis Fernando Núñez. 2009. "La carne pegada al hueso: Planteamiento sobre la concepción del cadáver en el Posclásico tardío, con énfasis en el México central." *Diario de Campo*: 40–48.

Mendieta, Gerónimo de. 1993 [1870]. *Historia eclesiástica indiana.* Mexico: Porrúa.

Mikulska, Katarzyna. 2008a. *El lenguaje enmascarado: Un acercamiento a las representaciones gráficas de deidades nahuas.* Mexico: Universidad Nacional Autónoma de México, Instituto de Investigaciones Antropológicas, Uniwersytet Warszawski, Polskiego Towarzystwa Studiów Latynoamerykanistycznych.

Mikulska, Katarzyna. 2008b. "El concepto de *ilhuicatl* en la cosmovisión nahua y sus representaciones gráficas en códices." *Revista Española de Antropología Americana* 38 (2): 151–171.

Mikulska, Katarzyna. 2015a. "Los cielos, los rumbos y los números: Aportes sobre la visión nawa del universo." In *Cielos e inframundos; una revisión a la Cosmología Mesoamericana,* ed. Ana Díaz, 109–173. México: Instituto de Investigaciones Históricas, UNAM, Fideicomiso Felipe Teixidor y Montserrat Alfau de Teixidor.

Mikulska, Katarzyna. 2015b. "Destronando a Ometeotl." *Latin American Literature Journal* 31 (1–2): 57–127.

Milbrath, Susan. 1999. *Star Gods of the Maya: Astronomy in Art, Folklore, and Calendars.* Austin: University of Texas Press.

Molina, Alonso de. 2005 [1571]. *Vocabulario en lengua castellana y mexicana y mexicana y castellana.* In *Gran Diccionario Náhuatl,* program by Marc Thouvenot. Centre National de la Recherche Scientifique/Instituto Nacional de Antropología e Historia/Universidad Nacional Autónoma de México/Université de Toulouse. http://www.sup-infor.com.

Montes de Oca, Mercedes. 2000. "Los difrasismos en el náhuatl del siglo XVI." PhD diss., Universidad Nacional Autónoma de México.

Montes de Oca, Mercedes. 2013. *Los difrasismos en el náhuatl de los siglos XVI y XVII*. Mexico: Universidad Nacional Autónoma de México-Instituto de Investigaciones Filológicas.

Montes de Oca, Mercedes. 2015. "Los difrasismos y la construcción de la identidad de la nobleza indígena." In *Identidad en palabras: Nobleza indígena novohispana*, ed. Patrick Lesbre and Katarzyna Mikulska, 249–266. Mexico: Universidad Nacional Autónoma de México–Instituto de Investigaciones Antropológicas, Uniwersytet Warszawski–Instituto de Estudios Ibéricos e Iberoamericanos, Université de Toulouse Le Mirail.

Nielsen, Jesper, and James E. Brady. 2006. "The Couple in the Cave: Origin Iconography on a Ceramic Vessel from Los Naranjos, Honduras." *Ancient Mesoamerica* 17: 203–217.

Nielsen, Jesper, and Toke Sellner Reunert. 2009. "Dante's Heritage. Questioning the Multi-Layered Model of the Mesoamerican Universe." *Antiquity* 83 (320): 399–413.

Olivier, Guilhem. 2008. "Las tres muertes simbólicas del nuevo rey mexica." In *Símbolos de poder en Mesoamérica*, ed. Guilhem Olivier, 263–291. Mexico: Universidad Nacional Autónoma de México, Instituto de Investigaciones Históricas, Instituto de Investigaciones Antropológicas.

Oudijk, Michel. 2011. "Elaboration and Abbreviation in Mexican Pictorial Manuscripts: Their Use in Literary Themes." In *Their Way of Writing: Scripts, Signs, and Pictographies in Pre-Columbian America*, ed. Elizabeth Hill Boone and Gary Urton, 149–174. Washington, DC: Dumbarton Oaks Research Library and Collection.

Pérez Castro, Ana Bella. 2012. "Los muertos en la vida social de la Huasteca." *Itinerarios* 15: 205–236.

Pitarch, Pedro. 2006 [1996]. *Ch'ulel: Una etnografía de las almas tzetzales*. Mexico: Fondo de Cultura Económica.

Ponce de Léon, Pedro. 1985. "Breve relación de los dioses y ritos de la gentilidad." In *Teogonía e historia de los mexicanos: Tres opúsculos del siglo XVI*, ed. Ángel María Garibay, 121–140. Mexico: Porrúa.

Preuss, Konrad Theodor. 1960. "La diosa de la tierra y de la luna de los antiguos mexicanos en el mito actual." *Boletín del Centro de Investigaciones Antropológicas de México* 10: 6–10.

Preuss, Konrad Theodor. 1982. *Mitos y cuentos nahuas de la Sierra Madre Occidental*. Mexico: Instituto Nacional Indigenista.

Ragot, Nathalie. 2000. *Les au-delàs aztèques*. International Series 881. Oxford: British Archaeology Reports.

Ramírez Castañeda, Elisa. 2002. "'Nuxi', trampeador de tropos." *Arqueología Mexicana* 10 (55): 17.
Real Academia Española. 2001. *Diccionario de la lengua española* (*DLE*). http://www.rae.es/recursos/diccionarios/drae.
Reyes Gómez, Laureano. 2008. "La visión zoque del inframundo." *Revista Española de Antropología Americana* 38 (2): 97–106.
Ricoeur, Paul. 1995 [1976]. *Teoría de la interpretación: Discurso y excedente de sentido.* Mexico: Universidad Iberoamericana, Siglo XXI Editores.
Ruiz de Alarcón, Hernando. 1953. *Tratado de las supersticiones y costumbres gentílicas que hoy viven entre los indios naturales de esta Nueva España*, vol. XX, ed. Francisco del Paso y Troncoso, 17–180. Mexico: Ediciones Fuente Cultural de la Librería Navarro.
Sahagún, Bernardino de. 1950–1982. *Florentine Codex: General History of the Things of New Spain by Fray Bernardino de Sahagún*, trans. Arthur J. O. Anderson, and Charles E. Dibble. Santa Fe, NM: University of Utah, School of American Research.
Sahagún, Bernardino de. 1997. *Primeros Memoriales* (*PM*). Translation and notes by Thelma D. Sullivan. Norman: University of Oklahoma Press, and Madrid: Patrimonio Nacional, Real Academia de la Historia.
Sandstrom, Alan R. 2010. *El maíz es nuestra sangre: Cultura e identidad étnica en un pueblo indio azteca contemporáneo.* Mexico: Centro de Investigaciones y Estudios Superiores en Antropología Social, Universidad Autónoma de San Luis Potosí, El Colegio de San Luis, Secretaría de Cultura del Estado de San Luis Potosí.
Schele, Linda, and David Freidel. 1990. *A Forest of Kings: Untold Stories of the Ancient Maya.* New York: William Morrow and Company.
Schwaller, John F. 2006. "The *Ilhuica* of the Nahua: Is Heaven Just a Place?" *The Americas* 62 (3): 391–412.
Seler, Eduard. 1990–2000. *Collected Works in Mesoamerican Linguistics and Archaeology*, 6 vols. Ed. J. Eric S. Thompson, Charles P. Bowditch, and Frank E. Comparato. Culver City, CA: Labyrinthos.
Serna, Jacinto de la. 1953. *Manual de ministros de indios para el conocimiento de sus idolatrías y extirpación de ellas*, vol. X. Ed. Francisco del Paso y Troncoso, 47–368. Mexico: Ediciones Fuente Cultural.
Siarkiewicz, Elżbieta. 1982. "La muerte y su significado ritual en la cultura nahua." PhD diss., Uniwersytet Warszawski, Instituto de Estudios Ibéricos e Iberoamericanos.
Taube, Karl Andreas. 1986. "The Teotihuacan Cave of Origin: The Iconography and Architectures of Emergence Mythology in Mesoamerican and the American Southwest." *Res: Anthropology and Aesthetics* 12: 51–82.

Taube, Karl Andreas. 1993. "The Bilimek Pulque Vessel." *Ancient Mesoamerica* 4 (1): 1–15.

Tavárez, David Eduardo. 2003. "Reproducción social de las prácticas rituales del Posclásico tardío en la época colonial temprana del centro de México." Fundación para el Avance de los Estudios Mesoamericanos. http://www.famsi.org/reports/96039es/index.html.

Thévet, André. n.d. *Fragments d'André Thévet surles Indes occidentales et sur le Mexique (Histoire du Mechique)* (*HM*). Ms. 19031, Fond Français. Paris: Bibliothèque nationale de France, Département des manuscrits.

Torquemada, Juan de. 1976. *Monarquía indiana*, 3 vols. Introduction by Miguel León-Portilla. Mexico: Porrúa.

Townsend, Richard F. 1987. "Coronation at Tenochtitlan." In *The Aztec Templo Mayor: A Symposium at Dumbarton Oaks, 8th and 9th October 1983*, ed. Elizabeth Hill Boone, 371–409. Washington, DC: Dumbarton Oaks Research Library and Collections.

Wright-Carr, David Charles. 2009. "Semasiografía y glotografía en las inscripciones de dos esculturas mexicas." In *Estudios acerca de las artes*, ed. Benjamín Valdivia, 226–253. Morelia: Universidad de Guanajuato, Azafrán y Cinabrio Ediciones.

9

Creating and Destroying the Upper Part of the Cosmos

A New Approach to the Study of Wixarika Cosmology

Johannes Neurath

RETHINKING COSMOLOGIES, CONTINUITIES AND CHANGE

To what extent have the Spanish conquest and subsequent events modified Amerindian cosmologies? Answers are rather varied. Some authors prefer to emphasize continuities. Against all odds, Amerindian people manage to reproduce what is often characterized as "their millenarian cosmovision." They do so because they are resisting the. Others reject such arguments as essentialist and tautologic: Do Indians have Mesoamerican cosmovisions because they still resist, or do they resist because they still have those cosmovisions? (see Navarrete 2012). Overemphasizing continuity is at odds with evident processes of change during more than 500 years of violent history. But the Indians are still here, and many Amerindian communities are actually thriving. So should we emphasize indigenous modernity, processes of hybridization, and *mestizaje*? Rather, we should try to restate the whole question.

My research is based on a long-term fieldwork project that started in 1992 and focuses on the ceremonial center of Keuruwit+a (Las Latas) that is part of the community of Tuapurie (Santa Catarina Cuexcomatitán), and the municipality of Mezquitic, Jalisco. I have always insisted on the contemporaneity of Wixarika (Huichol) culture. I believed my obligation as an anthropologist was to criticize exoticizing stereotypes, "imperialist nostalgia" (Rosaldo 1991, 71), and other misrepresentations of the Wixaritari (plural of *Wixarika*). So I made

DOI: 10.5876/9781607329534.c009

an effort to refute the idea the Wixaritari were something like a "cultural fossil" (see Seler 1901, 163) at the brink of extinction, or inevitably ruined by improved communications and technological progress. Instead, I emphasized that they are successful global players in a phase of demographic, cultural, and territorial expansion (Neurath 2002). The territories of the five traditional Wixarika communities (Tuapurie, Waut+a, Tutsipa, Tateikie, and Xatsitsarie) are located in the western Sierra Madre, in northern Jalisco and northeastern Nayarit, but many new communities have been founded in southern Durango, and other parts of Nayarit formerly settled by Coras or mestizos. According to a census from 2015, the *Wixarika* language has now 52,483 speakers (http://cuentame.inegi.org.mx/poblacion/lindigena.aspx?tema=P#uno). A century ago the Huichol were probably fewer than 10,000 people (Lumholtz 1904, 2, 100; Furst and Nahmad 1972, 7; Weigand 1981, 1985). Studying Wixarika ritual, my objective has been to describe it as an adequate expression of life in a complex world that is shared with non-Indians. Studying their art, I had to address a new stereotype: the Wixaritari as ruined by mass media. Somehow they are able to handle cultural industry, too (Neurath 2013).

In talking about Wixarika cosmology it is important to emphasize that we deal with a system of knowledge that is radically unlike Western ones, at least in some important aspects. The simple idea to reconstruct *the* Wixarika view of the world is problematic. Neither should cosmology be understood as a system of concepts *about* nature. Important aspects of the cosmos are *made* during ritual, so creation is an ongoing event that is not separable from ritual action. But there are limits to invention. There are parts of the cosmos that actually may be considered as naturally given. They were not made, neither by gods nor by men. Trying to homogenize those notions of cosmology in order to reconstruct something that might go for a comprehensive worldview, would be a failure. We'll never find models like the ones Europeans elaborated during the Middle Ages or the Renaissance.

In order to understand Wixarika cosmology, one good starting point is personhood. Following Marilyn Strathern (1988), Roy Wagner (1991), and other authors, we can characterize the Wixarika concept of the person as "dividual" (Neurath 2008). What I want to argue is that this idea can be extended to cosmology. In this context, shamanism may be understood as training in multiplication of personhood, which enables people to relate to animals, deities, and all kind of beings belonging to parts of an ontologically complex and differentiated cosmos. People do not always desire to be defined as individuals of one single place, they want to be functional in all kind of contexts. Apparently, this is a big advantage that non-Indigenous populations in Mexico cannot count on.

At this point it is worthwhile to take up the ideas of philosophers reflecting on modernity. According to Latour (1993), Western concepts of personhood—like the Cartesian subject or the bourgeois citizen of classical liberalism—are too simple to fit to the hybrid, often contradictory complexities of the contemporary world. As a consequence, Westerners have had problems coping with modernity and many still struggle transiting towards it. But it seems to be quite erroneous attributing those problems to Amerindian people like the Wixaritari. Rather one has to recognize that the Wixarika concept of the person has an affinity with avant garde concepts of decentered personhood, and this is why they are much better prepared to deal with complexity. In that sense, they never have been premodern.

In other words, the advantage Amerindian people like the Wixaritari have in the contemporary world has everything to do with their tradition. It is shamanism that enables Wixaritari to adapt successfully to all kind of circumstances.

COSMOLOGICAL ASYMMETRIES

In the study of Wixarika cosmology one may start from a geographic contrast between "above" (*hix+ata*) and "below" (*tat+ata*), between the eastern semidesert called Wirikuta and the fertile coastal plans in the west. The land of Wixarika communities is located between these, in the western Sierra Madre. The body–cosmos isomorphism known in many parts of Mesoamerica (see López Austin 1980; Monaghan 1995, 98; Galinier 2004) implies that Wirikuta (located in the state of San Luis Potosí) corresponds to the head and thus to authority, the Pacific coast to sexual organs, and the navel of the world is located on a ravine at the sierra. Moreover, Wirikuta corresponds to the day and the dry season, while the coast corresponds to night and rain. Autumn is "the break of day." Rain serpents are the world's vital soul or breath (*iyari*). The body is a microcosm; the world is a macrobody. The study of the body and the person cannot be separated from cosmology.

Many specialists have posited the dualist nature of Mesoamerican cosmologies, particularly Alfredo López Austin (1994) and Michel Graulich (1999). In the Wixarika case, however, important asymmetries stand out. Notably, dawn lasts almost the entire year—from the end of the rainy season (September/October) until May or June. In contrast, dusk takes place during a single ceremony associated with the summer solstice, the Namawita Neixa feast.

The temporality of the ritual calendar is inversely proportional to the duration of the time as it is experienced. Dawn is an elusive instant, while the realm of night has always existed and will always exist. The light of dawn is not naturally given, it must be "found" and envisioned. Wixarika initiation is the search for this elusive spot of time-space. A pilgrimage to the eastern desert must be organized, where participants—the *xukuri'ikate* "jicara carriers" (from *xukuri* "gourd bowl"; Span.: *jicareros*)—become *hikuritamete* "peyote-people" (from *hikuri*, "peyote"; Span.: *peyoteros*), attain *nierika* ("gift of sight" or "visionary ability") and are thus transformed into their own mythical ancestors who in their visionary experiences invent the world, particularly the luminous, orderly part of cosmos.

To find dawn one must refrain or walk away from anything having to do with the sea and darkness: one should not eat salt, must walk on the arid steppes of the Mexican highland, which are quite far from the coasts, must refrain from extramarital sex, and should sleep as little as possible. After days of purification, peyote (*Lophophora williamsii*) appears in front of the pilgrims like a deer that gives itself up to be hunted. The effect of the hallucinogenic is the light of dawn experimented when walking up the mountain named Under the Dawn (Paritek+a) or Burnt Hill (Reu'unari). Day (or daylight), and by extension knowledge and the sociocosmic authority of the Father Sun, are thus a vision and do not exist separately from ritual action and effort.

This same asymmetry is also seen in the architecture of Wixarika temples. Buildings of the *tuki* and *xiriki* types must be renewed every five years, but only thatch roofs are redone, while walls remain untouched. Roofs correspond to Wirikuta, the day sky and the Mountain of Dawn. The *tuki* roof posts are the *ocote* (pitch wood) torches or candles (*haurite*) that support the sky but are progressively worn out, so that they must be renewed periodically. The dark interior of the buildings corresponds to the sea and underworld. Central fires are replicas of the navel of the world, also known as Te'akata ("place of the earth oven").

The *wixarika* cosmos shows that only one part of the time-space is given; the celestial-sun half is artificial and ephemeral. It must be periodically recreated. As Roy Wagner (1981) points out, in many non-Western cultures, the artificial is not necessarily less true, less prestigious, or less important than the natural. Quite the opposite!

If the day sky belongs to the created part, itself a product of the initiate's visionary experience, the night sky is not just a simple opposite equivalent to the day sky. The light–darkness duality exists only during the day, so the day world is ontologically very different from the night world.

The lower half is an ancient world, paleontological or prehistoric, but at the same time is the urban modernity. Inhabited by monsters of Amerindian and Western tradition—cannibal giants, dinosaurs, mammoths, sea monsters, mermaids, goat-suckers (*chupacabras*), vampires—it is where filth accumulates and most sicknesses come from, but it is also the realm of non-indigenous populations, their advanced technology, and gadgets. Big cities, like Mexico City, are also part of this "underworld."

Wixaritari do not define themselves as indigenous but as the "younger brothers" of mestizo aborigines. They consider themselves as "those who arrived last" and as more evolved than mestizos. I have heard stories about the ancestors of the Wixarika as conquerors of mestizo territory. Apparently, the Wixaritari do not understand themselves as victims of history. They prefer seeing themselves as victorious warriors.

As descendants of cannibal monsters, mestizos show very rough social behaviors. They have lost their manners or never had them to begin with. Trusting their technology, they ignore the origin of things; for example, they think that electricity may be simply taken from an outlet without the need to give back to the god of fire. In contrast, the Wixaritari younger brothers know how to create their luminous world through initiation and the practice of the *yeiyari* tradition, "walking on the ancestors' footsteps" (see Kantor 2012; Lira 2014).

The technologically advanced underworld of the non-indigenous corresponds to a temporality where the most remote past and the hypertechnological present overlap—a phenomenon that has been observed in several Mexican indigenous communities (see Questa 2010; Romero 2010; Lorente 2011; Pitarch 2012). In contrast, the Wixarika solar world is, in some way, unreachable. "Walking on the ancestors' footsteps" is a deer hunt in which the hunter identifies with his own prey. The initiate almost dies and in this voluntary sacrificial quasi-death, he manages to experience dawn, awakening, to obtain knowledge. A perfect initiation is impossible to achieve. What is possible is but an infinitesimal approximation to Wirikuta.

Many never achieve any kind of initiation: unable to follow the right trail they get lost, they fall asleep or succumb to temptation. This is exactly what has happened to mestizos as well as many indigenous populations, the Coras (Náyari) and Tepehuanes (O'dam), and to animal species that are either nocturnal or inhabit the sea, the coasts, or the bottoms of canyons.

The initiation trip to Wirikuta (also known as "the peyote hunt") is a famous and relatively well-documented ritual (see, for example, Benítez 1984). Though much less ethnographied, it is noteworthy that the return from the desert to the community is a much more complex process than the other way around. It is difficult to be a common person and an ancestor at the same time. After dreaming of light, life, and peyote, pilgrims return triumphant to share these gifts with others, but commoners interpret it as an usurpation. At the peyote feast (*Hikuli Neixa*), celebrated as the culmination of the pilgrimage cycle, the uninitiated do not acknowledge *peyoteros* as special beings. When *peyoteros* form a snake of clouds and start distributing seeds and peyote, they are giving away themselves as ancestral beings transformed into precious gifts. However, they are obligated to accept payment in the form of cigarettes and tamales. That means that their divine gifts are not aknowledged as such. The reciprocal exchange nullifies the *peyoteros* as generous, living deities and therefore limits their power and special status (Neurath 2011a; 2011b; 2015; Rio 2007).

Soon after, the Namawita Neixa sowing feast is celebrated, and then the entire ordered cosmos created by the *peyoteros* collapses. What remains is the underworld; however, it is no longer a different world but simply *the* world. *Peyoteros* are replaced by the Nia'ariwamete group ("messengers" of the rain). In a wild dance they extinguish an *ocote* torch identified as the post holding up the sky: the top of the cosmos falls down. They also extinguish the central fire of the *tuki*, which is the god of fire, and at the end of the ritual the community's traditional authorities relinquish their power, turning over their staffs of authority (*its+*; Span.: *varas de mando*), the ropes to make arrests and to tie prisoners, and other emblems of office.

The Namawita Neixa celebration features a ritual song that mentions the sun visiting a place in the north known as El Bernalejo, Durango. At this (solstice) point it rests and is sexually seduced by a beautiful girl with a serpent tail. She devours him with her *vagina dentata* and once devoured, the Sun becomes that same female telluric monster that is his *nemesis* and *alter ego*. The earth monster named Takutsi Nakawé appears in the Namawita Neixa feast as the leader of the dance. She is the old goddess who, according to the myth, was the first shaman-singer, later overturned for being a despot, a drunkard, and/or a cannibal monster.

The return of Takusi Nakawé is much more than a carnivalesque rite of inversion, since in her kingdom, identified with the rainy season, a law different from the dry season sacrifice ideology applies. But it is a law nonetheless! Now they are all uninitiated humans and the "corn's wedding" is celebrated. According to mythology, a failed hunter turns into the first farmer. He attempts to buy corn from a mestizo lady living in the underworld or in a ranch located on the bottom of a deep canyon. Finally, he ritually marries the five maize girls, who are the daughters of that same mestizo lady or underworld goddess. Thus he establishes the first family, prepares the first cornfield (*waxa*; Span.: *milpa*), and settles down for a sedentary life. The corollary of this marriage alliance between the Wixarika farmer and gods from "below" is that now mestizos are not dismissed as filthy, lazy, or dumb subhuman beings, but acknowledged as persons, almost as equals (Neurath 2008, 2011b).

It is important not to underestimate the significance of the contrast between the heavenly solar world of the peyote pilgrims and the aquatic and dark world of Takutsi (illuminated only by the moon). The first world is the result of a transforming invention. The second world, or underworld, functions with a relatively simple logic of simple exchange or barter. As we have seen, according to mythology, the corn-maidens were acquired from mestizos. Consequently, during the rainy season, legitimate relationships contemplate and include non-indigenous populations. From a global perspective, there is a coming and going between dynamics that turn out to be incompatible. As a Wixarika, sometimes—especially during the time of the peyote-related rituals of the dry season—you are not supposed to engage too much with non-indigenous populations, but sometimes it is perfectly fine to do so.

Important aspects of Wixarika ritual can be understood as a continuous process of ethnogenesis, creating a luminous, ethnically exclusive space where non-indigenous people have no place. But other rituals are more about the importance of interacting in peaceful ways with others. Those rituals cannot be considered minor ceremonies. Corn itself is product of an alliance between Indians and mestizos, and some corn-related rituals celebrate just that. Seen in a global perspective, Wixarika ritual is about flexibility: an autonomous space is created and maintained, while the necessity to cooperate with non-Indians is also acknowledged.

In terms of mythology, one may talk about a first creation (when the ancestors emerged from the primordial sea, when sun and peyote were born, and in which shamanism and community organization had their origins), and a second cosmological era (focused on the origin of the family and corn agriculture), which began when the first had failed. But the two creations are imbricated, so it is never entirely clear which era came first. Often both exist simultaneously. What we can say is that one creation story highlights the importance of self-sacrifice and free gift, and characterizes non-Indians as lazy. The other talks about relations of exchange, including trade, with mestizos. The only structure that would possibly encompass the whole Wixarika world would be one resulting from the tension between free gift and reciprocal exchange. One scheme of practice (or regime of being) is not just the inversion of the other, but the articulation between both is always complicated and problematic (see Neurath 2015).

STAIRWAYS TO ZENITH

The first Cora and Wixarika researcher who consequently emphasized the importance of ritual practice to understand cosmogonies was Konrad Theodor Preuss. He did not think in terms of "invention," but writing on Cora Mitote dances in the 1930s, he said that according to his informants, every ritual occurs for the first time. Ritual is not a repetition, but a unique event (Preuss 1933). Such ideas were not appreciated by anthropologists of his time, but gained a renewed importance in the light of more contemporary authors, like Roy Wagner.

During his trips to the Wixarika sierra in 1906, Preuss collected a small wooden pyramid (figure 9.1). When documenting the ritual usage of this artifact he found evidence that shows how human action is indispensable to create the upper part of the world. The aforementioned piece is an object offered to the Father Sun (Tayau) to serve as a ladder or stair (*imumui*) for his ascent to the sky (figure 9.2). That is to say, the sky has to be built ritually. The miniature is a replica of the sky, which according to Preuss, was not conceived of as a dome but as a stepped pyramid, throughout ancient Mexico. The sun goes up one side and down the other. "When the Sun reaches the summit, it has reached the highest point of its trajectory" (Preuss 1998, 319). Ritual texts in indigenous languages documented by Preuss provide evidence for this interpretation. "Songs express clearly that the steep heights of the skies figure among the regions where gods perform their activities; that's why it is said that the gods request stairways . . . to ascend to and descend from heaven" (Preuss 1998, 246–247).

Preuss insisted on the importance of ritual action and rejected intellectualist trends that defined religions as results or expressions of human speculation on the cosmos. His main theoretical inspiration was Hermann Usener, a classical philologist, folklorist, and religion historian whose 1904 article "*Helige Handlung*" ("*The Sacred Act*") is considered one of the first anthropological texts that clearly states the primacy of ritual in the study of religion (Alcocer 2006, 2007). Today, this reasoning is a source of inspiration to develop more praxis-oriented models of the Mesoamerican cosmology.

Objects such as the *imumui* ladders or stairs are not understood through the reconstruction of an indigenous theory about the architecture of the cosmos, but through an analysis of ritual action, featuring the "sun person," personified by a specific human being who participates in the ritual, as making an effort to envision and to walk or to climb up the Mountain of Dawn. Wixarika cosmology is thus a type of metapragmatics, a reflection on their own ritual action and its effects.

FIGURE 9.1. *Wooden pyramid from Te'akata, replicating the shape of the sky that the sun climbs and descends. (14 cms, Martin Franken. Preuss Collection, Berlín-Stiftung Preussischer Kulturbesitz Ethnological Museum; after Artes de México 2007.)*

FIGURE 9.2. Imumui, *a stair for Father Sun's ascent. (Lumholtz Collection, American Museum of Natural History; after Lumholtz 1986.)*

Another very eloquent example already mentioned in Preuss' ethnography is that of the Wixarika's *nierika* discs. They are not cosmograms in the first place, but "instruments to see" (see Preuss 1998, 255, 292). In other words: "visions of the world" are the instruments to envision the world. What is captured is not as much a representation of the world, but rather what takes place during the visionary act of inventing it. This Wixarika metapragamitics of art also works for yarn tables like *The Vision of Tatutsi Xuweri Timaiweme* by José Benítez Sánchez that is at display at the Museo Nacional de Antropología and was analyzed in my book *La vida de las imágenes* (see Neurath 2013).

Preuss explains that the Wixarika and Cora arrows, gourd bowls, and other ritual objects are instruments that originally belonged to the gods. "The gods need these objects to keep the world functioning." Men's tasks consist in renewing them (Preuss 1998, 183). This way, the votive arrows "cannot be considered offerings, but are not prayers either. They are rather indispensable means to obtain life, health, rain and good harvests" (Preuss 1998, 107). They are even projectiles that are literally shot to the gods identified with sacrificial victims (Neurath 2013). It is difficult to distinguish between gods and their instruments. In ritual texts, ritual objects speak and act as persons, at the same level as other zoomorphic or anthropomorphic deities (see Preuss 1998, 268, 393–395).

Despite his emphasis on ritual and magical action, in Preuss's work we also find extensive information on indigenous speculation on the cosmos and its workings. However, the prevalence of semantic or symbolist approaches in anthropology has meant that until recently almost no one paid attention to the important nuances that Preuss introduced in his studies on Amerindian cosmologies. Today, in light of Alfred Gell's work (Gell 1998), his thoughts on the agency of ritual instruments and images turn out to be interesting.

I would like to clarify that, from my perspective, it is not about denying that there are native speculations on heaven or the underworld, but that the eagerness of many Mesoamericanists to rebuild cosmological models equivalent to those created by Europeans in the Middle Ages and the Renaissance should be conducive to understanding the Amerindian knowledge practices and life-worlds (see Nielsen and Sellner Reunert 2009; chapter 1, this volume; Botta, chapter 2, this volume; Díaz 2009; chapter 3, this volume; Vail, chapter 4, this volume).

Going back to the Wixarika case, it turns out that knowing, for example, how many levels are in heaven may not be that relevant. As I have described elsewhere (Neurath 2002), some of the Wixarika *xirikite* temples feature pyramidal foundations of up to eight levels, others have five steps, but the majority of those buildings feature just one. The number of steps corresponds, to a point, with the hierarchy of the temple. The *xirikite* of the communal ceremonial centers and in particular those identified with Paritek+a hill (Mountain of Sunrise) tend to be bigger, taller, and more elaborate architecturally than others.

FIGURE 9.3. *The Taimarita hill, with its natural tiers, as seen from Bule Hill. (Photography by Johannes Neurath.)*

The paradigm of the sun's stairways does consist of five steps. Some mountains are considered altars to the sun and, in the paths to the summit, five levels can be distinguished. The most important case is the Paritek+a hill in Wirikuta, also known as Reu'unari, Burnt Mountain (Cerro Quemado). The Tamairita hill (between Keuruwit+a and Xawiepa, in Tuapurie territory) is a particularly interesting case. It is a natural mountain that has five levels that are clearly distinguishable (figure 9.3).

Ritual dances and songs almost invariably consist of five parts, too, while the pilgrimage route from the sea to Wirikuta passes through five doors that must be crossed. These units and stages are associated with the steps of the *xirikite*, but, as we have seen, this in no way means that all temples must have exactly five steps.

It is worth mentioning that the sun's ladders or stairways are a ritual and iconographic element shared with Coras, the Wixarika neighbors. The altars of Cora churches usually feature five levels, and at the day of the winter solstice, Coras living at the border of the San Pedro River build a small pyramid of sand (Jáuregui and Magriñá 2007). Rock art in the region often shows emblems that can be identified as ladders and suns, and are interpreted as

"Sun Father's ladders" (Furst and Scott 1975). There are even indications that the sun's stairways to zenith may be part of a macroregional ritual complex associated with prehispanic cultures such as Aztatlán and Casas Grandes (Mathiowetz 2011).

Now, if we do not want to abandon the pragmatic perspective focused on the study of ritual action, all these stairs should not be seen as models but as steps to be effectively used. Ritual objects are not metaphors of cosmological ideas, but should be interpreted literally. The important thing is to climb up and down the ladder.

When we observe the ceremonies that take place in ceremonial centers, we actually see Father Sun walking up and down these steps. He is personified by the *jicarero* who lives in the temple of Father Sun. As we mentioned, the "*jícara* carriers" are the group of those in charge (Span.: *encargados*) of the *tukipa* ceremonial center. Their role is to personify the original family of the deities who are also the founders of the community. Each *jicarero* carries and protects an ancestral deity's gourd bowl. It could be said that their role is to "be born" from this *jícara* and to live the initiatory experience that enables one to effectively become an ancestor. Transforming into an ancestor implies to walk up the steps that lead from the underworld to the sky, from the primordial ocean in the west to the Place of Sunrise in the east. One *jicarero* is specifically in charge to be the Sun Father, but all participants perform the same set of ritual actions and also may become part of the original group of ancestors.

Emphasis on ritual action implies a negation of representation. In ritual action, the intention is to stage, not myths, but the real presence of the gods. But we must take into account that the "presentification" of the gods requires a certain effort and is not always achieved. The event that is the birth of the ancestors is the result of a complex initiatory process. Only when everything has been correctly executed does the *jicarero* personifying the sun become this deity and ascend the stairway to the summit.

From an epistemological perspective, a condition of possibility of this experience is the isomorphism between the micro- and the macrocosm. What happens on a ritual level is the same thing that happens in the world. Following Cassirer (1997), who largely based his theory of "magic" on observations made by Preuss among the Coras, ritual is based on a non-representational form of symbolization. But representation is not entirely foreign to the ritual. The aim is to overcome it, turning ritual representation into an event with an unmediated presence of powerful beings. However, in many cases, such an immediacy of ancestral deities is too dangerous, and more indirect ways to relate to those beings are preferred (Neurath 2013).

FIGURE 9.4. *A* xiriki *altar with thatched roof, Xawiepa ceremonial center. (Photograph by C. S. Lumholtz; Lumholtz 1904.)*

BUILDING ROOFS AND RENOVATING THE CANDLES OF LIFE

The importance of Father Sun's stairs must not make us ignore other ways of ritually (re)presenting or building the sky. By itself, each Wixarika temple or altar constitutes a microcosm and, within this context, the parts that are most closely identified with heaven are the thatched roofs made of *zacate* grass (figure 9.4).

As we have seen, *xiriki* temples are usually built above some type of pyramidal foundation. In other cases they are located on top of an actual mountain.

While those buildings are associated with the superior half of the cosmos, this is specifically valid for their roofs. In consequence, the most important objects are stored on elevated *tapancos* (attics) located on the eastern side of the building, the part that has the closest identification with the sky.

Circular *tuki* temples, within the group of temples which make up the ceremonial center, represent the space "below" (the underworld and the sea), but the roof of the grand temple is also associated with the sky. An *equipal* chair (Wixarika: *uweni*) hanging from the roof is the seat of the Sun.

The roofs of all Wixarika temples are renewed every five years, while other parts of the building are only restored occasionally. In cosmological terms, this corresponds to differentiated ontological statuses: the underworld is the preexisting part of the cosmos, the part that was never created. In contrast, the upper half of the cosmos is not a given, but a result of ritual action. As a consequence, it is mainly the roofs that must be periodically renewed. When this is done, some handfuls of grass gathered in Wirikuta are mixed with the rest of the *zacate* in order to establish an identification between the roofs of the temples and the eastern desert.

During the annual ritual cycle, the upper part of the cosmos is also rebuilt by erecting the pillars (*haurite*) that support the sky. The problem with *haurite* is that they are *ocote* torches or candles that burn away. At the Hikuli Neixa feast, the *haurite* are small pines placed in the corners of the *takwa* dance patio (somewhat similar to small Christmas trees). Dancers, personifying deer, visit the small pines and, according to the choreography, "scratch themselves" on them. Small trees are used in order to emphasize that they are just beginning to grow. The upper part of the world is just being built and an effort is made to separate heaven and earth.

THE EAGLE THAT NEVER SLEEPS

There is a third way to (re)present the sky: a sometimes two- or even four-headed eagle that is Our Mother the Young Eagle, Tatei Wierika Wimari (figure 9.5). She personifies the feathers of the *muwieri* sticks used by the *mara'akame* or shaman (figure 9.6). The ritual specialist is identified with his stick, and she with the feathers of this stick. He communicates with her who is up in the sky and "sees everything from above" in such a way that he can also communicate with other beings in the world.

As other ritual objects, the *muwieri* by itself is a person or deity. While the shaman is an intermediary between the uninitiated and the gods, the stick acts as an intermediary between the *mara'akame* and the other initiated persons or

FIGURE 9.5. *Front view and detail of a Huichol embroided textile with a motif combining the bicephalic eagle and peyote. (Digital Archive of the Collections of the Museo Nacional de Antropología Antropología. Reproduction authorized by the Instituto Nacional de Antropología e Historia, Secretaría de Cultura-INAH-MTM-MNA-CANON.)*

FIGURE 9.6. *A shaman's muwieri. (Digital Archive of the Collections of the Museo Nacional de Antropología. Reproduction authorized by the Instituto Nacional de Antropología e Historia, Secretaría de Cultura-INAH-MTM-MNA-CANON.)*

gods. The explanation that a *muwieri* works as a telephone and the goddess of heaven as a satellite or antenna is not uncommon. In the Cora Holy Week in Jesús María, some *borrados* ("erased ones") or *judíos* ("Jews") mock Wixarika *mara'akate* by using a cellphone as if it was a *muwieri*.

Again, we see the identification between the celestial realm and the "gift of vision" or the special power initiates possess. The characterization of the bird implies that it is a being with a highly advanced degree of initiation. For example, it is believed that the eagle never sleeps (Preuss 1993). One of the most important exercises for Wixaritari aiming to become shamans is sleep deprivation. This way, the eagle enjoys a privileged position to facilitate communication between the shaman and other beings. The equivalence between the goddess and panoptical telecommunication devices is not naive at all. The goddess of heaven is an extension of the shaman, an instrument at his service, while she herself is the most advanced *mara'akame*, so powerful and far from the night realm that she doesn't need to sleep at all.

There have been Wixarika rituals documented in which the entire group of participants transforms into a giant eagle (Neurath 2002; Pacheco 2016). On the other side, Tatei Wierika is also identified with non-indigenous symbols of power, particularly with the Virgin of Guadalupe who, as we know, wears a great starry mantle. The association between the Mexican eagle and the Virgin was very important during colonial times, and especially during the Independence period (Cuadriello 1995; Terán 1995). According to a Wixarika story documented in Tateikie (San Andrés Cohamiata) by Paul Liffman, a blood-red American anthropologist (Teiwari Xure, or Gringo Rojo) took photos of the Wixarika sky deity, sitting on the *nopal* cactus, devouring a serpent, and founding Mexico City. His photograph, obviously taken without permit, was used to produce peso coins, the Mexican coat-of-arms, and the Mexican flag (Liffman 2011, 2012). However, amongst Wixaritari of Tuapure, I also found that Tetei Wierika and the Mexican eagle are identified with the American bald eagle. When flying north, some *pieles rojas* ("redskins," Native Americans) shot at her and she captured the arrows with her talons. Then she sat down on the snow, thus her tail turned white.

THE DESTRUCTION OF THE UPPER HALF OF THE COSMOS

As we have mentioned, rites celebrated during the Namawita Neixa (summer solstice) sowing feast can be interpreted as the destruction of the celestial or solar world built over the rest of the year. Each beginning of the rainy season is a return to the almost undifferentiated world "below." The day sky is the ephemeral part of the cosmos whose existence depends on ritual action carried out by humans. In a way, it only exists during the dry season. During the rainy season, we find ourselves in a night (*t+karipa*), in a dark world. This world is not created by ritual action, but in order to return to this primordial cosmos where rain is constant, "where it is always rainy season" and "where everything grows," heaven must be destroyed first.

The crux of the Wixarika asymmetrical cosmological dualism lies in the contrast between two types of rituality. Heaven and daylight are *created* by vision seekers, who climb the Mountain of Dawn and identify themselves with the deer, the quintessential victim of sacrifice, or the eagle that never sleeps. The underworld is always there. It needs not be created since it is the realm where life originates spontaneously. This spontaneity may be recovered by destroying creation and suspending the solar authority system. But after Namawita Neixa the ritual labor of domesticating spontaneous fertility and creating the orderly, luminous realm of day and heaven is almost immediately taken up again.

One of the central rites in the Namawita Neixa feast is the burning of the post holding up the sky (see Neurath 2002). It should be noted that these rites are much more than an "inversion." During the rest of the year a solar order is created wherein the forces of darkness are repressed according to an ideology of (self-)sacrifice. The practice of sacrifice is suspended during the sowing feast, but now a law is fulfilled that is usually ignored and is the foundation for peaceful coexistence between beings "above" and those in the underworld. The result of this alliance is maize and humanity (Neurath 2008). The logic substituting initiation is exchange. It is no longer about differentiation from the uninitiated, but coexisting, exchanging, and trading peacefully with all beings of the world, including mestizos.

It is no exaggeration to state that the Namawita Neixa feast implies a change of worldview. During these rituals, that which according to solar ideology is considered "chaotic" suddenly becomes a cosmos in its own right. According to the solar worldview valid during the dry season, "chaos," "filth," and the underworld are located in the devalued extreme of the chronotope or time-space. Dark beings must be defeated and controlled ritually by forces associated with the sky, the rising sun, and the ancestors. However, from a non-solar viewpoint, dark powers are positive and must be released. The alternative worldview, valid during the rainy season, represents the viewpoint of the fertility deities. The supreme deity is no longer the god of fire (Tatewari) but the great goddess of fertility, Takutsi Nakawé. In their world, indigenous and mestizo people have few differences. Now sacrifice and shamanic invention are not as important.

Emphasis is mainly placed on exchange, reciprocity and commerce, cooperation, and alliance.

Considering its solar-celestial domain, the Wixarika tradition can be defined as visionary and focused on the creation of ephemeral, always radically new things. Shamanic visions are original events (Preuss 1933), or inventions in a poetic sense (Wagner 1981). On the other hand, the nocturnal realm of Wixarika tradition may be defined as a recovery of the past. To become involved with nocturnal, preshamanic life also entails establishing a constructive relationship with mestizos (Neurath 2011b).

As we have argued elsewhere, in every particular rite this coming and going between sacrifice and exchange is repeated. All ritual chants refer to trips to faraway places and a return to the place of celebration. The *mara'akame* travels to places where ancestral gods live and invites them to attend the feast. When men and gods are all reunited, animals are sacrificed. The shaman must convince the animals to be sacrificed. Thus, to "officially" kill animals is always an act of self-sacrifice on their part. The blood of agonizing animals is food for the gods and is distributed via different ceremonial objects. Some of these objects are offerings, which will be deposited in sacred places after the feast. When recovering the pilgrimage between ceremonial centers and sites in the landscape, the exchange rites that in principle oppose the rites of sacrifice are resumed.

In many indigenous communities in Mesoamerica and surrounding areas we can observe these systems of exchanging objects or substances between an altar or ceremonial center and places in the ritual landscape. A good example is Coyle's study of the Coras (Coyle 2000), where samples of water fetched at different sacred spots are ritually interchanged. But things are not quite that simple. Exchange is important, but deities are first identified with prey animals. And the relationship between predator and prey apparently precludes the possibility of relations of exchange. It seems to be rather complicated to exchange with a person you kill, but ritual acts may be both things at the same time: in ritual, noncompatible relations and contradictory actions can coexist and be performed simultaneously (Houseman and Severi 1998; also Humphrey and Laidlaw 1994).

The creation and destruction of the heavenly world occur in a simultaneity where the emphasis is always changing, but one process is never completely separate from the other. The typical Wixarika offering expresses this contradiction. A gourd *jícara* with blood is definitely an offering, something given as nourishment, but the votive arrow is a projectile shot at the deity as a prey (Neurath 2013).

Both ritual actions, exchanging and hunting, are essential. Those who stop placing offerings run the risk of becoming prey for gods and animals. However, the Wixarika gods always have the potential to become pathogenic agents, even the ones that are born from the visions of the initiates. Just as for the Mixtecos studied by Monaghan (1995, 103), "gods are diseases." Many rites describe how pathogenic beings from "below" are controlled by those "above." But the creation of the heavenly world only implies a somewhat lesser risk of producing incontrollable beings. Everything that relates to the solar realm is ephemeral, while beings from the underworld are eternal. Anyway, shamans are always confronted with mistrust. Often, they are accused of provoking sickness instead of promoting healing. It is for the same reason that *peyoteros* are received with such suspicion when returning to their communities. Non-initiated people may become sick just from their presence alone.

As a conclusion I would like to emphasize the importance of not trying to construct simple cosmological models. The complexity of ritual relationships produces differentiations and ambiguity on an ontological level that precludes the study of a single Wixarika *cosmovision*. This complexity is what really matters, because it allows Wixarika people to affirm a cosmological exclusiveness and ethnic particularity, but, at the same time, to live together with non-Indian populations. Such a cosmology may be characterized as an equivalent of what is known as dividual personhood.

REFERENCES

Alcocer, Paulina. 2006. "La forme interne de la conscience mythique: Apport de Konrad Theodor Preuss à la *Philosophie des formes symboliques* de Ernst Cassirer." *L'Homme* 180: 139–170.

Alcocer, Paulina. 2007. "Konrad Theodor Preuss: En busca de magia, ritos y cantos." *Artes de México* 85 (*Arte antiguo: cora y huichol*): 9–14.

Artes de México 2007. No. 85 (*Arte antiguo: cora y huichol*).

Benítez, Fernando. 1984. *Los indios de México*, vol. II: *Los huicholes*. Mexico: Era.

Cassirer, Ernst. 1997 [1925]. *Philosophie der Symbolischen Formen*, vol. 2: *Das mythische Denken*. Darmstadt: Primus Verlag.

Coyle, Philip E. 2000. "'To Join the Waters': Indexing Metonymies of Territoriality in Cora Ritual." *Journal of the Southwest* 42 (1, *Ritual and Historical Territoriality of the Náyari and Wixárika Peoples*): 119–128.

Cuadriello, Jaime. 1995. "Visiones en Patmos Tenochtitlán. La mujer águila." *Artes de México* 29 (*Visiones de Guadalupe*): 10–22.

Díaz, Ana. 2009. "La primera lámina del *Códice Vaticano A*: ¿Un modelo para justificar la topografía celestial de la antigüedad pagana indígena?" *Anales del Instituto de Investigaciones Estéticas* 31 (95): 5–44.
Furst, Peter T., and Salomón Nahmad, eds. 1972. *Mitos y arte huicholes*. Mexico: Secretaría de Educación Pública (SEP-Setentas 50).
Furst, Peter T., and Stuart D. Scott. 1975. "La escalera del Padre Sol: Un paralelo etnográfico-arqueológico desde el occidente de México." *Boletín INAH* 12: 13–20.
Galinier, Jacques. 2004. *The World Below: Body and Cosmos in Otomi Indian Ritual*. Boulder: University Press of Colorado.
Gell, Alfred. 1998. *Art and Agency: An Anthropological Theory*. Oxford: Clarendon Press.
Graulich, Michel. 1999. *Ritos aztecas: Las fiestas de las veintenas*. Mexico: Instituto Nacional Indigenista.
Houseman, Michael, and Carlo Severi. 1998. *Naven or the Other Self: A Relational Approach to Ritual Action*. Leiden: Brill.
Humphrey, Carolyn, and James Laidlaw. 1994. *The Archetypal Actions of Ritual: A Theory of Ritual Illustrated by the Jain Rite of Worship*. Oxford: Clarendon Press.
Jáuregui, Jesús, and Magriñá, Laura. 2007. "La escalera del Padre Sol en la Judea de los coras." *Arqueología Mexicana* 14 (85): 69–74.
Kantor, Lori. 2012. "El paisaje wixárika y su relación con la arqueología." Mexico: Museo Nacional de Antropología. http://www.mna.inah.gob.mx/contexto/el-paisaje-wixarika-l-kantor.html.
Latour, Bruno. 1993. *We Have Never Been Modern*. Cambridge, MA: Harvard University Press.
Liffman, Paul M. 2011. *Huichol Territory and the Mexican Nation: Indigenous Ritual, Land Conflict, and Sovereignty Claims*. Tucson: University of Arizona Press.
Liffman, Paul M. 2012. "Huichols and the Cosmopolitics of Mining in Mexico." Paper presented at the 12th European Association of Social Anthropologists Biennial Conference, Nanterre, France, July 10–13.
Lira, Regina. 2014. "L'alliance entre la Mère Maïs et le Frère Aîné Cerf: Action, chant et image dans un rituel wixárika (huichol) du Mexique." PhD diss., École des Hautes Etudes en Sciences Sociales, Paris.
López Austin, Alfredo. 1980. *Cuerpo humano e ideología*. Mexico: Universidad Nacional Autónoma de México, Instituto de Investigaciones Antropológicas.
López Austin, Alfredo. 1994. *Tamoanchan y Tlalocan*. Mexico: Fondo de Cultura Económica.
Lorente, David. 2011. *La razzia cósmica: Una concepción nahua sobre el clima, deidades del agua y graniceros en la sierra de Texcoco*. Mexico: Centro de Investigaciones y Estudios Superiores en Antropología Social, Universidad Iberoamericana.

Lumholtz, Carl S. 1900. "Symbolism of the Huichol Indians." *Memoirs of the American Museum of Natural History* 3 (1): 1–291.

Lumholtz, Carl S. 1904 [1902]. *El México desconocido: Cinco años de exploración entre las tribus de la Sierra Madre Occidental; en la tierra caliente de Tepic y Jalisco, y entre los tarascos de Michoacán*, 2 vols. Trans. Balbino Dávalos. New York: Charles Scribner's Sons.

Lumholtz, Carl S. 1986 [1900]. *El arte simbólico y decorativo de los huicholes*. Mexico: Instituto Nacional Indigenista.

Mathiowetz, Michael D. 2011. "The Diurnal Path of the Sun: Ideology and Interregional Interaction in Ancient Northwest Mesoamerica and the American Southwest." PhD diss., University of California, Riverside.

Monaghan, John. 1995. *The Covenants with Earth and Rain: Exchange, Sacrifice, and Revelation in Mixtec Sociality*. Norman: University of Oklahoma Press.

Navarrete, Federico. 2012. "¿Cómo pensar la historia indígena más allá de la aculturación y la continuidad?" Paper presented at the Seminario Permanente Taller "Signos de Mesoamérica," Mexico, Universidad Nacional Autónoma de México-Instituto de Investigaciones Antropológicas, September 21.

Neurath, Johannes. 2002. *Las fiestas de la Casa Grande: Procesos rituales, cosmovisión y estructura social en una comunidad huichola*. Mexico: Universidad de Guadalajara/Instituto Nacional de Antropología e Historia.

Neurath, Johannes. 2008. "Alteridad constituyente y relaciones de tránsito en el ritual huichol: iniciación, anti-iniciación y alianza." *Cuicuilco* 15 (42): 29–44.

Neurath, Johannes. 2011a. "Ambivalencias del poder y del don en el sistema político-ritual wixarika." In *Los pueblos amerindios más allá del Estado*, ed. Federico Navarrete and Berenice Alcántara, 115–143. Mexico: Universidad Nacional Autónoma de México, Instituto de Investigaciones Históricas.

Neurath, Johannes. 2011b. "Don e intercambio en los mundos rituales huicholes Una contribución a los debates sobre chamanismo y ontologías indígenas." In *Curanderismo y chamanismo: Nuevas perspectivas*, ed. Laura Romero, 21–41. Puebla: Benemérita Universidad Autónoma de Puebla.

Neurath, Johannes. 2013. *La vida de las imágenes: Arte huichol*. Mexico: Artes de México, Consejo Nacional para la Cultura y las Artes.

Neurath, Johannes. 2015. "Shifting Ontologies in Huichol Ritual and Art." *Anthropology and Humanism* 40 (1): 58–70.

Pacheco, Ricardo. 2016. "La navegación del tambor y el vuelo de los niños: complejidad ritual huichol." PhD diss., Universidad Nacional Autónoma de México.

Pitarch, Pedro. 2012. "La ciudad de los espíritus europeos: Notas sobre la modernidad de los mundos virtuales indígenas." In *Modernidades indígenas*, ed. Pedro Pitarch and Gemma Orobitg, 61–87. Madrid: Iberoamericana, Vervuert.

Preuss, Konrad T. 1933. *Der religiöse Gehalt der Mythen*. Tubinga: J. C. B. Mohr.
Preuss, Konrad T. 1993. "Dos cantos de los indios coras." In *Música y danzas del Gran Nayar*, ed. Jesús Jáuregui, 15–38. Mexico: Centro de Estudios Mexicanos y Centroamericanos, Instituto Nacional Indigenista.
Preuss, Konrad T. 1998. *Fiesta, literatura y magia en el Nayarit: Ensayos sobre coras, huicholes y mexicaneros de Konrad Theodor Preuss*, ed. Jesús Jáuregui and Johannes Neurath. Mexico: Instituto Nacional Indigenista, Centro de Estudios Mexicanos y Centroamericanos.
Questa Rebolledo, Alessandro. 2010. "Cambio de vista. Cambio de rostro: Parentesco ritual con no humanos entre los nahuas de Tepetzintla, Puebla." Master diss., Universidad Nacional Autónoma de México.
Rio, Knut. 2007. "Denying the Gift: Aspects of Ceremonial Exchange and Sacrifice on Ambrym Island, Vanuatu." *Anthropological Theory* 7 (4): 449–470.
Romero López, Laura. 2010. "Ser humano y hacer el mundo: La terapéutica nahua en la Sierra Negra de Puebla." PhD diss., Universidad Nacional Autónoma de México.
Rosaldo, Renato. 1991 [1989]. *Cultura y verdad: Nueva propuesta de análisis social*, trans. Wendy Gómez Togo. Mexico: Consejo Nacional para la Cultura y las Artes, Grijalbo (Los noventa).
Seler, Eduard. 1901. "Die Huichol-Indianer des Staates Jalisco in Mexico." *Mittheilungen der Anthropologischen Gesellschaft in Wien* 31: 137–163.
Strathern, Marilyn. 1988. *The Gender of the Gift: Problems with Women and Problems with Society in Melanesia*. Berkeley: University of California Press.
Terán, Marta. 1995. "La relación del águila mexicana con la Virgen de Guadalupe en los siglos XVII y XVIII." *Historias* 34: 30–50.
Usener, Hermann. 1904. "Heilige Handlung." In *Archiv für Religionswissenschaft* 7: 281–339. Leipzig: B. G. Teubner.
Wagner, Roy. 1981 [1975]. *The Invention of Culture*. Chicago, IL: University of Chicago Press.
Wagner, Roy. 1991. "The Fractal Person." In *Big Men and Great Men: Personifications of Power in Melanesia*, ed. Maurice Godelier and Marylin Strathern, 159–173. Cambridge, UK: Cambridge University Press.
Weigand, Phil C. 1981. "Differential Acculturation among the Huichol Indians." In *Themes of Indigenous Acculturation in Northwest Mexico*, ed. Thomas Hinton and Phil C. Weigand, 9–21. Anthropological Papers 38. Tucson: University of Arizona.
Weigand, Phil C. 1985. "Considerations on the Archaeology and Ethnohistory of the Mexicaneros, Tequales, Coras, Huicholes, and Caxcanes of Nayarit, Jalisco, and Zacatecas." In *Contributions to the Archaeology and Ethnohistory of Greater Mesoamerica*, ed. William J. Folan, 126–187. Carbondale: Southern Illinois University Press.

About the Authors

Sergio Botta (PhD) is associate professor in the History of religions at Sapienza University of Rome (Department of History, Culture, Religions) where he teaches the history of religions and history of the Americas. His main research interests are history and religions of indigenous cultures. His research focuses on a historical analysis of the missionary discourses in New Spain and on method and theory in the study of religions, religions and film, indigenous religions, and shamanism. He is the chief of the editorial committee of the journal *Studi e Materiali di Storia delle Religioni*.

Ana Díaz is associated with the Institute for Research in Aesthetics, at National Autonomous University of Mexico (UNAM). On 2019 she won the Cátedra Especial Miguel León-Portilla, a special chair offered by the UNAM. She holds a PhD in art history from UNAM (2011), for which her work was named the best dissertation in the humanities by the Mexican Academy of Sciences. She has also won the distinction of the best thesis on history and ethnohistory from the National Institute of Anthropology and History (INAH). From 2010 to 2012 she worked as assistant to the director of the National Museum of Anthropology of Mexico and as academic coordinator of this institution. In 2015 she won the prize for best article in cultural history from the Mexican Committee of Historical Sciences. She was a Fulbright fellow at Harvard University (2015) and is a lecturer in graduate and undergraduate programs at UNAM, ENAH, and the Iberoamerican University in Mexico City. Her research interests include Mesoamerican cosmologies, esthetical and environmental conceptions of sixteenth-century Nahuas, and calendrical concepts in the codices of the Borgia Group.

Kerry Hull holds a PhD in anthropology from the University of Texas, Austin, and is professor of ancient scripture at Brigham Young University. His research has been funded by the National Science Foundation (NSF) and the Foundation for the Advancement of Mesoamerican Studies, Inc. (FAMSI). He is the coeditor of *Parallel Worlds: Genre, Discourse, and Poetics in Contemporary, Colonial, and Classic Maya Literature* (with Michael D. Carrasco, 2012) and of *The Ch'orti' Maya Area: Past and Present* (with Brent Metz and Cameron L. McNeil, 2009). He recently published *A Dictionary of Ch'orti' Mayan-Spanish-English* (2016). He has authored or coauthored more than 20 peer-reviewed articles and chapters. His research interests include Mesoamerican languages, Ch'orti Maya language and culture, Mesoamerican epigraphy, Polynesian linguistics, and Mesoamerican ethno-ornithology.

Katarzyna Mikulska holds a PhD in human studies (literature), Faculty of Modern Languages, University of Warsaw (2005), where she is now an assistant professor at the Institute for Iberian and Iberoamerican Studies. She received an award for the paper "Tlazolteotl: Una Nueva Diosa del Maguey" in the Competition of Young Americanists (at the 50th International Congress of Americanists, Warsaw), and a grant for the project "The Graphic Communication System of Pre-Hispanic Central Mexico as Utilized in Calendar-Religious Books." She is a member of the scientific board of the peer-reviewed journal *Anales de Antropología* (National Autonomous University of Mexico, Institute for Anthropological Research) and a member of the editorial board of the peer-reviewed journal *Itinerarios* (published by the Institute for Iberian and Iberoamerican Studies, University of Warsaw). She has coordinated five international symposia and authored more than twenty publications. Her research interests include Mesoamerican cosmologies and beliefs; Mexican divinatory codices, especially the codices *Borgia*, *Vaticanus B*, and *Cospi*; central Mexican graphic communication systems; and theory of writing.

Johannes Neurath holds a PhD in anthropology from the National Autonomous University of Mexico (1998). Since 1992 he has conducted fieldwork in northwestern Mexico among the Cora and Huichol. He is a researcher at the National Institute of Anthropology and History (INAH), distinguished with the "Titular C" category. He is associated with the Ethnography Department of the National Museum of Anthropology and a member of the National Researchers System (SNI) of the CONACYT. He is lecturer in the graduate program of Mesoamerican studies at the Faculty of Philosophy and Letters of the UNAM. He is coordinator of a Franco-Mexican project in the anthropology of art (since 2006). From 2010 to 2011 he participated as invited professor at the Laboratory of Social Anthropology of the Collège de France. He has authored two books and chapters in collective works and specialized journals and has served as coordinator of four books. His research interests include ritual, cosmology, and anthropology of art.

Jesper Nielsen holds a PhD in American Indian languages and cultures from the University of Copenhagen (2003). He was awarded the University's gold medal for his master's thesis (1998) on the dedication rituals of the ancient Maya. Currently he is associate professor and head of the Department of American Indian Languages and Cultures, University of Copenhagen, and epigrapher with the "Pintura Mural Prehispánica"-project at the Instituto de Investigaciones Estéticas at UNAM (National Autonomous University of Mexico). He has organized several international symposia and the sixteenth European Maya Conference (Copenhagen, 2011). He has published extensively in peer-reviewed journals (e.g., *Antiquity, Latin American Antiquity, Ancient Mesoamerica*) as well as books and edited volumes on various subjects related to Mesoamerica. His most recent research focuses on the writing systems of Teotihuacan and the Epiclassic cultures, Nahuatl writing, Teotihuacan iconography and religion, Teotihuacan's imperial presence in north-central and western Mexico, and the religious and visual syncretism of the early colonial period in New Spain.

Toke Sellner Reunert † held an MA in philosophy and American Indian languages and cultures from the University of Copenhagen (2008). He was coauthor of a comprehensive biography on the Danish Maya archaeologist Frans Blom (with Jesper Nielsen and Tore Leifer) and author of peer-reviewed articles on Maya shamanism and worldview, and on religious and visual syncretism of the early colonial period in New Spain.

David Tavárez holds a combined PhD in history and anthropology, University of Chicago (2000). His research (1997–2017) has been supported by fellowships and grants from the John Simon Guggenheim Memorial Foundation, the National Science Foundation, the Mellon Foundation, the École des Hautes Études en Sciences Sociales, the Foundation for the Advancement of Mesoamerican Studies, the John Carter Brown Library, and the National Endowment for the Humanities (twice). He is professor of anthropology at Vassar College. He is the author of *The Invisible War: Indigenous Devotions, Discipline, and Dissent in Colonial Mexico* (2011, paperback 2013), and coauthor of *Chimalpahin's Conquest: A Nahua Historian's Rewriting of Francisco López de Gómara's* La conquista de México (with Susan Schroeder, Anne Cruz, and Cristián Roa, 2010). Both works appeared in Spanish in 2012. More recently, he coauthored *Painted Words: Nahua Catholicism, Politics, and Memory in the Atzaqualco Pictorial Catechism* (with Elizabeth Hill Boone and Louise Burkhart, 2017), and edited *Words and Worlds Turned Around: Indigenous Christianities in Latin America* (2017). He has also published more than 40 peer-reviewed articles and book chapters. He has served as editorial board member for *Ethnohistory* and *Studies in Medieval and Renaissance History*, and as councilor of the American Society for Ethnohistory. His research interests include Mesoamerican religion and calendars, Nahua and Zapotec social history and philology, and colonial Latin American history.

Alexandre Tokovinine is a Maya epigrapher and archaeologist. He participated in several projects in Guatemala including the Holmul Archaeological Project and Proyecto Arqueológico de Investigación y Rescate Naranjo. He received a doctoral degree in anthropology from Harvard University in 2008. His doctoral research centered on Classic Maya place names and was supported by a Junior Fellowship at Dumbarton Oaks. Its results were published as a monograph, *Place and Identity in Classic Maya Narratives*. Other projects include 3D documentation of Classic Maya monuments, buildings, and artifacts, and contributions to *Ancient Maya Art at Dumbarton Oaks* (CAA Alfred H. Barr Jr. Award of 2013). Tokovinine currently holds the position of assistant professor in the department of anthropology, University of Alabama, Tuscaloosa.

Gabrielle Vail holds a PhD in Anthropology from Tulane University. She was awarded an NSF Research Grant, "Maya Hieroglyphic Database Project: The Codices," 1997–1999. She has codirected (with Martha Macri) both an NSF Research Grant, "Discourse Analysis and Grammatical Structure in the Maya Codices," 1999–2001, and an NEH Collaborative Research Grant, "Commentary and On-Line Database of the Maya Madrid Codex," July 2001–December 2002. She held a fellowship in precolumbian studies at Dumbarton Oaks in summer 2004 and was awarded the Colorado Endowment for the Humanities Publication Prize for *The Madrid Codex: New Approaches to Understanding an Ancient Maya Manuscript* (coedited with A. Aveni), 2004. She was also awarded an NEH Collaborative Research Grant, "Mesoamerican Codices Database Project: An On-Line Tool for Researching the Maya Codices," July 2004–April 2007. She is the author of books, chapters, and articles in peer-reviewed books and journals, and has participated as editor and coeditor of collective volumes. She specializes in precolumbian studies with an emphasis on the iconography and hieroglyphic texts of the screenfold codices painted by the prehispanic Maya. Her recent collaborative research emphasizes the Borgia Group of codices from central Mexico and Postclassic murals from the Maya area, as well as ethnohistoric documents from the Maya region. These studies contribute to our understanding of the ideology and religion of Mesoamerica and of interactions among Maya and central Mexican cultures during the Postclassic and colonial periods.

Index

A

Above / up, 7–8, 16, 18, 36, 44, 54, 109–110, 120, 127, 181, 190(fig.), 190–191, 199–201, 215, 217, 237, 239n, 252, 258, 282–283, 291–294, 296, 298–299, 305, 321, 331, 332n; above us, 43, 287, 289, 292, 300–301

Altar, 45, 64, 82, 118(fig.), 119, 125, 151, 159, 215, 216(fig.), 233(fig.), 241n, 253n, 254, 260(fig.), 261, 265, 283, 285(fig.), 329, 331, 336

Alternative: antiquities, 76, 90; cosmographic alternatives, 5; dimensions, 129; spaces, 104, 129; worlds / otherworld, 5–6, 8, 17–18, 32, 215, 217, 219–220, 258, 283, 292, 299, 300–301, 305, 309

Ancestors, 5, 7, 18, 119–120, 125, 129, 132, 182, 185, 200–202, 226, 229, 236, 253, 261, 268, 289, 299, 301–305, 307n, 322–323, 330, 335

Ángel, 16, 141, 142, 149–151, 152(fig.), 153; cangel / archangel, 153–156, 169, 172

Arrangements of cosmos / Cosmological arrangements, 5, 35, 110; horizontal, 5, 39, 42, 44, 297, 301; horizontal layers, 3–36, 53–54(fig.), 58–59, 81; horizontal structure, 38, 45; vertical, 7, 37, 44, 80, 82, 101, 201, 266; vertical layers, 5, 8, 33, 36, 38–39, 100, 130n, 251, 258; vertical structure, 35, 37, 46, 58, 59

ascend, to (verb), 105, 130n, 131n, 168, 193, 253 264, 270–271, 326

B

Bacab, 36, 149, 153, 156, 158, 173, 215

Below, 3, 7–8, 16, 18, 52, 166, 173, 181, 190, 190(fig.), 191, 199, 200–201, 215, 217, 237, 239n, 258, 282–283, 288, 291–292, 294, 301, 304–305, 321, 325, 232, 334

Beyond, 31, 43, 251, 255, 263, 292–293, 304, 306; world beyond, the, 292; sea, beyond the, 258, 259

Biyé, 8, 10, 55, 183, 185, 189, 202

Blood, 32, 120, 149, 186, 223–224, 231, 309, 336; sea of blood, 218

Boccacio, Giovanni, 73

Bolon Tz'akab, 149, 150–151, 153, 169

Borders / Frames, 121, 122(fig.), 123, 147, 155(fig.) 161, 165

Borgia Group, 181, 283

C

Calendar, 16, 104–105, 112, 142, 174n, 182, 184–185, 189, 191, 193–194, 197, 199, 201, 202n, 203n, 264, 321; cosmogonic discourse, 129; graphic discourse, 103, 109, 110, 113, 124, 126; signs, 5, 132n, 186, 232n, 253; wheel, 111, 231, 253

Cangel, 153–156, 169, 172. *See also* Ángel
Cardinal: directions, 5, 35, 38, 39(fig.), 44–46, 152(fig.), 186, 187(fig.); points, 5, 153, 186, 201
Cartari, Vincenzo, 15, 72–76, 79, 80(fig.), 82, 83(fig.), 84(fig.), 86(fig.), 87–88, 89(fig.), 90(fig.), 92n, 93n, 94n
Cave, 5, 32, 37, 45, 54(fig.), 57, 114, 161, 162, 220, 252, 262, 289; earth cave, 161–162, 176n
Chac/Chaak, 144–145, 148(fig.), 149, 152(fig.), 153–162, 155(fig.), 163(fig.), 169, 175n, 176n
Ch'orti', 16, 209–232, 234–238, 241n, 242n
Chicomoztoc, 57, 125, 304
Chilam Balam, Books of, 11, 16, 36, 44, 141–143, 150, 153, 156, 158–160, 170, 175n, 176n, 177n, 182, 253
Citlalinicue, 15, 84(fig.), 103, 105, 107, 109, 110, 112, 120, 123, 124–125, 127–129, 132n
Citlallatonac, 77, 83–84, 93n, 295–296
Clouds, 55, 117, 129, 154–155, 175n, 221, 227–228, 230, 232, 233(fig.), 238, 239n, 283–286, 288(fig.), 290, 224
Clusters, iconographic, 117–118, 120, 124, 126, 129
Codex Borgia, 38, 41, 100, 121–122, 124, 131n, 132n, 202, 283, 284(fig.), 284–286, 287(fig.), 288(fig.), 289(fig.), 294, 303
Codex Fejérváry-Mayer, 39(fig.), 186, 189, 297, 301, 305
Codex Vaticanus A / Codex Vaticanus 3738 / Codex Ríos, 3–4, 6, 16, 32–34, 34(fig.), 35–36, 42, 45–46, 53, 55–58, 70–77, 79–88, 90, 91n, 93n, 94n, 100–101, 101(fig.), 102, 107, 108, 110, 112, 118, 120, 124, 130n, 200, 283, 296–297
Codex Vindobonensis, 114, 115(fig.), 117(fig.), 118, 131n
Colors, 13, 81, 105, 118, 129, 154–156, 215, 238
Cosmography, 15, 32, 34–35, 42, 46, 101–103, 106–109, 112, 130n, 282, 306n; Eurasian, 4; Mesoamerican, 3, 8, 18, 31–32, 34, 42, 46–47; Roman, 46, 54–55, 56(fig.), 72, 81, 309n
Cosmology, 4, 6, 11–12, 15, 18, 32–33, 35, 37, 60n, 70, 72, 76–77, 91, 114, 199, 201, 319, 321, 342; Christian, 73, 89; Dante's, 46, 51, 53, 58; generative, 125, 127–128, 145; hybrid, 72, 79; Mesoamerican, 6–7, 9, 11, 33, 43, 75, 77–79, 81, 88, 91, 326; Wixarika, 319–321, 326
Cosmological, 8, 18, 19n, 51, 60n, 77, 108–109, 111–115, 125, 127, 129, 130; architecture, 10, 37, 59, 265, 322, 326; arrangements, 5, 35, 110; asymmetry, 18, 322, 335; discourses, 110, 124,

126, 130n; eras, 116; flexible conception, 13–17, 52, 103; geography, 7; image, 82, 113; model, 3, 5, 107, 112–113; mutable composition, 6, 8, 127; narratives, 7, 15, 17, 59, 103–104, 112–116, 122, 122(fig.), 123, 125, 129, 151, 253, 255, 258, 259, 262–271, 269(fig.); pattern, 5, 7–8, 13, 37, 40, 53, 56, 71, 159, 191, 198, 214, 255, 283, 291, 298; pyramidal structure, 100; representation, 3–4, 12–13, 15, 18, 33, 38, 40–42, 47, 53, 54(fig.), 57, 58, 71, 80–81, 81(fig.), 85, 103, 117–120, 130n, 131n, 147, 160, 162, 186, 190–191, 252, 254, 259, 262, 267(fig.), 282–286, 291, 297, 327, 330; structure, 3, 4, 6, 14–17, 31, 33–39, 34(fig.), 41, 44–46, 48, 58, 100, 101(fig.), 102(fig.), 103, 105, 107, 110, 112–114, 127–128, 130n, 166(fig.), 190, 201, 251, 297, 306n, 325
Cosmos, 3–8, 14–18, 31–39(fig.), 40–48, 51–53, 58–59, 71, 80–82, 100–116, 101(fig.), 123–127, 129, 130n, 131n, 132n, 180–181, 189–193, 197, 199–201; three-tiered cosmos, 16, 191, 200; Aristotelian Metaphysics, 110; body, 8, 120–121, 127, 132n, 321; cardinal directions/regions, 35, 39(fig.), 44–46, 153, 186, 187(fig.), 201, 257, 271, 297; center, 4, 5, 38, 39(fig.), 45–46, 52–53, 154, 157(fig.), 158, 161, 164(fig.), 169, 174, 175n, 180, 186, 187(fig.), 190, 196, 213–214, 217, 227, 239n, 301, 328, 330, 331(fig.), 332, 336; Ch'orti', 16, 209–232, 210(fig.), 211(fig.), 212(fig.), 216(fig.), 234–239, 241n, 242n; Chalcan, 107, 112, 125; Christian, 11–12, 14, 31, 46–48, 58, 70–73, 76, 88, 108, 111–112, 132, 169; living entity, 15, 103; Greco-Roman, 46; houses, 8, 191, 193, 197, 199, 201; Mayan, 16, 46, 235, 272; Mesoamerican, 5–7, 9, 31–32, 37–38, 42–46, 48, 58, 70, 75–78, 81, 88–91, 101, 113, 131n, 181, 199–200, 319, 321, 326, 328; Medieval, 52, 282, 307n; Mexica, 3–4, 8, 15, 31–33, 35–37, 40, 76–77, 79, 82, 100, 101(fig.), 102–104, 107, 108–110, 112–114, 118–119, 122, 124, 129, 131n, 168, 175n, 186, 200–201, 203n, 283, 322, 334; multilayered, 5, 31, 39, 41(fig.), 42, 44, 45, 59, 60n, 200; Otomi, 93n, 108–110, 112, 292, 297; Tetzcocan, 106–107, 112, 125; Wixarika, 8, 319, 320–323, 325–329, 331, 332, 334–337; Zapotec, 6, 10, 16, 34, 44, 180–188, 187(fig.), 190, 190(fig.), 193–197, 199–202, 202n, 203n, 204n, 205n

346 INDEX

Creativeness/to create, 12–13, 17, 74, 105, 107–108, 110, 127, 156, 163, 168, 213–214, 225, 234, 288, 293–294, 300, 303, 322–325, 328, 335

D

Dante, 14, 32–33, 46, 48, 51–53, 55–58, 60n, 201, 307n

Darkness, 106–107, 157(fig.), 232, 234, 236, 285, 304, 322, 335

Dawn, 160, 170, 184, 235–236, 321–323, 326, 335

Deep-time, 258, 259–263, 268, 270–271

Descend, to (verb), 35–36, 40, 109, 114–115, 120, 140n, 150–151, 171, 186–187, 193, 196, 253, 289, 295, 323, 326, 327(fig.)

Diablo, 16, 141, 225

Dimensions, alternative, 129

Directions/Regions, 35, 38–39, 44–46, 120, 139, 153, 154, 161, 176n, 186, 187(fig.), 194, 201, 257(fig.), 271, 297, 305; cardinal, 35, 39(fig.), 45; gods, 7, 15, 36–39, 42, 44–47, 54–57, 72–80, 82–83, 83(fig.), 100, 104–107, 109, 111–113, 119–120, 122, 125, 129, 153, 155–156, 158–159, 165, 173, 175–176, 212–213, 216, 228, 232, 233(fig.), 239n, 241, 253, 258, 261–263, 265, 289–290, 296, 301, 320, 325–326, 328, 330, 332, 334, 336–337

Discourse, 5, 6, 11, 15, 70, 73, 76, 79, 86, 88, 90, 91n, 103–104, 109–110, 113, 124, 126, 129, 130n, 293, 294; cosmogonic discourse, 129; discoursive productions/configurations, 112, 126, 130; graphic discourse, 103, 109–110, 113, 124–126; visual discourse, 11, 15, 113

Distance, 7, 17, 74, 78, 251, 253, 255, 258, 263, 266, 271

Divine Comedy, 51–52, 57

Down, 3, 17, 43, 170, 194, 197, 200, 308n, 326, 330

Dreams, 18, 104, 114, 129, 229, 300, 305; place of dreams, 114, 129; space of dreams, 7, 104, 229

Dresden Codex, 144, 146, 1458(fig.), 149, 150, 154, 156, 157(fig.), 158, 160(fig.), 161, 167(fig.)

E

Earth, 3–8, 10, 16–18, 31–35, 38–40, 42, 44–46, 48, 50(fig.), 52–53, 56, 100, 101(fig.), 106, 109, 114–118, 115(fig.), 116(fig.), 120–125, 121(fig.), 129, 150, 154, 169, 173, 176n, 180–181, 191–193, 195, 197–201, 213–221, 224, 227, 228, 234–235, 237–238, 239n, 240n, 242, 285–287, 289(fig.), 294, 296, 298–300, 303–307, 322, 332; crocodilian earth god/earth god, 39, 54 (fig.), 57, 145, 157(fig.), 160; earth cave, 5, 161–162, 176n; earth monster/Tlaltecuhtli, 40, 40(fig.), 285, 324; earthly creatures, 123, 124, 144

Egypt/Egyptian 6, 74–79, 82, 83(fig.)

Ethnogenesis, 325

Ephemeral, 18, 322, 335–337

F

Far, 4, 7, 38, 217, 253, 334

First Cause, 52, 81–82, 110. *See also Primum mobile*

Five levels, 329

Five partitions, 7

Five steps, 328, 329

Flood, 78, 82, 108, 157–159

Florentine Codex, 11–12, 42, 47, 51, 111–113, 127, 128(fig.), 288–289, 292, 296

Foreigner, 255, 256(fig.). *See also tz'ul*

Four cardinal directions, 5, 35, 38, 39(fig.), 44–46, 152(fig.), 186–187(fig.)

Four directional points, 180

Four directions, 297

Four quadrants, 169

Four quarters, 158

Four world directions, 154, 161

Frames/Borders, 121, 122(fig.), 123, 147, 155(fig.), 161, 165

G

Galilei, Galileo, 74

Genesis, 72, 306

Generative cosmology, 125, 127–128, 145. *See also* Cosmology

H

Heaven/heavenly, 7, 18, 31, 33–44, 41(fig.), 48, 50(fig.), 51, 109–111, 120, 124, 127–128, 156, 164, 168, 199, 212–219, 231, 235, 238, 251, 258, 262, 282, 287, 291, 295–296, 300, 305, 307n, 309n, 310n, 325–326, 328, 331–337; heavenly bands/heavenly layers, 24, 34(fig.), 35, 53–58; heavenly spheres/celestial spheres, 3, 35, 52, 111. *See also* Sky

Hell, 3, 18, 35–36, 48, 50(fig.), 51, 52–53, 57, 166, 212–213, 217, 307n

Histoire du Mechique, 42, 100, 102, 106–107, 112, 120, 125, 302, 309n

Historia de los mexicanos por sus pinturas, 42, 102, 104–105, 295

Historia sacra, 79, 82, 85, 88, 91

Homeyoca, 80, 108, 110, 288–290, 292, 294–296, 298, 303–305. *See also* Omeyocan

Horizontal, 5, 39, 42, 44, 297, 301; layers, 35–36, 53, 54(fig.), 58–59, 81; structure, 38, 45

Houses of the cosmos, 8, 191, 193, 197, 199, 201; earth, 193, 195, 199; goddess, 106; jaguar, 32, 33; Mictlan, 106, 287; sky/heaven, 196, 199, 287, 294; underworld, 37, 45, 193, 195–196, 199

Hun Ajaw, 161–162, 166, 167(fig.), 168–169

I

Idol, 15, 74–79, 82, 89, 92n, 93, 106, 141, 144, 147, 150

Images: cosmos, 8–12, 18, 46–48, 49(fig.), 56, 58, 72, 74–79, 82–83, 83(fig.) 85(fig.), 86, 102–105, 108–115, 119, 124, 142, 146–147, 149, 162, 166, 232, 282, 284, 286, 294, 328; new, 8, 11, 15, 17–18, 71–72, 110, 126, 200

Illusio, 71

Imago Mundi, 17, 282

K

K'iche', 32–33, 44, 166, 240n, 241n, 292

K'uh, 141, 144–149, 146(fig.), 154–155, 159–162, 169, 174n, 259, 271n, 272n, 273n

L

Landscape, 12, 42, 217, 251–252, 258, 261–263, 266, 268, 270–271, 336; deep-time, 258–263, 268, 270–271

Lápida de los cielos, 3, 40–42, 57, 58, 131n

Layers/Levels, 3–5, 8, 14, 16, 31–40, 34(fig.), 43–47, 50, 50(fig.), 52–53, 53(fig.), 55–59, 55(fig.), 56(fig.), 60n, 79, 80, 81, 81(fig.), 100, 101(fig.), 106–107, 109–110, 120–121, 130n, 161, 190(fig.), 191, 200, 201n, 228, 240, 251, 262, 296, 297

Leyenda de los Soles, 43, 290, 295, 310n

Location qualifiers, 251

M

Madrid Codex, 38, 144, 147(fig.), 148, 150, 151(fig.), 152(fig.), 154, 155(fig.), 158–159, 161, 163, 164(fig.), 165, 166(fig.), 174n, 175n

Maya, 6–17, 19n, 31, 33–39, 41, 44–46, 115, 117, 141–147, 146(fig.), 151, 153–159, 161–162, 165–170, 174n, 175n, 176n, 182, 209, 212–213, 222, 224–225, 227, 229–230, 232, 233, 235, 237, 239n, 240n, 241n, 242, 251–256, 252n, 258–259, 261, 263–264, 266, 270–271, 292, 298, 307, 308n

Metnal, 16, 141, 165, 176n. *See also* Xib'alb'a

Mictlan (see Underworld), 3, 42–43, 57, 100, 122–123, 287, 288–295, 298–306, 307n, 310n

Mixtec, 13, 34, 40, 42, 57, 114–120, 115(fig.), 121(fig.), 123–124, 129, 303, 337

Moon, 35, 41, 52, 54, 56–58, 106–107, 109–113, 120, 125, 160, 168, 176n, 213, 221–226, 231–238, 240n, 241n, 283, 284(fig.), 286, 307n, 325; Moon goddess, 162, 227, 308n. *See also* Underworld

Mother, 84, 107, 109, 125, 128–129, 176n, 213, 223–224, 229, 236–237, 288–289, 332

Mountain, 5, 40, 42, 52, 57, 114–115, 118, 129, 161, 215, 220, 230, 253, 258–259, 261–263, 270–271, 302, 305, 307, 322, 326, 328–329, 331, 335

Multilayered Universe, 5, 31, 39, 41(fig.), 42, 44, 45, 59, 60n, 200

Multiple, 7, 14–15, 33, 48, 54, 103, 114, 116, 120, 123, 127, 169, 181, 185, 222; Positions, 130, 187, 240n, 297

Muwieri, 332, 333(fig.), 334

Mythology, comparative, 72

N

Nahua, 6, 7, 9–10, 13, 18, 19n, 38, 45, 47, 49(fig.), 60n, 81, 103–108, 111, 114–115, 123–125, 127, 129, 131n, 132n, 176n, 183, 203n, 282, 294, 298, 299, 301–302, 307n, 308n, 309n, 310n

Narrative, 7, 15, 17, 32, 51, 54(fig.), 59, 78, 88, 103–109, 114, 116, 121, 123, 125, 129, 131n, 161, 170, 181, 185, 214, 251–255, 258–259, 261–271

Nature, 8, 18, 38, 48, 75–76, 113, 115, 121, 123, 125, 153, 162, 181, 184–185, 212, 261, 271, 306, 320–321

Near, 57, 195, 219, 220–221, 225, 238, 251, 253, 256, 261–262, 289, 307n

Nepanihui, 43, 309n

New Spain, 10, 51, 52, 55, 71, 74, 78, 90, 107–108, 110, 112, 130, 182

New World, 12, 46, 48, 71, 88, 92n

Nine (number), 16, 18, 45, 30

Nine divisions, 37–39, 52, 58, 107, 131n
Nine gods, 37, 38, 42, 44–45
Nine levels/layers, 3–5, 8, 16, 43–44, 52, 58, 106, 109–110, 190(fig.), 191
Nine lords, 295–304
Nine places/regions, 7, 35, 301

O

Oaxaca, 182, 198, 202n, 230
Ometeotl, 106, 108, 110, 309n
Omeyocan, 53, 57, 58, 60n, 108, 110, 288–290, 292, 294–296, 298, 303–305. *See also* Homeyoca
Otherworld/Otherworldly spaces, 6, 18, 32, 215, 217, 219–220, 258, 282, 292, 301, 305, 309n
Origin, 17, 33, 44, 45, 57, 76, 79, 82, 83(fig.), 91n, 104–105, 141, 180, 183, 197, 200, 218, 258, 259, 262–263, 264(fig.), 270, 298, 300–306, 323, 325

P

Pauahtun/Pawahtun, 36, 153
Philosophia perennis, 78, 88, 91
Pignoria, Lorenzo, 15, 70–90, 91n, 92n, 93n, 94n
Pilgrimage, 17, 262–266, 268–271, 272n, 322, 324, 329, 336
Plato, 74; Neoplatonic, 78, 85, 88
Popol Vuh, 32, 37, 44–45, 57, 166, 168, 253, 259, 285, 290
Portal, 17, 144, 180, 215, 219
Primum mobile, 52, 111(fig.). *See also* First Cause
Pyramid, 35, 37, 52, 130n, 264, 326, 327(fig.), 328, 329; double-pyramid, 100, 102(fig.)

R

Rain, 17, 144–145, 153–156, 162, 169, 172, 175n, 210, 214–215, 220–224, 227–233, 238–239, 239n, 284, 287, 290, 321, 324, 328, 335
Realms, 16–18, 51, 101, 103, 114, 121, 128–129, 131n, 159, 161, 169, 181, 189, 219, 221, 229, 238, 239n, 304–305
Relación de las cosas de Yucatán, 36, 141–142, 149
Regions/Directions, 35, 38–39, 44–46, 120, 139, 153, 154, 161, 176n, 186, 187(fig.), 194, 201, 257(fig.), 271, 297, 305
Renaissance, 6, 15, 47, 53, 73, 77, 297, 320, 328
Rhetoric, 71, 78, 88
Rhetorica Christiana, 47, 48(fig.), 50(fig.)

Ritual, 7, 16, 18, 45–46, 60n, 114, 115(fig.), 119, 123, 126–130, 131n, 142, 144–148, 148(fig.), 149–150, 153, 165, 180–187, 191, 193, 197, 199–200, 202n

S

Sacrifice, 38–39, 122, 129, 146, 150, 165, 166(fig.), 176n, 290, 307n, 324–325, 335–336
Sea of blood, 218
Season, 161, 214, 230, 238, 286–287, 304, 321, 324–325, 334–335; Rainy, 230, 286, 287, 304, 321, 324–325, 334–335
Selden Roll, 42, 53, 54(fig.), 57–59, 120–121, 131n
Seler, Eduard, 4, 6, 19n, 34–35, 38, 57, 100, 102(fig.), 107, 118, 121–123, 130n, 131n, 201, 203n, 283, 320
Seven (number), 7, 14, 37, 40, 54, 60n, 221–222, 272n, 273n
Seven bodies, 130
Seven caves, 54, 57, 304
Seven days, 202
Seven Divisions, 263
Seven lords, 222
Seven planets, 109
Seven terraces, 52
Sickness, 42, 220, 225, 234, 238–239, 323, 337
Sky, 7, 18, 31, 33–44, 41(fig.), 48, 50(fig.), 51–53, 54(fig.), 55(fig.), 56–58, 109–111, 120, 124, 127–128, 156, 164, 168, 199, 212–219, 231, 235, 238, 251, 258, 262, 282, 287, 291, 295–296, 300, 305, 307n, 309n, 310n, 325–326, 328, 331–337; iconographic clusters, 118, 120, 124, 127; night sky, 18, 42, 106–107, 157(fig.), 232–236, 283, 284(fig.), 285–286, 285(fig.), 289(fig.), 291, 294, 300, 304–306, 322, 335; representations, 103, 117(fig.), 284, 285, 286; segmentation/compartmentalization, 5; spheres, 3, 17, 33, 35, 44, 52–55, 56(fig.), 110–112, 130n. *See also* Heaven
Six Sky Mountain, 262
Stair/Ladder, 301, 326, 327(fig.), 329–331
Spiritual, 88, 90(fig.), 111(fig.), 219–220, 227, 229, 238
Stars, 40–42, 44, 52, 57–58, 107, 110, 111(fig.), 117, 120–121, 123–125, 127, 129, 131n, 132n, 213, 233–235, 238, 241n, 242n, 283–286, 290, 305, 307n, 308n
Sun, 35, 38–41, 52, 57, 105–116, 120, 124–125, 155, 157, 159–162, 166–169, 167(fig.), 175n, 176n, 188, 198, 221–227, 231–236, 238, 240n, 241n,

INDEX 349

264, 282–285, 287, 287(fig.), 292, 294–295, 305–306, 307n, 310n, 322, 324–325, 327, 329–332, 335; night Sun, 38–39; Sun god, 43, 145, 148 (fig.), 159, 160, 213, 222, 227
Strategies, 8, 13, 15, 70, 103, 104, 108, 110, 111, 112, 113, 251, 258, 268, 270

T

Talocan, 45–46, 298. *See also* Tlalocan
Thirteen divisions, 255
Thirteen gods, 44
Thirteen levels/layers, 4, 80
Thirteen lords, 100, 255
Threshold, 15, 104, 116–117, 120–121, 123, 127
Time: deep time, 258, 259, 260, 261, 262, 263, 268, 270, 271; time and space, 18, 31, 104, 180–181, 185–186, 253, 271, 299, 304–306, 322, 335
Tlalocan, 109, 287, 290, 294
Tlatelolco, Colegio de, 47, 52
Transformation, 10–12, 16, 73, 82–83, 103, 128, 141–142, 149, 168–169
Travel, 8, 17, 42, 43, 51, 114, 131n, 180, 199, 201, 219, 221, 236, 251, 253, 263, 266, 268, 269
260-day count, 8, 16, 142, 174n, 180–183, 185–187, 197, 199, 200, 202, 260, 263; *Biyé*, 8, 10, 55, 183, 185, 189, 202; *Tzolk'in*, 142, 168, 263
Tzitzimime, 123, 125, 290, 291, 308n
tz'ul, 255, 256(fig.). *See also* Foreigner

U

Underworld, 3–8, 14, 16, 31–40, 42–45, 51–54, 57–59, 60n, 77, 80, 83, 100, 101(fig.), 104, 106, 109, 114, 116(fig.), 123–125, 129, 130n, 131n, 132n, 140, 155, 158–161, 165, 166(fig.), 168–169, 176n, 180–181, 191, 193, 195–199, 217, 240n, 251, 261–262, 282, 285, 287–288, 291, 305, 322–325, 328, 330, 332, 335, 337. *See also* Layers; Mictlan; Xib'alb'a

V

Valadés, Diego de, 47–48, 49(fig.), 50(fig.)
Venus, 40, 41(fig.), 117–118, 118(fig.), 123, 131n, 155, 157(fig.), 158–162, 167–169, 176n, 235–238, 242n, 283, 285(fig.), 291n
Vertical, 7, 37, 44, 80, 82, 101, 201, 266; layers, 5, 8, 33, 36, 38, 39, 100, 130n, 251, 258; structure, 35, 37, 46, 58, 59
Vision, 7, 14, 31–32, 36, 38, 51, 58–59, 101, 110, 122, 211, 262, 282, 305, 307n, 322, 327–328, 334–337
Visual, 10–12, 15, 48, 74–75, 79–82, 86–88, 100, 102–103, 108, 110–111, 113–114, 117, 118, 122(fig.), 127, 129, 130n, 227, 252, 258, 266; investigation/research, 72, 74–75, 251

W

Water, 53, 107, 117–118, 121, 129, 147, 147(fig.), 149, 157(fig.), 161–164, 169, 176n, 188, 209, 213–221, 223–225, 227, 231–232, 238–240, 252–254, 259, 272, 286, 289–290, 300, 302, 304–305, 307, 309n, 336
Wind, 57, 85, 105–106, 112, 145, 146(fig.), 153–154, 162, 188, 221, 259, 290; 9 Wind, 57, 114, 115(fig.), 120, 121(fig.), 129
Wixarika/Wixáritari, 8, 319, 320, 321, 322–329, 331–332, 334–337
World, 8, 18, 38, 43, 52, 56, 73, 76, 89, 104–105, 107, 111(fig.), 112, 114, 116, 120, 126, 129, 130n, 142, 144, 149, 154, 156, 158–159, 164(fig.), 170, 173–174, 176n, 214–217, 222, 231, 239n, 253, 255–256, 263, 271, 282, 285–287, 289(fig.), 290, 293–295, 300, 301, 304–306, 309n, 320, 321–328, 330, 332, 334–337; composition/configuration, 5, 101, 102, 104, 105, 107–108, 110, 113–115, 117, 130n, 297; Dantean, 14, 32–33, 46, 51, 53, 55–58, 107, 111(fig.), 112, 114, 116, 120, 126, 129, 130n, primordial, 18, 122–123, 288, 305–306, 325, 330, 335; solar/sun, 222, 323, 325, 334–337; view (visions of world), 32, 36, 38; watery, 161–162, 214–217, 220, 229, 231–233, 233(fig.), 239. *See also* Cosmos

X

Xib'alb'a, 16, 32. *See also* Underworld

Z

Zapotec, 6, 8, 10, 16, 34, 44, 180–201, 201n, 202n, 203n, 204n, 205n